An Introduction to
Digital Signal Processing

An Introduction to Digital Signal Processing

John H. Karl
Department of Physics and Astronomy
The University of Wisconsin-Oshkosh
Oshkosh, Wisconsin

Academic Press
San Diego New York Boston
London Sydney Tokyo Toronto

Copyright © 1989 by Academic Press, Inc.
All Rights Reserved.
No part of this publication may be reproduced or transmitted in any form or by any means, electronic or mechanical, including photocopy, recording, or any information storage and retrieval system, without permission in writing from the publisher.

Academic Press, Inc.
A Division of Harcourt Brace & Company
525 B Street, Suite 1900, San Diego, California 92101-4495

United Kingdom Edition published by
Academic Press Limited
24–28 Oval Road, London NW1 7DX

Library of Congress Cataloging in Publication Data

Karl, John H.
    An introduction to digital signal processing / by John H. Karl.
    p. cm.
    Includes index.
    ISBN 0-12-398420-3 (alk. paper)
    1. Signal processing--Digital techniques. I. Title.
TK5102.5.K352 1989
621.38'043--dc19   88-26822
                                                                                  CIP

PRINTED IN THE UNITED STATES OF AMERICA
95 96 97 QW 9 8 7 6 5 4 3

To Karen

Contents

Preface   xi

1 / Signals and Systems

Sampling and Aliasing   4
Linear Time-Invariant Systems and the Convolution
    Operation   8
Constant-Coefficient Difference Equations   18
System Block Diagrams and Flow Graphs   21
Problems   24

2 / Sampled Data and the Z Transform

The Z Transform, Polynomial Multiplication, and
    Convolution   27
Factoring Z Transforms into Couplets   29
Inverse Operators: Stability, Causality, and Minimum Phase   30
Problems   37

3 / Sinusoidal Response of LSI Systems

Sinusoidal Signals as Eigenfunctions and the Spectrum
    as Eigenvalues   42
Frequency Response of Some Simple LSI Systems   44

Frequency Response of Digital Differential
and Integral Operators   50
The Bilinear Transform and Its Application
to Differential Equations   53
Problems   60

4 / Couplets and Elementary Filters

The Single-Zero Couplet   65
The Single-Pole Couplet   69
The Single-Zero, Single-Pole, Allpass Filter   71
Elementary Filters Classified by Their Poles and Zeros   74
Problems   76

5 / The Discrete Fourier Transform

Sampling the System Response in the Frequency Domain   79
Properties of the DFT   82
Special Values of the DFT   86
The Phase-Shift Theorem   87
The Convolution Theorem   88
Cross-Correlation and Autocorrelation   91
Problems   93

6 / The Continuous Fourier Integral Transform

The Fourier Integral Transform Developed from the DFT   98
Properties of the Fourier Integral Transform   102
The Wiener–Khintchine Theorem   104
The Time-Limited Band-Limited Theorem   106
A Repertoire of Transforms and Their Importance   107
Problems   124

7 / Application of the Fourier Transform to Digital Signal Processing

Continuous Time, Discrete Frequency: The Fourier Series   128
The Least-Squares Convergence of the Fourier Series   131
The Sampling Function   133
The Relationship between the FT and the DFT:
Resolution and Leakage   135
Interpolation, Decimation, and Multiplexing   143
Digital Control Systems   149
Problems   160

Contents

8 / Digital Filter Design

Digital Filter Design—The Problem   167
Designing FIR Filters Using Windows   171
Frequency Sampling and the Parks-McClellan Algorithm   175
Recursive Filters   182
Digitizing Rational Functions of ω 185
DFT Frequency Filtering   190
Problems   192

9 / Inverse Filtering and Deconvolution

Exact Inverses via the DFT   198
Linear Deconvolution—The Problem   203
Wiener Least-Squares Filters   206
Application to Inverses   208
Filter Delay Properties   211
Applications to Prediction Filters   213
Matched Filters and Output Energy Filters   217
The Levinson Recursion   225
Appendix: The Least-Squares Property of Wiener Filters   228
Problems   230

10 / Spectral Factorization

The Root Method   236
The Spectrum of a Real Causal Function   238
Kolmogoroff Factorization   242
Least-Squares Zero-Delay Factorization   245
Iterative Least-Squares Factorization   246
Applications to IIR Filter Design   250
Problems   253

11 / Power Spectral Estimation

Signal-Like and Noise-Like Processes—The PSD Problem   261
The MA Model: The Approach of Blackman and Tukey   267
The AR Model: The Approach of Yule-Walker and Burg   271
The Maximum Entropy Principle   279
Other Data Models   284
Problems   287

12 / Multidimensional DSP

Multidimensional Difference Equations   294
Fourier Transform Methods in Multidimensions   310
Two-Dimensional FIR Frequency Filter Design   320
Problems   326

References   333

Index   335

Preface

This book is written to provide a fast track to the high priesthood of digital signal processing. It is for those who need to understand and use digital signal processing and yet do not wish to wade through a three- or four-semester course sequence. It is intended for use either as a one-semester upper-level text or for self-study by practicing professionals.

The number of professionals requiring knowledge of digital signal processing is rapidly expanding because the amazing electronic revolution has made convenient collection and processing of digital data available to many disciplines. The fields of interest are impossible to enumerate. They range widely from astrophysics, meteorology, geophysics, and computer science to very large scale integrated circuit design, control theory, communications, radar, speech analysis, medical technology, and economics.

Much of what is presented here can be regarded as the digital version of the general topic *Extrapolation, Interpolation, and Smoothing of Stationary Time Series*, a title published in 1942 by the famous Massachusetts Institute of Technology mathematician Norbert Wiener. This original classified Department of War report was called "the yellow peril" by those working on the practical exploitation of this theoretical document. The nickname came from the report's frightening mathematics, lurking within a yellow book binding. Since the report became available in the open literature in 1949, many works have followed, including textbooks on many facets of the subject. Yet topics important for the digital implementation of these ideas have not been made available at an introductory level. This book fills that gap.

To minimize the required mathematical paraphernalia, I first take the viewpoint that the signals of interest are deterministic, thereby avoiding a heavy reliance on the theory of random variables and stochastic processes. Only after these applications are fully discussed do I turn attention to the nondeterministic applications. Throughout, only minimal mathematical skills are required. The reader need only have a familiarity with differential and integral calculus, complex numbers, and simple matrix algebra. Super proficiency is not required in any of these areas.

The book assumes no previous knowledge of signal processing but builds rapidly to its final chapters on advanced signal processing techniques. I have strived to present a natural development of fundamentals in parallel with practical applications. At every stage, the development is motivated by the desire for practical digital computing schemes. Thus Chapter 2 introduces the Z transform early, first as a simple device to represent advance and delay operators and then as an indispensable tool for investigating the stability and invertibility of computational schemes. Likewise, the discrete Fourier transform is not introduced as an independent mathematical entity but is developed quite naturally as a tool for computing the frequency response of digital operators. The discrete Fourier transform then becomes a major tool of digital signal processing, connecting the time and frequency domains by an efficient computing algorithm, the fast Fourier transform.

The idea of a data model is threaded throughout much of the discussion. Since continuous functions are an underlying model for many sampled data, Chapter 6 is devoted to the continuous Fourier transform and Chapter 7 addresses its all-important relationship to discrete data.

Using the fundamental concepts developed in these early chapters, each of the last five chapters covers a topic in digital signal processing that is rich in important applications. The treatment of each is sufficiently detailed to guide readers to practical solutions of their own signal processing problems and to enable them to advance to the research literature relevant to their applications.

Much effort has been spent in making the text as complete as possible, avoiding distracting detours to references. Consequently, few references are included. Those that are included, however, provide the tip of a rapidly expanding pyramid of references for readers desiring further study.

Problems at the end of each chapter reinforce the material presented in the text. Because many of these problems include practical computing applications, it would be best if the book were read with at least a small computer available to the reader. Computer routines are presented in a psuedo-FORTRAN code for all of the fundamental processing schemes discussed. My purpose is not to provide commercial-grade digital signal processing algorithms; rather it is to lay open the utter simplicity of these little gems. Then, armed with a sound understanding of concepts, readers with a working knowledge of computer coding will be able to quickly adapt these central ideas to any application.

I greatly appreciate the patience of my students, who suffered through using the manuscript in its developmental stages. Sue Birch is gratefully acknowledged for helping draft the figures and for being a pleasure to work with. I especially extend my greatest appreciation to Joan Beck for typing the manuscript, which required a multitude of corrections and changes, and for cheerfully tolerating all my ineptness. Any remaining errors are my responsibility, not hers.

John H. Karl

1

Signals and Systems

The feeling of pride and satisfaction swelled in his heart every time he reached inside the green, felt-lined mahogany box. At last the clouds were breaking up, so now Captain Cook reached for the instrument, exquisitely fashioned from ebony with an engraved ivory scale and hand-fitted optics mounted in brass to resist the corrosive environment of the sea. With mature sea legs, he made his way up to the poop deck of the Resolution. There the sun was brightening an aging sea—one whose swells still remained after the quieting of gale force winds. James Cook steadied himself against the port taffrail. His body became divided. Below the waist he moved with the roll of the ship; above it he hung suspended in inertial space. Squinting through his new Galilean ocular, he skillfully brought the lower limb of the noon sun into a gentle kiss with the horizon.

As always, Master Bligh dutifully recorded the numbers slowly recited by the navigator as he studied the vernier scale under the magnifying glass. At first the numbers slowly increased, then seemed to hover around 73 degrees and 24 minutes. When it was certain that the readings were decreasing, Cook called a halt to the procedure, took Bligh's record, and returned to his cabin below deck.

After first returning his prized octant to its mahogany case, the navigator sat at his chart table, carefully plotting Bligh's numbers in search for the sun's maximum altitude for use in computing the Resolution's latitude. Next course and estimated speed were applied to the ship's previous position to produce a revised position of 16° 42' north latitude and 135° 46' west longitude. A line drawn on the chart with a quick measurement of its azimuth produced a beckon to Bligh: "tell the helmsman to steer west northwest one-quarter west."

This one example from the cruise of the *Resolution* includes analog to digital conversion, interpolation, extrapolation, estimation, and feedback of digital data. Digital signal processing certainly extends back into history well before Cook's second voyage in 1772–1775 and was the primary form of data analysis available before the development of calculus by Newton and Leibnitz in the middle of the seventeenth century. But now, after about 300 years of reigning supreme, the classical analytical methods of continuous mathematics are giving way to the earlier discrete approaches. The reason, of course, is electronic digital computers. In recent years, their remarkable computing power has been exceeded only by their amazing low cost. The applications are wide ranging. In only seconds, large-scale super computers of today carry out computations that could not have been even seriously entertained just decades ago. At the other end of the scale, small, special-purpose microprocessors perform limited hard-wired computations perhaps even in disposable environments—such as children's toys and men's missiles.

The computations and the data they act on are of a wide variety, pertaining to many different fields of interest: astrophysics, meteorology, geophysics, computer science, control theory, communications, medical technology, and (of course) navigation—fundamentally not unlike that of James Cook. For example, a modern navigation system might acquire satellite fixes to refine dead reckoning computations derived from the integration of accelerometer outputs. In modern devices, the computations would act on digital data, just as in Cook's day.

In the many examples given above, the data involved can have different characteristics, fundamentally classified by four properties: analog, digital, deterministic, and innovational. These properties are not all-inclusive, mutually exclusive, nor (since all classification schemes contain arbitrary elements) are they necessarily easy to apply to every signal.

We will now discuss these four properties of signals, considering that the independent variable is time. In fact, for convenience throughout most of the book, we will take this variable to be time. But, it could be most anything. Our signals could be functions of spatial coordinates x, y, or z; temperature, volume, or pressure; or a whole host of other possibilities. The independent variable need not be a physical quantity. It could be population density, stock market price, or welfare support dollars per dependent family. Mostly we will consider functions of a single variable.

An *analog* signal is one that is defined over a continuous range of time. Likewise its amplitude is a continuous function of time. Examples are mathematical functions such as $a + bt^2$ and $\sin(\omega t)$. Others are measured physical quantities, such as atmospheric pressure. We believe that a device designed to measure atmospheric pressure (such as a mercurial barometer) will have a measurement output defined over a continuous

1/ Signals and Systems

range of times, and that the values of this output (the height of the mercury column) will have continuous values. When sighting the barometer's column, any number of centimeters is presumed to be a possible reading.

A meteorologist might read the barometer at regular periods (for example, every four hours) and write the result down, accurate to, for example, four decimal digits. He has *digitized* the analog signal. This digital data is defined only over a discrete set of times. Furthermore, its amplitude has been quantized; in this example, the digital data can only have values that are multiples of 0.01 cm. A digital signal specified at equally spaced time intervals is called a *discrete-time sequence*.

We see that our definitions of analog and digital signals include two separate and independent attributes: (1) when the signal is defined on the time scale and (2) how its amplitude is defined. Both of these attributes could have the continuous analog behavior or the quantized discrete behavior, giving rise to four possibilities. For example, we might record values of a continuous-time function on an analog tape recorder every millisecond. Then the resulting record has analog amplitudes defined only at discrete-time intervals. Such a record is called *sampled data*. Another combination is data that are defined over continuous time, but whose amplitude values only take on discrete possibilities. An example is the state of all the logic circuits in a digital computer. They can be measured at any time with an oscilloscope or a logic probe, but the result can only take on one of two possible values.

Generally speaking, in our work we will either be considering continuous signals (continuous values over a continuous range of time) or digital signals (discrete values over a discrete set of time).

The other aspect of signals that we wish to consider is their statistical nature. Some signals, such as $\sin(\omega t)$, are highly *deterministic*. That is, they are easily predictable under reasonably simple circumstances. For example, $\sin(\omega t)$ is exactly predicated ahead one time step Δt from the equation

$$U_{t+2\Delta t} = aU_{t+\Delta t} - U_t \qquad (1.1)$$

if we know the frequency ω_0 of the sinusoid and only two past values at $t + \Delta t$ and t. You can easily verify that Eq. (1.1) is a trigonometric identity if $a = 2\cos(\omega_0 \Delta t)$ and $U = A\sin(\omega_0 t + \phi)$. An interesting and significant property of Eq. (1.1) is that its predictive power is independent of a knowledge of the origin of time (or equivalently, the phase ϕ) and the amplitude A. Hence it seems reasonable to claim that sinusoids are very predictable, or deterministic.

On the other hand, some signals seem to defy predication, no matter how hard we may try. A notorious example is the Dow Jones stock market indicator. When a noted analyst was once asked what he thought the market would do, he replied, "It will fluctuate." When these fluctuations defy prediction or understanding, we call their behavior random, stochastic, or innovative. Some call it *noise-like* as opposed to deterministic *signal-like* behavior. There are many other examples of random signals: noise caused by electron motion in thermionic emission, in semiconductor currents, and in atmospheric processes; backscatter from a Doppler radar beam; and results from experiments specifically designed to produce random results, such as the spin of a roulette wheel. Many of these random signals contain significant information that can be extracted using a statistical approach: temperature from electron processes, velocities from Doppler radar, and statistical bias (fairness) from the roulette wheel.

It is not universally recognized that it is not necessarily the recorded data that determines whether a certain process is random. But rather, we usually have a choice of two basically different approaches to treating and interpreting observed data, deterministic and statistical. If our understanding of the observed process is good and the quantity of data is relatively small, we may well select a deterministic approach. If our understanding of the process is poor and the quantity of data is large, we may prefer a statistical approach to analyzing the data. Some see these approaches as fundamentally opposing methods, but either approach contains the potential for separating the deterministic component from the nondeterministic component of a given process, given our level of understanding of the process.

For example, the quantum mechanical description of molecules, atoms, nuclei, and elementary particles contains both deterministic and statistical behavior. One question is, can the behavior which appears statistical be shown, in fact, to be deterministic via a deeper understanding of the process? (This is the so-called hidden variable problem in quantum mechanics.) Some, like Albert Einstein even to his death, believe that the statistical nature of quantum mechanics can be removed with a deeper understanding.

Because it requires less mathematical machinery, and hence it is more appropriate for an initial study of digital signal processing, we will use primarily a deterministic approach to our subject. Only in later chapters when we discuss concepts such as prediction operators and the power spectrum of noise-like processes will we use the statistical approach.

Sampling and Aliasing

Our immediate attention turns to deterministic digital signals. Frequently these digital signals come from sampling an analog signal, such as Captain Cook's shooting of the noonday sun. More commonly today, this sampling

Sampling and Aliasing

is done with an electronic analog-to-digital (A/D) data acquisition system. Sampling rates can vary widely, from one sample per day, or less, to more than ten million samples per second.

There are, however, signals that are inherently digital, such as those that occur in demographic studies. The deer kill in the Wisconsin woods during the gun season is inherently a discrete variable that is a function of discrete, equally spaced time samples, measured in the number of deer per year. It has no continuous function of time from which it was digitized. Others perhaps are not so clear. Some economists say that a commodity's price is determined only at the time of sale. What then is the price of beans between sales, and is it a continuous function of time or a discrete function (which would not be equally spaced in time)?

If we consider the case where a clearly continuous function is digitized by an A/D conversion, a fundamentally important phenomenon arises. As an example, sinusoids are the most instructive because, as it turns out, they will play a basic role in our studies. Thus, let us consider the familiar complex sinusoid

$$f(t) = \cos(\omega t) + i \sin(\omega t) = e^{i\omega t} \tag{1.2}$$

and form a time-sampled version of this signal by restricting time to only the discrete equally spaced values $t = n \Delta t$, where Δt is the time between samples. The sampled signal is now defined for integer values of n:

$$f_n = e^{i\omega n \Delta t} \tag{1.3}$$

Now, f is not a digital signal in the sense used in the introduction; its amplitude values are not quantized. Indeed, it is a continuous function of ω, a property that we now wish to explore.

We might imagine that our sinusoid of Eq. (1.2) is at the input of an A/D converter as shown in Fig. 1.1(a). (Such an instrument would produce a discrete amplitude as well, but that is not a point of interest at the moment.) Next, we consider ω starting as a low frequency near zero and increasing. We wish to study the output of the A/D converter as a function of the continuous parameter ω. As ω increases, f behaves as expected until $\omega = \pi/\Delta t$. Then

$$f_n = e^{i(\pi/\Delta t)n \Delta t} = e^{i\pi n} = (-1)^n$$

which suddenly does not look like a digitized sinusoid at all. A similar thing happens when $\omega = -\pi/\Delta t$. To explore this phenomenon further, let us compare the output of the A/D converter at frequencies that differ by multiples of $2\pi/\Delta t$:

$$\omega \rightarrow \omega \pm 2\pi M/\Delta t \quad M = 1, 2, \ldots$$

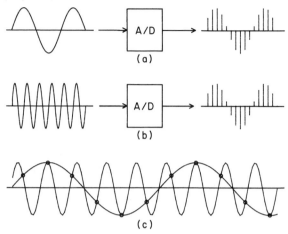

Figure 1.1 (a) Analog-to-digital conversion of a sinusoid of a certain frequency. (b) At higher input frequencies, related in a special way to the original input frequency, exactly the same digitized output occurs. (c) Shows how the A/D conversion of the two sinusoids can produce the same result. This aliasing phenomenon is discussed in the text.

Then Eq. (1.3) becomes

$$f_n \to e^{i(\omega \pm 2\pi M/\Delta t)n\Delta t} = e^{\pm i2\pi nM} e^{i\omega n\Delta t}$$

$$f_n \to e^{i\omega n\Delta t}$$

which is identical to Eq. (1.3). This is to say, the digitized result is periodic in frequency, with a fundamental period of $2\pi/\Delta t$. This is the sampling frequency ω_s of the A/D converter. Those readers who are familiar with Fourier synthesis will immediately recognize that since the output of the A/D converter is periodic for sinusoids, it will also be periodic for all signals because they can be synthesized from the sinusoids. This periodicity of the output from digitizing is of fundamental importance to digital signal processing. We will gain a deeper insight into how it arises and its significance in the coming chapters.

This periodicity means that the output of the A/D conversion is not unique; $\sin \omega t$ and $\sin(\omega + \omega_s)t$ produce the same output. Additionally there is yet more ambiguity. We next digitize sinusoids of frequencies

$$\omega \to 2\pi/\Delta t - \omega$$

The output becomes

$$f_n = e^{i\omega n\Delta t} \to e^{i(2\pi/\Delta t - \omega)n\Delta t} = e^{i2\pi n} e^{-i\omega n\Delta t}$$

$$f_n \to e^{-i\omega n\Delta t} = \cos \omega t - i \sin \omega t$$

Sampling and Aliasing

Again we get another output of the same lower original frequency ω. This time there is a sign change for the odd functions, the sines, but not for the even ones, the cosines. This is a general symmetry property, but it is not of prime concern here.

The main point that we have discovered in relating sinusoids digitized from analog sine waves of frequencies ω, $2\pi/\Delta t + \omega$ and $2\pi/\Delta t - \omega$ is summarized in Fig. 1.2. In all cases, the result is a digitized version of a sinusoid. But the results are not unique. The output frequency only follows the input frequency up to $\pi/\Delta t$; then it decreases, looking like the digitized version of a lower frequency sine-wave input. This bogus low-frequency output is called an *alias* because it is masquerading as a frequency not really present in the input. Aliasing is avoided only at input frequencies below one-half the sampling frequency. This frequency $\omega_s/2$, obviously of prime importance, is called the *Nyquist frequency* ω_N. Another name sometimes used for $\omega_s/2$ is the *folding frequency*, because it folds higher frequencies back down into the range $-\omega_s/2$ to $\omega_s/2$.

Aliasing is familiar to all of us from western movies where we see wagon wheels obviously misbehaving by going too slow for the speed of the

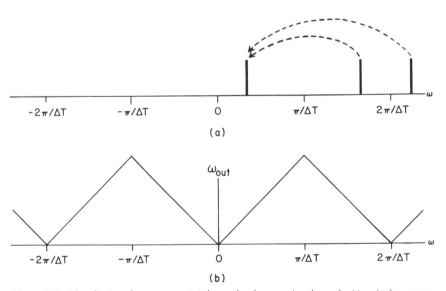

Figure 1.2 The aliasing phenomenon: (a) shows that frequencies above the Nyquist frequency of $\pi/\Delta t$ are folded back down into the Nyquist interval; (b) shows an alternate representation of the aliasing showing a plot of the output frequency of an A/D converter versus the input frequency. The frequency of the digitized output is equal to the frequency of the analog input up to Nyquist. At input frequencies above Nyquist, the output frequencies fall back below the Nyquist frequency, producing aliasing.

wagon, or worst yet by going backward when the wagon is moving forward. Modern motion pictures show a still-frame sample every $\frac{1}{24}$ of a second, making the Nyquist frequency 12 samples/second. This Nyquist frequency is well above the possible rotation speed of any horse-drawn wagon wheel, but the multiple spokes of the wheel produce much higher periodicities that do exceed the 12 samples/second Nyquist frequency.

Linear Time-Invariant Systems and the Convolution Operation

We will think of systems as entities that generate, measure, or process signals. Systems can be categorized in many different ways using many different attributes. Of course, no categorization is complete or always easy to apply. The range of complexity of systems varies greatly—from the ultrasimple, like a toggle switch, to the ultracomplex, like the human body.

The four properties discussed in the previous section—analog, digital, deterministic, and innovation—can be applied to systems. The implementation medium (i.e., the physical structure) of systems includes a vast number of possibilities: biological, chemical, electrical, mechanical, pneumatic, hydraulic, social, and economic, to name a few. They may be natural, like a river or a thunderstorm; they may be manufactured, like a hydraulic control system or a wind tunnel. Systems can contain subsystems upon subsystems. The subsystems can be quite different from one another; one might be a deterministic digital computational scheme, another might be a naturally occurring random analog component. Figure 1.3 shows some simple examples of various systems.

Of all of the possible properties of systems, there are five that are of particular importance for our study of digital signal processing: linearity, time invariance, stability, causality, and invertibility.

Linearity is the grand premiere property of systems that allows for great simplification in understanding and treating systems that would otherwise be quite forbidding. Linearity allows effects to be superimposed. The sum of two effects is equal to the effect of their sums. To be more precise, let us think of some operator S. The operation S could be anything, as long as it is defined. Additionally, the operations of summation and multiplication must be defined over a scalar number field and an operand field. Then we can put the concept of linearity on a mathematical level, defining a linear operator S to be one which satisfies

$$S(ax_1 + bx_2) = aSx_1 + bSx_2 \qquad (1.4)$$

Linear Time-Invariant Systems

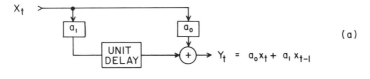

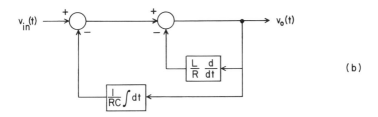

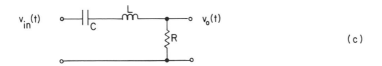

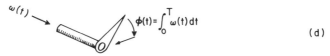

Figure 1.3 Some simple systems: (a) shows a digital system employing one unit delay and the multiplication by two constant coefficients; (b) shows a second-order analog system and one possible implementation of it, using electrical elements in (c). (d) The rotation of a shaft can be thought of as a mechanical system whose system output (the needle's position) is the time integral of the systems input (the angular velocity applied to the shaft).

Here, S operates on x_1 and x_2, while a and b are scalar numbers. One of the most common examples of a linear operator is integration:

$$\int [af(x) + bg(x)] \, dx = a \int f(x) \, dx + b \int g(x) \, dx$$

Another example is matrix multiplication:

$$\begin{pmatrix} 1 & 2 \\ 7 & 5 \end{pmatrix} \left[a \begin{pmatrix} 1 \\ 0 \end{pmatrix} + b \begin{pmatrix} 3 \\ 4 \end{pmatrix} \right] = a \begin{pmatrix} 1 & 2 \\ 7 & 5 \end{pmatrix} \begin{pmatrix} 1 \\ 0 \end{pmatrix} + b \begin{pmatrix} 1 & 2 \\ 7 & 5 \end{pmatrix} \begin{pmatrix} 3 \\ 4 \end{pmatrix}$$

The linearity condition defined by Eq. (1.4) has two parts: superposition and proportionality. Not only does superposition hold, but the output of the operation [the right side of Eq. (1.4)] contains contributions from x_1 and x_2 exactly in proportion to their content in the input (the left side of the equation). This means that if a linear system has only one input, and then that input is increased by a certain factor, the output will then increase by exactly the same factor.

It is easy to construct linear mathematical operators, but there are really very few physical systems that are exactly linear. Additionally, we must always be aware that while a system may be linear in one pair of variables, it can be highly nonlinear in another pair. For example, Maxwell's equations of electromagnetic theory are linear in the fields, but the forces vary as the inverse square of distance from point sources, giving rise to all kinds of nonlinear effects in the behavior of macroscopic matter. At first sight, it may seem that a certain system is composed of only linear components, but it may have hidden nonlinear ones. A simple toggle switch, for example, might contain just one linear spring. Yet when combined with the geometry of the toggle action, extreme nonlinear behavior occurs.

In spite of ubiquitous nonlinear systems, a study of linear systems is still very much worth our time. The reason is twofold: (1) a powerful theory of systems can be developed giving us great insight, and (2) many real systems can be treated as linear over some region giving us very useful results. Many examples of this last point can be given; one will do. Acoustic waves in the solid earth (or in the oceans) are so weak that the medium supporting their propagation undergoes extremely small variations in pressure, velocity, and displacement. Over these small excursions, the medium's behavior is very nearly linear. This is a manifestation of a general principle; most any system can be treated as though it were linear for sufficiently small excitation. This point alone is enough to justify a thorough study of linear systems.

Returning then to the mathematical implications of linearity, let us consider a discrete linear operator S that acts on a discrete sequence of data x to produce an output y as diagrammed in Fig. 1.4(a). Symbolically we express this action by writing

$$y = Sx \qquad (1.5)$$

We call y and x *time sequences* because each is a sequence of digital data equally spaced in time. The operator S is a time domain operator acting on various x's to produce an output at time t. Next, for reasons to be seen shortly, we introduce the so-called Kronecker delta $\delta_{t,k}$, which is a time sequence that contains just one unit pulse at time k as shown in Fig. 1.4(b). This allows us to write

$$x_t = \sum_k x_k \delta_{t,k} \qquad (1.6)$$

Linear Time-Invariant Systems

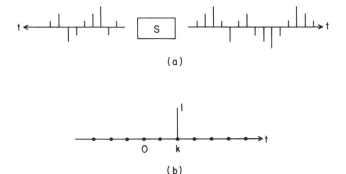

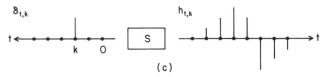

Figure 1.4 Digital systems and the unit impulse. (a) A general digital system showing output due to an arbitrary input. (b) A unit pulse at time k, sometimes called the Kronecker delta $\delta_{t,k}$. (c) The response of the system due to input δ is $S\delta_{t,k}$, the system impulse response function. Note that time increases to the left for the input, pictorially representing data fed into the system.

because, when summed over k, the result is zero until $k = t$, returning the value x_t. Equation (1.5) then becomes

$$y_t = S \sum_k x_k \delta_{t,k} \tag{1.7}$$

$$y_t = \sum x_k S \delta_{t,k} \tag{1.8}$$

In the second step, we have moved the operator over to the unit impulse because S operates on functions of t, using the linearity property with the x_k appearing as multiplicative factors. In this last step, we have invoked the linear property of S; the result of operating on a sum [Eq. (1.7)] is the same as the sum of the effects [Eq. (1.8)].

Now, $S\delta_{t,k}$ represents the output of the system at time t due to a unit impulse input at time k. This is the *impulse response* (IR) of the system S, as depicted in Fig. 1.4(c), where it is called $h_{t,k}$. Using this notation $h_{t,k}$ for the impulse response, Eq. (1.8) becomes

$$y_t = \sum h_{t,k} x_k \tag{1.9}$$

This equation can be viewed as matrix multiplication of the column vector **x** by the IR matrix **h** to give the output of the system at each time in the output column vector **y**. We shall see that this relationship takes on special simplicity when the system S does not change its properties with time.

Time invariance means that an impulse applied to the input of a system will produce the same output independent of when the input occurs, except of course for the corresponding time shifts. Examples of time-invariant systems are a mechanical gear train and an electrical circuit with fixed

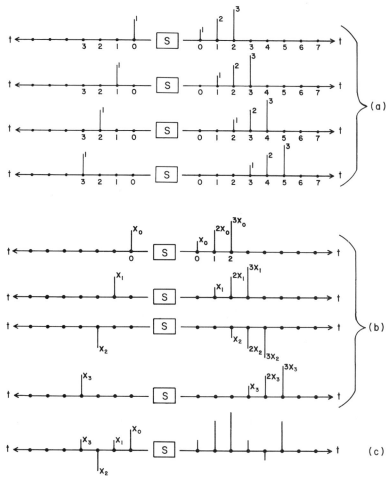

Figure 1.5 Pictorial representation of the action of a digital system whose impulse response is (1, 2, 3). (a) The input of a unit pulse successively delayed in time results in identical outputs, except for corresponding delays. (b) The unit impulse input is now scaled by factors (x_0, x_1, x_3) each of which simply scales the output by the same factor. (c) The result of the input (x_0, x_1, x_3) is the sum of the outputs in (b).

Linear Time-Invariant Systems

components. If the gear train has a gear shifting feature or if the electrical network has variable components, such as potentiometers, then these systems have the ability to change with time. Most of the systems that we will consider will be time-invariant (TI) systems. Because the independent variable in our discussions need not always be time, but could be anything, such as a spatial variable, an uncommitted term like *shift invariance* (SI) is preferred, avoiding any reference to time. Nonetheless, we will continue the use of time invariance in keeping with some authors and in keeping with our exemplary use of time as the independent variable. When it is desired to emphasize that the independent variable might not be time, we shall refer to LSI systems, linear shift-invariant systems.

It may seem that time invariance is a severe restriction on systems that we will discuss, and indeed it is. However, it is the proper starting point for an introduction to digital signal processing with many important applications. Additionally, simple time varying or adaptive processes can be developed after an understanding of time-invariant processing is achieved.

Thus, we will mostly be considering linear time-invariant systems, *LTI systems*. These two properties are exhibited in Fig. 1.5(a), which shows a digital LTI system with unit impulse inputs occurring at successively later times. Because of time invariance, the impulse response is the same in all four cases, except for time delays. That is, the system does not treat the impulse at $t = 0$ any differently than it does at $t = 2$.

Next, let the inputs be scaled by multiplicative factors (x_0, x_1, x_2, x_3) as shown in Fig. 1.5(b). Because of the linearity, two statements can be made: (1) the output due to each impulsive input is simply scaled by the x factor as shown in Fig. 1.5(b), and (2) the output due to the sum of the inputs is the sum of the outputs due to each input acting alone, as shown at the bottom of the figure. Thus, we see that, in general, the output of an LTI system is the sum of the impulse responses at different times, scaled according to the magnitude of the inputs at each input time.

This situation is better expressed by writing Eq. (1.9) to explicitly display the time invariance. In the example of Fig. 1.5, the systems impulse response is $(1, 2, 3)$. The matrix **h** has this same impulse response in every row, only shifted in time. Thus, in this example, Eq. (1.9) would read

$$\begin{pmatrix} y_0 \\ y_1 \\ y_2 \\ y_3 \\ y_4 \\ y_5 \end{pmatrix} = \begin{pmatrix} 1 & 0 & 0 & 0 \\ 2 & 1 & 0 & 0 \\ 3 & 2 & 1 & 0 \\ 0 & 3 & 2 & 1 \\ 0 & 0 & 3 & 2 \\ 0 & 0 & 0 & 3 \end{pmatrix} \begin{pmatrix} x_0 \\ x_1 \\ x_2 \\ x_3 \end{pmatrix} \qquad (1.10)$$

Note that elements in the **h** matrix appear reversed from those in Fig. 1.5 because time increases to the left in the rows of the matrix, while it increases to the right in the figure. The **h** matrix has the number of columns to match the length of the input and sufficient rows to include the last element of the impulse response arising from the last input element. It can be seen that if the impulse response is n long and the input is N long, the output will be $n + N - 1$ long. This matrix equation simply represents the time shifted IRs of Eq. (1.9). The column vector of inputs scales the IR to the value dependent on the input at that time. For example, by matrix multiplication in Eq. (1.10), at time $t = 2$ the output is

$$y_2 = 3x_0 + 2x_1 + x_2$$

The first term of this output, $3x_0$, is the combined effect of the input at time $t = 0$ and the IR at time $t = 2$. The second term $2x_1$ is the combined effect of the input at $t = 1$ and the resulting part of the IR that occurs at $t = 2$, and so on.

Clearly, it is the linearity of the system that allows us to express the system's input–output relation using matrix algebra. Additionally, the shift-invariant property of the system means that the **h** matrix has a very special form; each row is just a shifted version of the preceding row. That is, the elements are the same along each upper left to lower right diagonal. A matrix with this structure is called a *Toeplitz matrix*. We can express this structure mathematically by observing that the elements of $h_{t,k}$ are not independent functions of their indices, but rather are a function solely of their difference:

$$h_{t,k} = h_{t-k} \tag{1.11}$$

Thus, Eq. (1.9) can be written

$$y_t = \sum_{k=0}^{n} h_{t-k} x_k \tag{1.12}$$

where h_{t-k} is now just one dimensional. Equation (1.12) is called the *convolution* of h with x and is symbolically written

$$y = h * x \tag{1.13}$$

By writing out terms in Eq. (1.12), it can be seen that it implements the same shifting, multiplying, and addition of the matrix operation in Eq. (1.10). A convenient computational scheme for doing convolutions with pencil and paper is to invert the order of one factor, write it above the other, then shift, multiply, and add. Thus,

$$(3 \quad 2 \quad 1) \rightarrow$$
$$(x_0 \quad x_1 \quad x_2 \quad x_3)$$

Linear Time-Invariant Systems

```
    DIM X(M), Y(N), C(M + N − 1)
    C = 0
    DO 10 I = 1, M
    DO 10 J = 1, N
10  C(I + J − 1) = C(I + J − 1)  +  X(I) * Y(J)
```

Figure 1.6 A computer program to perform the convolution C = X * Y. All possible products of X and Y are required in the convolutional sum. Basically, it takes only three lines of computer code to perform the most general LSI operation possible.

yields the convolution $(1, 2, 3) * (x_0, x_1, x_2, x_3)$ when products are summed for each shift of $(3, 2, 1)$. Note that the convolution commutes:

$$h * x = x * h$$

and remember that the length of the convolution is $n + N - 1$, where n and N are the lengths of each factor.

From inspection of Eq. (1.12), it can be seen that the convolution operation requires all the possible products of the two factors and sums them into the output when the sum of the indices of h and x are equal to t [i.e., $(t - k) + k = t$]. This recognition leads to a computing algorithm for doing convolutions, shown in Fig. 1.6, which is the epitome of simplicity; it contains only three primary statements.

It is worth reflecting on what we have discussed. For LTI systems, the most general relationship between the input and the output is the convolution operation of Eq. (1.12). The impulse response h gives the output for any input; hence, an LTI system is completely and uniquely specified by its impulse response. Although we used digital systems in our discusssion, the same conclusions apply to analog systems, see Problem 1.12.

The convolution operation uses only shifting, multiplication, and addition. These operations are well suited to digital computing implementation in a wide variety of environments, from microcoding of integrated circuits to the three-line code of a high-level computer language.

Stability is an important property of both physical systems and computing schemes. One way of defining stability is in terms of the impulse response of a system. A digital system is defined to be stable if and only if its impulse response is absolutely summable:

$$\sum_{-\infty}^{\infty} |h_n| < \infty \tag{1.14}$$

Since the impulse response of a system uniquely defines that system, its properties can be used for system classification. Clearly Eq. (1.14) allows for an infinitely long impulse response. Such systems are called *IIR systems*. These IIR systems may be stable or unstable according to whether Eq. (1.14) is satisfied or not. Obviously, all finite impulse response systems, *FIR systems*, are stable.

Causality is a property that we expect of all passive systems. A system with no energy source, such as an electrical filter circuit consisting of only resistors, capacitors, and inductors, would not have an output before it is excited with an input. It is causal. For a digital system, this same idea can be expressed through the impulse response

$$h_t = 0 \quad \text{for} \quad t < 0 \quad (1.15)$$

where $t = 0$ is the time that the impulsive input occurs. Thus, the example of Fig. 1.5 is a causal system because there is no output until there is an input. This condition of Eq. (1.15) does not preclude the possibility that $h = 0$ for some positive times also, which would only imply a delayed output.

Acausal systems are not uncommon. An electronic signal generator has an output with no input (the ac power line does not count; we are only considering signal lines). In the digital signal processing world, the simple moving average is usually employed in an acausal manner. A simple example is the operator $(h_{-1}, h_0, h_1) = (\frac{1}{2}, 1, \frac{1}{2})$. This averaging operation replaces each value of a time sequence with the current value, averaged in with one past and one future value, weighted one-half as much. Because it uses future values, it is acausal.

Causality is an important consideration in many digital signal processing schemes. In real-time computing, for example, future data is not available without a delay in the computing process. In other applications, physical considerations may dictate that a certain operator be causal. Finally, in recursive computing schemes the output is used in the calculation itself; obviously, only past computed values can be made available for this type of calculation.

Invertibility of systems is a topic that occupies much of the interesting and useful concepts of digital signal processing. An LTI system is invertible if Eq. (1.12) can be solved for x in terms of y. The convolutional operator that does this is denoted h^{-1}, recovering x:

$$x = h^{-1} * y = h^{-1} * h * x$$

where $h^{-1} * h = 1$ is the unit impulse at $t = 0$. This unit impulse could occur at some time other than $t = 0$, giving rise to inverses with various delay properties. The discussion of these delay properties of inverses will be postponed until Chapter 9.

Inverse operators are important because they provide a means of undoing or *deconvolving* an operation that has been applied to a signal. For example, an input transducer for a particular measurement will have some impulse response that adulterates the measurement by a convolution (if the transducer is an LTI instrument). The effects of this transducer can be

Linear Time-Invariant Systems 17

removed by using an inverse operator—a deconvolution operator—to recover the original input signal that was applied to the transducer. Usually, the deconvolution operation is limited in accuracy, and thus it can only recover an approximate version of the original signal. This property, coupled with stability and causality considerations, provides the study of inverse operators with considerable substance.

An example of a linear operator and its exact inverse is provided by rotations about an axis in three dimensions. A vector **V** with components x, y, and z in one coordinate system has components x', y', and z' in a rotated coordinate system given by

$$\mathbf{V}' = \mathbf{R}\mathbf{V} \qquad (1.16)$$

$$\begin{pmatrix} x' \\ y' \\ z' \end{pmatrix} = \begin{pmatrix} \cos\theta & \sin\theta & 0 \\ -\sin\theta & \cos\theta & 0 \\ 0 & 0 & 1 \end{pmatrix} \begin{pmatrix} x \\ y \\ z \end{pmatrix}$$

The rotation is about the z axis by an angle θ. Note that this is one example of a linear operator that is not shift invariant. The inverse operator that recovers x, y, and z is just

$$\mathbf{R}^{-1} = \begin{pmatrix} \cos\theta & \sin\theta & 0 \\ -\sin\theta & \cos\theta & 0 \\ 0 & 0 & 1 \end{pmatrix}^{-1} = \begin{pmatrix} \cos\theta & -\sin\theta & 0 \\ \sin\theta & \cos\theta & 0 \\ 0 & 0 & 1 \end{pmatrix}$$

as can be readily verified by matrix multiplication. Systems whose inputs and outputs are related by a square matrix as in Eq. (1.16) can always be inverted exactly to recover (x, y, z) providing that the determinate of the coefficients in the matrix is not zero. On the other hand, if we wished to find the nine rotational parameters in the matrix given $\mathbf{V}'$, and $\mathbf{V}$, there are only three equations and nine unknowns; the plot thickens. A similar situation occurs in deconvolution. Imagine, for example, that we wished to invert Eq. (1.10) to find the convolutional factor represented in the matrix. If nothing is known about this factor, it represents nine unknowns [the three leading and the three trailing zeros bracketing the sequence (1,2,3) are also unknowns]. Given the six equations of y versus x, it would be impossible to solve for these nine elements. On the other hand, if somehow we knew that the zeros were indeed zero, there would be only three unknowns with six equations, producing an overconstrained set of equations. The question then would be to find the best inverse according to some criterion. Thus, we see that the subject of inverses of LTI systems is going to be an interesting one indeed.

The five properties of systems—linearity, time invariance, stability, causality, and invertibility—play a central role in our discussion of digital signal processing. The last three properties—stability, causality, and invertibility—can also be applied to sequences as well as systems. The sequence under discussion need not represent the impulse response of some system, but it could be anything, such as data. Additionally, it is straightforward to adapt the past discussion to analog systems. See Problems 1.12, 1.13, and 1.14.

Constant-Coefficient Difference Equations

Applications in many fields of science require the solutions of constant-coefficient differential equations. Frequently it is preferred to solve these equations numerically, which brings us to an important connection with digital signal processing. The numerical solutions of these constant-coefficient differential equations produces digital computing schemes that we wish to relate to our previous discussion.

In the preceding section, we made it a point to emphasize that any LTI system can be represented by the operation of just one convolution; see Fig. 1.7(a). Even though this is the most general operation possible, it is instructive to consider other possibilities. One very powerful scheme—found in computing systems, control systems, and many natural systems—is *feedback*. A system with feedback feeds the output back into the input. Such a system is also called *recursive* or *autoregressive* (AR) because in recures, or regresses, back onto itself. Figure 1.7(b) shows the simplest such system where a convolutional operator F acts on the output and adds it to the input. The system in Fig. 1.7(a) is called a *moving average* (MA)

X ⟶ [S] ⟶ Y = S ∗ X   (a)

X ⟶⊕⟶ Y = X + F ∗ Y   (b)
     ↑___[F]←___|

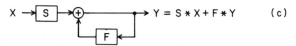

Figure 1.7 Three different digital schemes. (a) The simplest convolves the system response function with the input, called a moving average system. (b) The output is operated on by an LSI system and fed back into the input. This system is called autoregressive. (c) The above two schemes are combined into an autoregressive moving average (ARMA) scheme.

Constant-Coefficient Difference Equations

system because the convolution acting on the input is the familiar averaging operation that moves along for each input data point. These two schemes can be combined as shown in Fig. 1.7(c) to produce an *autoregressive moving average* or *ARMA system*. Now, in any computing scheme designed to be used in a real physical computer, the MA and AR convolutional sums must only have a finite number of terms:

$$y_t = \sum_{k=t-m}^{t+n} S_{t-k} x_k + \sum_{k=t-N}^{t-1} F_{t-k} y_k \qquad (1.17)$$

The MA operator has $m + n + 1$ terms and, in general, may utilize both future and past values of the input. That is, S may be acausal. On the other hand, F must be a causal operator using only previously computed values of the output. In Eq. (1.17), F has N terms operating on values of y from time $t - N$ to time $t - 1$.

A simple example of a computing scheme of the form of Eq. (1.17), that is, a digital ARMA system, is the trapezoid rule of integration:

$$y_t = y_{t-1} + \tfrac{1}{2}(x_t + x_{t-1}) \qquad (1.18)$$

This rule uses trapezoids to approximate the area under a continuous curve. In a similar spirit, we can approximate slopes of continuous curves with chords between digitized values to translate differential equations into difference equations. As an example, imagine that we wish to solve

$$\frac{dy}{dt} + 2y = x(t) \qquad (1.19)$$

At time t, we will approximate the derivative with the slope and let the sampling interval Δt be equal to unity. Thus, one difference equation that approximates Eq. (1.19) is

$$\tfrac{1}{2}(y_{t+1} - y_{t-1}) + 2y_t = x_t$$

or

$$y_{t+1} = 2x_t - 4y_t + y_{t-1}$$

Because the system is time invariant, this equation likewise must be true at any other time, such as $t = t - 1$, allowing us to rewrite the equation as

$$y_t = 2x_{t-1} - 4y_{t-1} + y_{t-2} \qquad (1.20)$$

This equation represents a digital ARMA computational system, having the form of Eq. (1.17).

Several major points can now be stressed. (1) By using simple rules, integrals and derivatives in any constant-coefficient differential-integral equation can be approximated by difference operators, producing a digital

ARMA system. (2) Such difference equations will be useful in solving initial value problems where the starting values of y are given. (3) The steady-state solutions are also useful for implementing many analog devices in digital fashion. We have only begun our study of digitizing analog systems; there is much more to be said on topics such as approximating derivatives that will be addressed in the following chapters.

Last, what about the contention that any LSI system can be represented by just one convolutional sum? Does the ARMA system contradict this? To see the answer, let us write the ARMA system of Eq. (1.17) in simpler notation:

$$y = S*x + F*y \qquad (1.21)$$

and solve for y in terms of x:

$$(1 - F)*y = S*x$$
$$y = (1 - F)^{-1} * S * x \qquad (1.22)$$
$$y = E*x$$

So, providing that $(1 - F)^{-1}$ exists, there is an equivalent MA operator E that does relate the output to the input by just one convolution. As we shall see in the next chapter, this inverse can always be found. However, under some circumstances, it can exhibit very nasty behavior. Although F has a finite number of terms, the required inverse will always have an infinite number. If these terms rapidly converge, a truncated version of the inverse can be useful. Additionally, these terms may have strong acausal components, requiring many future values of x. Using future values of x is not a limitation in many types of off-line data processing where all the input data x may be available, for example, stored on magnetic computer media. But the requirement for a large number of future values of the input can be disastrous in some real-time applications where a large delay in computing the output would be intolerable.

To summarize, any large number of convolutional operations with both AR and MA components can be reduced to one AR and one MA operator as expressed in Eq. (1.21). Furthermore, because the convolution is the most general operation of an LSI system, the ARMA model can be performed, in principle, with one MA operator. Thus computationally there is always a choice; if there is an AR component in the computational scheme, it can be performed in that *AR mode* or it can be performed in the *MA mode* by using an inverse operator. In either case the AR operator has an infinite impulse response and hence is always an IIR system. Clearly, the AR system has great computing power; in effect, it has the influence of an infinite number of terms by using only a finite number of feedback coefficients.

System Block Diagrams and Flow Graphs

Because systems, in general, can be very complicated, some means of analyzing them is required. The digital ARMA system diagrammed in Fig. 1.7(c) is quite simple, containing but one MA operator and one AR operator. However, many systems contain numerous components including analog devices; digital devices or operators; and signal paths, including feedback loops. However complicated, we know that any system containing only LTI components is equivalent to the simplest system in Fig. 1.7(a), containing just one MA operation. Frequently, we would like to determine that equivalent single operator. Analyzing block diagrams can answer this question as well as other questions relating to various operators and subsystems.

For an example of this concept look at Fig. 1.8, where four LTI subsystems are combined to relate one output to one input. By simply labeling the signal on each line in terms of the most recent upstream operator, we can write the output as

$$y = G * [S * (x + G * F * y) + F * y]$$

$$y = \left(\frac{G * S}{1 - G * S * G * F - G * F} \right) * x$$

This last operator in parentheses is the equivalent one that relates the output to the input by a single MA convolution. This equivalent operator is called the system transfer function because it provides the overall relationship between the input and the output; it transfers the input to the output. Clearly, the system transfer function is essentially the same as the system's impulse response; when x is the unit impulse, y is equal to the system impulse response. Besides computing system transfer functions, block diagrams also can be useful for increasing our understanding of systems; for example, Problem 1.21 asks you to show the equivalence of two different feedback schemes. Similar block diagrams are also used for analog functions and devices in their analysis.

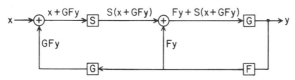

Figure 1.8 A system containing four LSI operators. By labeling signals on each line, it is possible to write an equation for the output that can then be solved for the overall system transfer function.

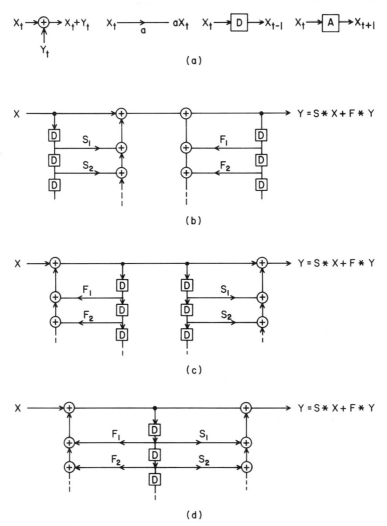

Figure 1.9 Flow graphs of digital systems. (a) The four basic digital operations of addition (or subtraction), multiplication by a constant coefficient, delay, and advance. (b) The direct form I realization of an ARMA system. (c) The system redrawn by interchanging the MA and AR portions, and (d) using common delays to form the direct form II realization (sometimes called the canonical form).

The use of diagrams can be extended down one lower level to display the fundamental operations used in convolutions: shifting, multiplication, and addition. Figure 1.9(a) shows symbols for these operations that enjoy somewhat standard use, and Fig. 1.9(b) shows how they are used to represent the ARMA model of Fig. 1.7(c).

Diagrams of these elemental operations are sometimes called *flow graphs* and can be exceedingly useful in designing computational algorithms for use in digital signal processing. As an illustration, Problem 1.21 states that the MA and the AR components of Fig. 1.9 can be switched in order, resulting in the flow graph of Fig. 1.9(c). Now, the delay elements are used in common for both the MA and AR components, which are sometimes called *feedforward* and *feedback* operations because of the direction of signal flow through them. With this common usage, one whole set of delay elements can be eliminated as shown in Fig. 1.9(d), resulting in quite a saving. Depending on the application, this saving could be significant. If the algorithm were implemented in hardware, perhaps burned into microchips, a whole shift register would be eliminated; or, if the computation were done with software, perhaps in a high-level language on a general purpose digital computer, memory would be economized.

Using concepts from topology and matrix algebra, the subject of flow graphs can be cast into well-known mathematical forms. The version in Fig. 1.9(b) is called the *direct form I* realization of the computing algorithm in question, while Fig. 1.9(d) is called the *direct form II* realization or the *canonical form*. Principles from conventional network theory such as Tellegen's theorem, are used to develop a framework for studying the topology of these flow graphs for more complicated systems than we are interested in here. See Oppenheim and Schafer (1975) for more details on the subject of flow-graph theory.

Thus ends our introductory chapter. We have discussed fundamental properties of digital systems. One of these—aliasing—arises because of the sampling in equally spaced time intervals. It is unique to sampled systems, and as we shall continually see, it is one of their most important characteristics. The limitation of linearity and time invariance dictates that the most general operation possible for an LTI system is convolution. Using simple approximations, constant-coefficient integral-differential equations can be converted to difference equations, providing a computational algorithm using convolutions with both the input and the previously computed outputs. These MA and AR processes can be represented in block diagrams and reduced to other equivalent forms. In particular, using inverses allows an AR process to be implemented in the MA mode. Clearly, the reverse is also true; an MA process can be implemented in the AR mode, taking advantage of the power of only a few autoregressive coefficients. There are prices to be paid, however, for this convenience—prices that will become more clear in following discussions. The block diagram concept can be carried down one further level to the elemental operations used in the convolution, providing insights and efficiencies of system design. Many of the concepts discussed apply to analog systems as well and will be familiar to readers that have studied continuous system theory. The good news is that discrete system theory, in many ways, is simpler than analog theory. The powerful mathematical machinery of the next chapter exploits this simplicity.

Problems

1.1 We want to digitize high-fidelity program material that spans a frequency range from 20 Hz to 20,000 Hz. What should our minimum sampling frequency be to avoid aliasing?

1.2 A certain A/D converter samples at a 10 kHz rate. List the frequencies of at least four different input sinusoids that will produce a sinusoidal output from this A/D converter at 1 kHz.

1.3 Write a graphics computer program to display the results of digitizing a series of nine sinusoids with equally spaced frequencies from zero to, and including, the sampling frequency.

1.4 Write a computer program that calculates convolutions and displays the results graphically. Experiment with a variety of sequences, such as boxcars, triangular, single spike, two spikes, etc.

1.5 Write a matrix equation showing the convolution of $(1, 3, 2, -1)$ with $(i, -1, 1, 2)$ as a matrix multiplication.

1.6 Give three practical examples of the convolution operation.

1.7 Prove that convolutions obey the associative rule $A * (B * C) = (A * B) * C$.

1.8 Prove the distributive addition rule $A * (B + C) = A * B + A * C$.

1.9 By finding a counter example, show that the operation of point-by-point multiplication of two time sequences and the convolution operation are not associative [i.e., $A(B * C) \neq (AB) * (C)$].

1.10 Give an example of linear operators that are not time invariant.

1.11 In Eq. (1.16), the transformation is given for components of a three-dimensional vector under a rotation by an angle θ about the z axis, $\mathbf{V}' = \mathbf{R}\mathbf{V}$. Show that the inverse operator $\mathbf{R}^{-1}$ given in that discussion recovers the original components via $\mathbf{V} = \mathbf{R}^{-1}\mathbf{V}'$. Furthermore, construct a matrix that describes a rotation about the X axis by an angle ϕ, and show that it does not commute with $\mathbf{R}$ [i.e., $\mathbf{R}(\phi)\mathbf{R}(\theta) \neq \mathbf{R}(\theta)\mathbf{R}(\phi)$].

1.12 By analogy, extend the five concepts of linearity, time invariance, stability, causality, and invertibility to continuous systems. Assume that the linear operation is integration, and convince yourself that the operation corresponding to the discrete convolutional sum is the convolutional integral

$$y(t) = \int_{-\infty}^{\infty} x(\tau)h(t - \tau)\,d\tau$$

1.13 The convolution integral of Problem 1.12 computes the output from a linear shift-invariant effect of inputs due to a continuously infinite application of unit impulses. The same can be done using a superposition of unit step function responses (the unit step function is unity for all $\tau > 0$ and zero for $\tau < 0$). Using this approach, derive Duhamel's integral:

$$y(t) = x(0+)U(t) + \int_{0+}^{t} x'(\tau) U(t - \tau) d\tau$$

where U is the response of the system due to a unit step function.

1.14 Show that

$$\int_{-\infty}^{\infty} f(x)[g(x) * h(x)] \, dx = \int_{-\infty}^{\infty} g(x)[(f(x) * h(-x)] \, dx$$

and that

$$\int_{-\infty}^{\infty} g(x)[f(x) * h(-x)] \, dx = \int_{-\infty}^{\infty} h(x)[f(x) * g(-x)] \, dx$$

Finally, cast the above two results into the appropriate form for discrete variables.

1.15 The derivative can be approximated by the 1st forward difference operator

$$\frac{dx}{dt} = \frac{x_{t+1} - x_t}{\Delta t}$$

By using this 1st forward difference operator twice in succession, find the 2nd forward difference operator.

1.16 Use the 2nd forward difference operator from Problem 1.15 to write a difference equation for the differential equation

$$\frac{d^2 y}{dt^2} + 3y = x(t)$$

1.17 Draw a flow graph for implementing the difference equation in Problem 1.16.
1.18 Draw a flow graph of an ARMA filter.
1.19 Draw a flow graph for the equation

$$y_t = 2y_{t-1} + 5y_{t-2} - 3x_t + 4x_{t-1}$$

1.20 Find the system transfer function for the system shown here.

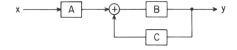

1.21 Show that these two ARMA systems have the same transfer function.

1.22 Following the procedure used in Problem 1.15, compute the 4th forward difference operator.

1.23 Any interpolation scheme must make some assumption about the missing data. If we assume that the data behaves locally like a 3rd-degree polynomial, then the 4th difference operator should annihilate the data at every point. Use the 4th forward difference to find an operator that interpolates the center point from its four neighboring points, two on each side.

1.24 Give an example of an unstable sequence.

1.25 Give an example of a stable causal sequence.

1.26 We would like to make a digital recording of a 3-minute-long stereo music performance. Assume that frequencies exist up to 20 kHz but there are none above that limit. We would like a 96-decibel(db) dynamic range (i.e., each sample will require two 8-bit bytes). What is the minimum sampling rate that we could use? How many bytes of storage will be required to record the program at this minimum sampling rate?

1.27 By expanding $y(x_i + h)$ and $y(x_i - h)$ in a power series about x_i, show that $\Delta^2 y = (y_{i+1} - 2y_i + y_{i-1})/h^2$ does equal the second derivative of y at (x_i) plus correction terms of order of h^2.

2

Sampled Data and the Z Transform

Every piece of mathematical machinery must be weighted for the power and insight it produces versus the overhead required to grease and turn its wheels through a given problem area. For studying digital signals and systems, the Z transform offers an outstanding price/performance ratio. This amazingly simple transform, with its intimate connection with the all-important convolution operation, requires only a slight mathematical investment in return for a powerful insight into digital, linear shift-invariant systems. Furthermore, it turns out that the Z transform is a generalization of the Fourier transform of periodic signals. In this chapter, we will first introduce the Z transform and then exploit it in the discussion of the properties of systems and sequencies.

The Z Transform, Polynomial Multiplication, and Convolution

In the previous chapter, we have represented sampled data by writing the values, corresponding to the magnitude of the signal at each time, as an array of numbers. The sequence

$$x(t) = (1, 3, 0, -1, -5)$$

for example, has a value of one at $t = 0$, a value of three at the second time increment, and so on. This sequence may represent a stream of data or the impulse response of an LSI system. The sequence may be infinitely

27

long, as is the case for the impulse response of an IIR system, but for our opening discussion we have a finite sequence in mind, such as the data gathered from any real experiment.

Since the limitation of equally spaced data is fundamental to our discussion, we have introduced in the previous chapter the idea of delay and advance operators. These operators are an important component both of the convolution operation and of computing hardware implemented by shift registers. For this reason, it will behoove us to cast these delay and advance operators into a mathematical scheme. As we will see shortly, by introducing a variable called Z, with its powers corresponding to the number of delays or advances, an attractive picture emerges. That is, our example above would be written

$$X(Z) = 1 + 3Z - Z^3 - 5Z^4$$

This polynomial in Z is called the Z *transform* of the sequence $x(t)$. Each power of Z produces one unit of delay. The last term is delayed four time increments from the first term at $t = 0$. Advances are represented by inverse powers of Z. If our example sequence were advanced one time increment, the resulting new sequence would become

$$Z^{-1}X(Z) = Z^{-1} + 3 - Z^2 - 5Z^3$$

We must immediately pause to explain that, unfortunately, there are two conventions used for the Z transform. The one we are using was introduced by Laplace and is generally used by physicists and other scientists. Engineers commonly use the negative powers of Z for the delay operators and positive powers for the advance operators. Using Laplace's convention results in writing fewer negative powers because advance operators are in the minority. Switching conventions is clearly a trivial but annoying matter, resulting in sort of an occupational hazard when reading the literature using the opposite definition.

One of the chief properties of Z transforms is that their multiplication is equivalent to the convolution of their corresponding time sequences. This is so because polynomial multiplication automatically collects contributions from factors having a common delay (i.e., the same powers of Z). To demonstrate, consider that

$$x_1 = (1, 5)$$
$$x_2 = (1, 2, 3)$$

so that

$$x_1 * x_2 = (1, 7, 13, 15)$$

while the Z transforms of the two sequences are
$$X_1(Z) = 1 + 5Z$$
$$X_2(Z) = 1 + 2Z + 3Z^2$$
giving for their product
$$X_1(Z)X_2(Z) = 1 + 7Z + 13Z^2 + 15Z^3$$
which is the Z transform of the convolution of $x_1(t)$ with $x_2(t)$. We call this equivalence of convolution in the time domain with multiplication in the Z domain the *convolution theorem*. It affords two ways of looking at the same operation, which we will see is extremely valuable. For example, we can now use known properties of polynomials to study LSI systems and sampled time data.

Factoring Z Transforms into Couplets

Perhaps the most important property of a polynomial is stated in the fundamental theorem of algebra: Any polynomial of degree n can be factored into exactly n factors. The importance of this fact, for LSI systems, is that any FIR system can be broken down into a series of simpler systems. For example, if we were studying a system with an impulse response
$$h(t) = (2, -1, -1)$$
whose Z transform is
$$H(Z) = 2 - Z - Z^2$$
the result of this system acting on a sampled data sequence could be accomplished in an equivalent fashion by factoring the polynomial:
$$H(Z) = 2 - Z - Z^2 = (2 + Z)(1 - Z)$$
Since multiplication of these two factors corresponds to convolution in the time domain, these two factors lead to a series action of two, 2-term, impulse response systems as shown in Fig. 2.1. Thus, any sequence with an nth-degree Z transform may be written as a product of n factors, which we shall call *couplets*:
$$X(Z) = a_0 + a_1 Z + \cdots + a_n Z^n = a_n(Z - r_1)(Z - r_2) \cdots (Z - r_n)$$
displaying the n roots, r_i, some of which may be complex. Naturally, in most cases, both a system's impulse response and a data sequence will be real. In this situation, the roots of the polynomial in Z will occur in

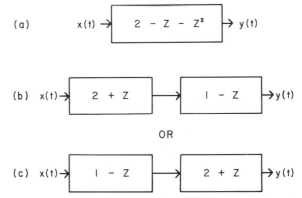

Figure 2.1 The system with a 3-term response function in (a) can be viewed equivalently as two systems in series, each with a 2-term system response function, as shown in (b) and (c). Since both convolution and multiplication commute, the order is unimportant.

complex pairs. However, remember from our examples in Chapter 1 that there are some cases where we would find a complex data sequence to be convenient. Now we see that the decomposition of a system into its 2-term components will, in general, also lead to complex impulse response functions. Such systems have complex outputs for real inputs.

Inverse Operators: Stability, Causality, and Minimum Phase

In the previous chapter, we have seen how important inverses are to "undoing" the effects of an LSI operator and how inverses play a critical role in recursive filters and constant-coefficient difference equations. Recall that the meaning of the inverse of an LSI operator is an operator such that

$$A(t) * A(t)^{-1} = \delta$$

where δ is the spike sequence; it has all zeros except at $t = 0$, where the value is unity. The Z transform of the above equation is simply

$$A(Z)A(Z)^{-1} = 1$$

because of the convolution theorem. Thus, the inverse operator is just

$$A^{-1}(Z) = \frac{1}{A(Z)}$$

So, in the Z domain, writing the inverse of an LSI operator is child's play. But, there are complications. Since the Z transform of any finite sequence

Inverse Operators

can be written in terms of couplets,

$$A^{-1}(Z) = \frac{1}{a_n(Z - r_1)(Z - r_2) \cdots (Z - r_n)}$$

it seems logical to start our study of inverses by looking at the inverse of a single couplet. Then, we can expand the discussion to include Z transforms of any degree.

Instead of writing $(Z - r)$ for the couplet, we will use $(1 - aZ)$, which only differs by an overall constant. Thus, we are led to consider the inverse

$$(1 - aZ)^{-1} = \frac{1}{1 - aZ}$$

What is the meaning of such an inverse Z transform? Specifically, how is it implemented in a computational scheme? In our discussion on flow graphs we saw one possibility: feedback, that is, recursive schemes involved inverses. The other approach, which we wish to pursue now, is the conversion of this inverse to operations requiring only the input data, avoiding the use of the system's output. This is possible by virtue of the expansion

$$\frac{1}{1 - aZ} = 1 + aZ + (aZ)^2 + (aZ)^3 + \cdots \qquad (2.1)$$

which is called the geometric series. It converges absolutely for all aZ such that

$$|aZ| < 1$$

This is a fundamental power series studied in complex analysis; for our purpose we take the result on faith. It follows that aZ is a variable that may take on complex values. Previously, we observed that, in general, we would expect the coefficient a to be some complex number that arises from the roots of a polynomial. This variable Z, which was originally designed only to represent delays and advances via its exponent, now demands more serious attention: in order to locate the roots of polynomials in Z, we must allow Z to take on complex values.

Returning now to the geometric series expansion, we see that the expression for the inverse will converge on the unit circle if

$$|aZ| < 1 \quad \text{or} \quad |a| < 1$$

Under this condition, the inverse can be implemented in a nonrecursive manner, using only the input data. The price paid for avoiding feedback is the requirement for an infinite number of coefficients: an FIR sequence has an IIR inverse. In practice, the terms are truncated when they have diminished to an acceptable level.

Before proceeding, we will belabor the convergence requirement of Eq. (2.1) a bit further. The couplet $(1 - aZ)$ has a zero at $Z = Z_0 = 1/a$. So, we can think of the convergence requirement in three ways: (1) $|a| < 1$, that is, (2) the couplet must have the magnitude of the second term less than the magnitude of its first term or (3) the couplet must have its zero outside of the unit circle in the Z plane

$$\left(Z_0 = \frac{1}{a}, \quad |Z_0| > 1\right)$$

Such a couplet is called a *minimum phase couplet*. A sequence whose every factor is a minimum phase couplet is a *minimum phase sequence*.

A couplet whose first coefficient has less magnitude than its second is called a *maximum phase couplet*; its zeros are inside the unit circle. A sequence composed of all maximum phase couplets is a *maximum phase sequence*. Sequences containing both minimum and maximum phase couplets are called *mixed phase sequences*. A couplet whose zero lies on the unit circle is a special case, being neither minimum nor maximum phase. Later, when we study the frequency response of couplets, these names will seem more appropriate.

Let us now look at the maximum phase couplet that causes the geometric series of Eq. (2.1) to diverge. Does this mean that it does not have a stable inverse? Actually, it is possible to form a stable inverse by writing

$$\frac{1}{1 - aZ} = -\frac{1}{aZ}\left[\frac{1}{1 - (1/aZ)}\right]$$

$$= -\frac{1}{aZ}[1 + (aZ)^{-1} + (aZ)^{-2} + (aZ)^{-3} + \cdots]$$

where we have used convergence of the geometric series in powers of $1/aZ$ for

$$|aZ| > 1 \quad \text{or} \quad \frac{1}{|aZ|} < 1$$

Accordingly, the maximum phase couplet has a perfectly stable IIR inverse, but note that it requires future values of the input. This is completely unacceptable in some real-time computing situations. For this reason, such an inverse is called physically unrealizable. This term, however, is somewhat misleading because in other applications all the data may be available, such as on a magnetic tape record, and the inverse can be readily applied to future values. Even in the real-time computing environment, the nonrecursive maximum phase inverse may still be useable if a delay in output is tolerable until the series converges to an acceptable level.

Inverse Operators

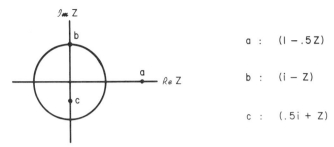

Figure 2.2 An example of three couplets: a has its zero outside of the unit circle (minimum phase), b has its zero right on the unit circle, and c has its zero inside the unit circle (maximum phase). The real part of Z is plotted on the x axis and its imaginary part is plotted on the y axis.

So we have found that, in general, $1/(1 - aZ)$ may always be expanded into powers of Z in two ways, into a convergent series or a divergent one. The value of $|a|$ determines which powers of Z will belong to which series. If $|a| < 1$, we can expand $1/(1 - aZ)$ into either a convergent series in positive powers of Z or a divergent one in negative powers. If $|a| > 1$, the reverse is true. The coefficients of these series represent the time domain terms, that is, the inverse Z transform. Because there are always two choices for this expansion, we see that the inverse Z transform is not unique. More often than not we are interested in the convergent, or stable, Z transforms. Figure 2.2 shows examples of three different couplets. The minimum phase couplet has a stable *causal* inverse and the maximum phase couplet has a stable *acausal* (i.e., requires anticipatory data) inverse.

Next, we need to expand the discussion to include Z-transform polynomials of degree greater than one. We proceed by example. Consider the sequence $(4, 0, -1)$ with Z transform:

$$X(Z) = 4 - Z^2$$

It has an inverse

$$(4 - Z^2)^{-1} = \frac{1}{4 - Z^2} = \frac{1}{(2 + Z)(2 - Z)} = \frac{1/4}{2 + Z} + \frac{1/4}{2 - Z}$$

The last step is called a partial fraction expansion, which you can verify by adding the last two fractions together. This partial fraction technique allows the geometric series to be used for each inverse couplet. Summing the results for each fraction then gives the expansion of the inverse:

$$(4 - Z^2)^{-1} = \frac{1}{8} \left[2 + \frac{Z^2}{2} + \frac{Z^4}{8} + \frac{Z^6}{32} + \cdots \right]$$

This partial fraction technique can be applied, in general, by writing unknowns for the numerators of each couplet and solving the resulting equations by standard techniques discussed in algebra books. Likewise, another straightforward method of computing the inverse is to use polynomial division before factoring into couplets. Special treatment is required for multiple poles. Luckily, our concern here is not to develop proficiency in these algebraic gymnastics, but rather we wish to arrive at some important conclusions concerning properties of sequences. Note that the sequence discussed above, $(4, 0, -1)$, factors into two minimum phase couplets. By the partial fraction form of the inverse, we see that the stable inverse of any such minimum phase polynomial will only contain positive powers of Z, that is, the inverse will be causal. On the other hand, if the sequence is mixed phase, couplets will occur in the partial fraction sum with both minimum and maximum phase, generating both positive and negative powers of Z. Since these terms require anticipatory data, it is acausal: only the pure minimum phase combination (the minimum phase polynomial) has a stable, causal inverse.

Now, we are in a position to consider the most general form of a rational, stable sequence. Its Z transform may be written as a rational fraction of two causal polynomials. Within an overall scale factor, it appears in factored form as

$$S = \frac{(Z - a_1)(Z - a_2) \cdots (Z - a_n)}{(Z - b_1)(Z - b_2) \cdots (Z - b_m)} \tag{2.2}$$

In general, S may have a mixed phase denominator. Stability is assured by expanding the appropriate terms infinitely far into the future and into the past. Thus, with no further restrictions, S represents the most general stable sequence possible. Now, we pursue a classification scheme. If the denominator is pure minimum phase, S is causal and stable. If the numerator is also minimum phase, then we define S to be a minimum phase sequence. This minimum phase sequence will clearly have a causal, stable inverse. This classification scheme is summarized in Fig. 2.3. It is worth emphasizing that not all causal, stable sequences are minimum phase. If S has a minimum phase denominator, but not a minimum phase numerator, S is causal and stable but not minimum phase.

Zeros in the denominator of a fraction cause the fraction to "blow up," that is, to go to infinity. These zeros are called poles of the fraction. They give an alternate succinct statement for the minimum phase condition: a causal stable sequence is minimum phase if its Z transform has no poles or zeros inside the unit circle. This is simply a slightly fancier way of saying the Z transform has no zeros of either its numerator or its denominator inside of the unit circle.

Inverse Operators

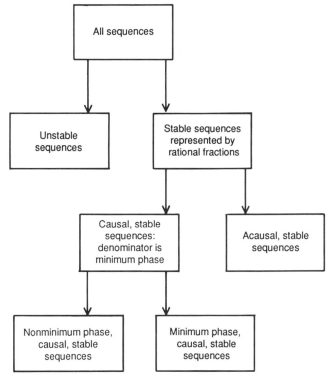

Figure 2.3 A classification of sequences and systems. The minimum phase sequence is a subclass of causal, stable sequences.

The inverse of $S(Z)$ above is

$$S^{-1} = \frac{(Z-b_1)(Z-b_2)\cdots(Z-b_m)}{(Z-a_1)(Z-a_2)\cdots(Z-a_n)} \qquad (2.3)$$

Clearly, if S is minimum phase, so is its inverse. Likewise, S^{-1} is causal and stable. Furthermore, it is only the minimum phase subclass of causal stable sequences that have causal stable inverses. This very important property can be seen easily by comparing Eq. (2.2) and Eq. (2.3): if S^{-1} is to be causal, its denominator, and hence the numerator of S, must be minimum phase. Since the denominator of S is already minimum phase (because S is causal), S itself must be minimum phase.

Why have we bothered to define this minimum phase condition? The answer has two parts: the first we will present now, the second will have to wait until we discuss spectra and Fourier transforms. The first important aspect of the minimum phase condition is its relationship to real, physical systems. Consider an example of a one-port, passive, linear, electrical

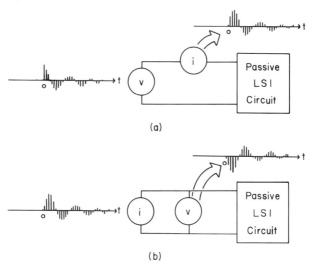

Figure 2.4 A digitized passive LSI electrical circuit driven in (a) by a voltage and in (b) by a current source. In the first case, the system output is the measured current into the circuit. In the second case, the output is the measured voltage across the circuit.

network. We can choose to excite this network with a voltage source and measure the resulting current, or we can drive it with a current source and measure the resulting voltage. The two situations are depicted in Fig. 2.4. In both cases, the system is causal, that is, no response occurs until there is an excitation present. Normally, you may think of continuous signals in an electrical example like this, but we are primarily interested in sampled data here. Therefore, for the first case, when the excitation is a voltage source, we relate the output of the circuit i to the input by writing

$$i(Z) = v(Z)y(Z)$$

That is, because the system is linear and shift invariant, the output i is related to the input v by a convolution. In the Z domain, the convolution becomes multiplication by the Z transform of y, the circuit's admittance. In the second case, when a current source is the excitation, we would likewise write for the output

$$v(Z) = i(Z)z(Z)$$

where z is the Z transform of the circuit's impedance. Clearly, since the circuit is assumed to be passive, both y and z are causal, stable sequences, and furthermore,

$$z(Z) = 1/y(Z)$$

We have seen that the only causal, stable Z transform that has a causal, stable inverse is the minimum phase one. Hence, we arrive at a significant conclusion: a certain real analog system may have a causal inverse demanded by physical considerations. In order to preserve this causal invertibility property when we represent the system digitally, the discrete version must have a minimum phase Z transform. Herein lies one important characteristic of the minimum phase condition—it limits the expected impulse behavior of many systems according to their known physical properties. Generalizing, we can say this: excite any causal LSI system in any way. If (and this is sometimes a big "if") we can argue that the relationship between the input excitation and the output signal is invertible, then the system's response function must be minimum phase.

A very unfortunate, very interesting, and very profound point in digital signal processing is that the minimum phase condition is not necessarily preserved when an analog system is written in discrete form. Furthermore, the minimum phase condition can be erroneously incorporated in the system's response function when it does not belong—one has to be careful. In Problems 2.11 and 2.12, we address a simple analog RC circuit example. Two approaches to finding the digital version of the system's response function are taken; and two different results are obtained. The relationship among the minimum phase condition, physical aspects, and stability (which includes numerical stability of computing schemes) means that the minimum phase condition is going to be an important part of our study in digital signal processing.

This short, but wise, investment in the Z transform will pay out well in the remainder of the book. This is because in any digital application of signal processing, the sequences, systems, and constant-coefficient difference equations can always be represented by a rational fraction of polynomials in Z. The Z transform, via the simple geometric series expansion, has shown us how to convert the recursive action of a denominator couplet into an equivalent, although infinitely long, moving average operator. Consequences of this expansion relate the minimum phase condition to a physical aspect—causality. In the next chapters, where we study the frequency response of sequences and systems, the Z transform will again serve as an indispensable framework for our ideas.

Problems

2.1 Write the Z transforms of the following sequences and factor them into couplets: $(2, -3, -2)$, $(2, -5, 2)$, $(1, -2, 4)$, and $(2, -2, 1)$. Which are minimum phase sequences?

2.2 Express the inverse of $(1, -2)$ as a stable sequence in powers of Z.

2.3 Classify the time series $(4, 0, -1)$ and then find its inverse.

2.4 Show that multiplication by $(1 - Z)$ is analogous to differentiation in continuous time. Show that dividing by $(1 - Z)$ likewise corresponds to integration. What are the limits of this integration?

2.5 A recursive LSI system representing the trapezoid rule of integration can be written

$$y_t = y_{t-1} + \tfrac{1}{2}[x_t + x_{t-1}]$$

Find the Z transform of the impulse response function of this system. Classify this system according to the location of its poles and zeros in the Z plane.

2.6 A certain nonrecursive LSI system, describing a kind of differentiation, can be written

$$y_t = \tfrac{1}{2}[x_{t+1} - x_{t-1}]$$

Find its impulse response function in the Z domain and classify it according to its poles and zeros.

2.7 Find the Z transform of the system function for the feedback arrangement shown here in terms of $F(Z)$ and $G(Z)$.

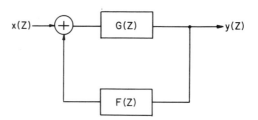

2.8 Specialize Problem 2.7 by setting $G(Z) = 1$. Under what conditions on $F(Z)$ does the system now have a stable output?

2.9 A moving vehicle receives 1-D navigation information at equal time intervals. We would like to smooth this data so as to get a better estimate of the vehicle's true position by using a time domain filter.

received data $x(t) \longrightarrow \boxed{F} \longrightarrow y(t)$ better estimate of true position at time t

We design a recursive filter by reasoning

$$y_t = y_{t-1} + v_{t-1}\Delta t$$

That is, at time t, our improved estimate of position [over simply using the current received information x_t] is the previous estimate

Problems

y_{t-1} plus the velocity from $t-1$ to t multiplied by the interval Δt. Estimating

$$v_{t-1} = (x_t - y_{t-2})/2 \Delta t$$

gives us the ARMA filter

$$y_t = y_{t-1} - \tfrac{1}{2}y_{t-2} + \tfrac{1}{2}x_t$$

(a) Show that the above ARMA filter does follow from the reasoning given.

(b) Write down the Z transform of this filter function and classify this digital processing scheme according to the poles and zeros of $F(Z)$.

2.10 Exponentials occur frequently as continuous functions of time in many applications, such as the simple decay of current into a resistive load from a charged capacitor. Sample the continuous exponential

$$x(t) = \begin{cases} 0 & t < 0 \\ e^{-t} & t > 0 \end{cases}$$

at equal time intervals Δ to get

$$x_n = (e^{-\Delta})^n$$

and use the geometric series to write the Z transform of $x(\Delta n)$.

2.11 The simple RC circuit shown below is excited with a decaying exponential voltage source. The output voltage is the difference of two exponentials as shown. Compute the Z transform of the input voltage, the output voltage, and the transfer function of the circuit. Is the transfer function causal and stable? Is the transfer function minimum phase and, for that matter, would we expect it to be?

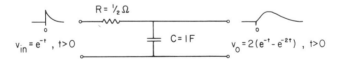

2.12 Analog theory tells us that the impulse response of the RC circuit in Problem 2.11 is

$$h(t) = e^{-2t}$$

Digitize this impulse response directly, using the same method of Problem 2.11, showing that this procedure gives a discrete version $H(Z)$ with minimum phase. Which result is better? Is there a correct result?

2.13 Find the inverse of $A = (1 - aZ)$ by expanding it in the geometric series for $|a| < 1$. Then, write a computer program that convolves A with a truncated version of A^{-1}. Compute $A * A^{-1}$ for various values of a and for various lengths of A^{-1}.

2.14 Repeat Problem 2.13 for values of $|a| > 1$.

2.15 Another way of digitizing an analog system is called step invariance. Even though the Z transform of the step function does not converge, we can still write

$$1/(1 - Z) = 1 + Z + Z^2 + \cdots$$

for the digital form of the step function. Given that a certain analog system has a step response of

$$y(t) = 1 - e^{-t/\tau}$$

digitize the system by computing the ratio of the digitized step function response to the digitized step function input. Where are the poles and zeros of the resulting $H(Z)$?

2.16 Show that the in-place, time-reversed version of an $(n + 1)$-long sequence $A(Z)$ is $Z^n A(1/Z)$.

2.17 Find the Z transform of $\sin \omega t$ and $\cos \omega t$. Comment on the location of their poles and zeros.

2.18 Find the Z transform of $f(t) = e^{-at} \sin \omega t$ for $t > 0$ where $f(t) = 0$ for $t < 0$. Comment on the location of its poles and zeros.

2.19 Sometimes the method of finite differences can be applied to nonlinear differential equations in a straightforward manner. Convert the nonlinear differential equation below to a difference equation using the forward, backward, and central difference operators. What happens to our pole–zero stability analysis now?

$$\frac{dy}{dt} + ay^2 = x(t)$$

3
Sinusoidal Response of LSI Systems

All students of the physical sciences learn very early in their studies that sinusoidal functions play a special role in what may seem like the natural scheme of things. Yet, no one has ever seen an actual sine wave; its infinite length precludes the possibility. Of course, we may view a portion of a sine wave, but that is not a pure sine wave; it is too short. Nonetheless, the concept of sinusoidal waves, including their impossible infinite length, gives us valuable insight to certain problems. This insight comes from the fact that sinusoids are the eigenfunctions of LSI systems. Everyone who has studied electrical circuits knows that any arrangement of ideal linear components such as resistors, capacitors, and inductors is an LSI system. In practice, many of these real components approximate an ideal LSI system quite well. If such a system is excited by a sinusoidal source of voltage or current, all resulting voltages and currents vary in a sinusoidal fashion. The frequency of the response is the same as that of the excitation, leaving only the magnitudes and phases as variables for study. This chapter begins our study of this response of LSI systems under sinusoidal excitation; advantages and limitations accompanying this approach will occupy most of the remainder of the book.

The fact that the excitation must be turned on and off means that the source is not a pure sine wave. In some cases, this limitation is satisfactorily dealt with by making measurements after the transient response has died to an acceptable level. On the other hand, the resulting finite data stream creates severe problems in many important cases. This situation applies equally to digital LSI systems. Finite data and operator length will lie at the root of our concern in our study of digital filters, inverse filtering, and spectral estimation.

The use of sinusoids in describing LSI systems will lead to a rich and beautiful picture consisting of complementary views in time and frequency called Fourier, or spectral, analysis. The development of this technique has a long history dating back to the Babylonians who used similar ideas in their study of astronomical events. Modern treatment of the subject began with L. Euler in 1748 and was applied to heat diffusion problems by Joseph Fourier in 1807, but was not accepted by the scientific community until Fourier's 1822 publication—15 years later. In 1829, P. L. Dirichlet provided the convergence conditions for Fourier series that placed the theory on a firm mathematic basis. Our rather pragmatic treatment of this subject, extending over five chapters, will concentrate on those aspects central to digital signal processing. In this short chapter, we introduce the concept of eigenfunctions with the frequency responses of LSI systems as their eigenvalues. Then, we study the frequency response of some important digital LSI operators.

Sinusoidal Signals as Eigenfunctions and the Spectrum as Eigenvalues

Eigenfunctions are special functions associated with a given type system (or operator) such that, except for a possible scale factor, the output of the system has the same functional form as the input. The scale factor is called the eigenvalue. For example, the derivative operator has exponential functions for their eigenfunctions because

$$\frac{d}{dx} e^{-\alpha x} = -\alpha e^{-\alpha x}$$

with the decay constant α as the eigenvalue. (The word "eigen" comes from German meaning *self*.) Now, we show that digitized sinusoidal functions are eigenfunctions of digital LSI systems.

We have seen in the last chapter that both FIR and IIR digital systems can be implemented in the moving average mode. The output of the most general LSI system is thus obtained by convolving its impulse response function h with the system's input x:

$$y_n = \sum_k x_k h_{n-k} = \sum_k x_{n-k} h_k$$

where the sum extends to infinity for the IIR system. If the system is excited by a sampled sinusoidal input

$$x_n = e^{i\omega n}$$

Sinusoidal Signals as Eigenfunctions

the output becomes

$$y_n = \sum_k e^{i\omega(n-k)} h_k$$

$$= e^{i\omega n} \sum_k e^{-i\omega k} h_k$$

This result shows that for a sampled sinusoidal input, the output is likewise a sampled sinusoidal function modified only by the factor

$$H(\omega) = \sum_k e^{-i\omega k} h_k \qquad (3.1)$$

That is to say, sampled sinusoids are eigenfunctions of digital LSI systems and $H(\omega)$ is the eigenvalue. In general, $H(\omega)$ may be complex, which means the output sampled sinusoid may be phase shifted from the input. If H turns out to be real in a particular instance, the output is in phase with the input. Because the eigenvalue $H(\omega)$ is a function of frequency, it is called the system's *frequency response, spectral response*, or simply *spectrum*. Two fundamental properties of the frequency response must be immediately emphasized from Eq. (3.1): $H(\omega)$ is a continuous function of frequency and it is completely determined by the system's impulse response function. Furthermore, in Chapter 7, we will see that Eq. (3.1) can be uniquely inverted to give the impulse response directly from the frequency response; the system can be completely specified by either. These are the two pictures promised in the introduction: any LSI system (or sequence) can be completely described by its time representation h_k or equivalently by its frequency representation $H(\omega)$.

Before we continue with examples, we pause to show the relationship between the frequency response and the Z transform of h_k. By writing Eq. (3.1) as

$$H(\omega) = \sum_k e^{-i\omega k} h_k = \sum_k h_k (e^{-i\omega})^k$$

and remembering that the Z transform of h is

$$H(Z) = \sum_k h_k Z^k$$

we can see that when $Z = e^{-i\omega}$, then

$$H(\omega) = H(Z) \qquad (3.2)$$

In words: the frequency response of a system is the Z transform of its impulse response evaluated on the unit circle. Like the Z transform, Eq. (3.1) is also a transform. This transform, from the time domain to the frequency domain, is one kind of *Fourier transform* that we see from Eq. (3.2) in a subset of the Z transform, occurring for the special case when

Z is on the unit circle. It transforms an equally spaced sampled signal (which may be infinitely long) into a continuous but periodic function of frequency. This transform of Eq. (3.1) is our first encounter with one of four types of Fourier transforms that we will consider in subsequent chapters. (In fact, if you imagined switching, for a moment, the roles of time and frequency, you may recognize our first Fourier transform as the Fourier series having continuous, periodic time functions and discrete, aperiodic frequency components.)

Frequency Response of Some Simple LSI Systems

Some examples will not only help to illuminate these important ideas, but will yield a significant insight to some traditional data manipulations. To some (those deprived of the frequency domain picture), a common 3-point moving average may seem like a very reasonable way to smooth data; after all, if one point is good, wouldn't the average of three be better? The 3-point operator that we are considering is

$$y_t = (\tfrac{1}{3})(x_{t-1} + x_t + x_{t+1})$$

and its Z transform is

$$Y(Z) = (\tfrac{1}{3})(Z + 1 + Z^{-1})X(Z)$$

The operator's frequency response is given by simply evaluating $Y(Z)$ on the unit circle

$$Z = e^{-i\omega}$$

or

$$H(\omega) = \frac{Y(\omega)}{X(\omega)} = (\tfrac{1}{3})(e^{-i\omega} + 1 + e^{i\omega})$$

$$H(\omega) = (\tfrac{1}{3})(1 + 2\cos\omega) \tag{3.3}$$

Now, we have a new view of the operator's action. Because the 3-point MA filter is an LSI operator, we know that when it acts on a sinusoidal signal, the result is a sinusoidal output. Equation (3.3) gives us the magnitude of that output for every ω. [The phase is zero since $H(\omega)$ is real.] A plot of $H(\omega)$ in Fig. 3.1 raises objections to this operation that are difficult to overcome. If we simply desire to smooth the data, to remove small irregularities and jitter for example, with no other objectives in mind, why would we want to treat sinusoid inputs as shown in Fig. 3.1? It seems impossible to justify, for example, completely rejecting a sine wave signal

Frequency Response of Some Simple LSI Systems 45

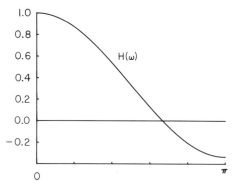

Figure 3.1 The frequency response of the equal-weighted, 3-point moving average operator. The higher frequencies are admitted with a sign inversion.

at $\omega = 2\pi/3$ while accepting sinusoids at higher frequency with reduced amplitude and, of all things, with inverted values. Intuitively, I think you will agree that a much more reasonable smoothing operator should have a spectrum as shown in Fig. 3.2, displaying a monotonically decreasing response. The higher frequencies, which contribute to the irregularities and jitter in the original data, are rejected in an unbiased fashion. The frequency domain gives us an insight in the operator's performance that would be otherwise unavailable. Simply enough, by tapering the weights of the 3-point smoothing operator to

$$y_t = (\tfrac{1}{4})(x_{t-1} + 2x_t + x_{t+1})$$

we get the monotonically decreasing frequency response of Fig. 3.2:

$$H(\omega) = (\tfrac{1}{2})(1 + \cos \omega)$$

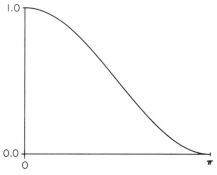

Figure 3.2 The frequency response of the tapered, 3-point operator. It monotonically rejects higher frequencies.

This achievement of the desirable frequency response via the simple modification of the MA weights suggests a tack in a direction opposite to our discussion. The desired frequency response of the operator may be known from the properties of the data reduction problem. The problem would then be the inverse of our discussion here: find the time domain operator that has a specified frequency response. The solution of this problem, thought not particularly simple, is indeed possible and is addressed in Chapter 8, Digital Filter Design.

We will continue our present course of investigating the frequency response of familiar operators. But we next pause to emphasize two technical points in the development of Eq. (3.3). First, we can see that it was the symmetry of delays and advances in the time domain operator that led to equal coefficients of the complex exponentials, making $H(\omega)$ real—symmetric operators always have real frequency responses. That is, symmetric operators will have zero phase—no delay—outputs. Second, Eq. (3.3) is valid for all values of ω, exhibiting the frequency folding or aliasing inherent in all digital signal processing. Extending the plot in Fig. 3.1 with Eq. (3.3) verifies that the frequency response is periodic and that it is folded about the Nyquist frequency $\omega = \pi$ in the same manner discussed in Chapter 1.

Returning now to the equal-weighted 3-point moving average, we note that such an operation results from fitting the best straight line through three points using a least-squares criterion. We will next use this class of symmetric smoothing operators to add to our examples of frequency response analysis. We start by seeking a straight line passing through $t = 0$ that minimizes the sum squared differences from an odd number of data points, $2N + 1$. That is, we wish to find a and b such that

$$\sum_{m=-N}^{N} [x_m - (a + bm)]^2$$

is a minimum. Many come to believe that this least-squares criterion is based on some fundamental notion. In truth, it is generally used (as here) only for the reason that it leads to simple results. Proceeding then, we minimize the above expression by setting

$$\frac{\partial}{\partial a} \sum_{m=-N}^{N} [x_m - (a + bm)]^2 = -2 \sum_{m=-N}^{N} [x_m - (a + bm)] = 0$$

or

$$\sum_{-N}^{N} x_m - \sum_{-N}^{N} a - \sum_{-N}^{N} mb = 0$$

giving

$$a = \frac{1}{2N + 1} \sum_{-N}^{N} x_m$$

Frequency Response of Some Simple LSI Systems

since $\quad \sum_{-N}^{N} a = a(2N+1) \quad$ and $\quad b \sum_{-N}^{N} m = 0$

This is the familiar result that the average is the number, among a collection of values, that minimizes the sum squared differences between those values and that number. (We do not need b, the slope of the line, for our discussion.) To smooth a data stream by replacing each point with a least-squares, straight-line fit to N points on either side, we simply replace that point with the equally weighted average

$$y_t = \frac{1}{(2N+1)} (x_{-N} + \cdots + x_{t-1} + x_0 + x_{t+1} + \cdots + x_N) \quad (3.4)$$

So, we see that the 3-point MA operator discussed earlier is just a special case of Eq. (3.4) corresponding to a straight-line, least-squares fit among three data points. If, for example, it were desired to make the fit with seven points, the appropriate operator would be

$$\tfrac{1}{7}(1,1,1,1,1,1,1)$$

The 3-point operator was sufficiently simple that we could see how to correct its disappointing frequency response by tapering the coefficients to form a new operator. Now, we can inquire what the effect of lengthening the equal-weight operator has on its frequency response. Lengthening the operator just corresponds to fitting more points of the data to a straight line. At first sight, it may seem that this smoothing of data by a least-squares fit to a straight line is a reasonable treatment. But, look at the frequency response of Eq. (3.4). Its Z transform is

$$Y(Z) = \frac{1}{(2N+1)} (Z^N + \cdots + Z^1 + 1 + Z^{-1} + \cdots + Z^{-N}) X(Z)$$

giving a frequency response of

$$H(\omega) = \frac{1}{(2N+1)} (e^{-i\omega N} + \cdots + e^{-i\omega} + 1 + e^{i\omega} + \cdots + e^{i\omega N})$$

$$H(\omega) = \frac{1}{(2N+1)} \left[1 + 2 \sum_{n=1}^{N} \cos(n\omega) \right] \quad (3.5)$$

A plot of this function is shown in Fig. 3.3 for 5, 7, and 9 points. Lengthening of the equal-weight operator does give it more smoothing action because it rejects a wider band of high-frequency components, but the behavior of the spectral response in the reject region is still quite objectionable. It is again difficult to imagine a justification for smoothing data in this fashion: why would one wish to admit, reject, and even invert sine wave inputs according to the undulations observed in the frequency response of

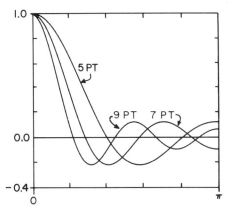

Figure 3.3 The frequency response of 5-, 7-, and 9-point equally weighted moving averages.

these operators at high frequencies? This frequency domain picture has enabled us to view the action of MA operators in a different light, a light that casts a suspicious shadow on their behavior.

Perhaps we have been too simplistic in forcing the data to fit a straight line at each data point; possibly polynomials would be more realistic. The same least-squares fitting procedure can be carried out for polynomials of any degree, and the frequency response of the corresponding operator computed. See Hamming (1983), for example. The results for second- and fourth-order polynomials, shown in Figs. 3.4 and 3.5, show no signs of improvement in the higher frequency behavior with either increasing the degree of the polynomial or increasing the number of data points included in the fit. It seems that modeling our data by polynomials will never give

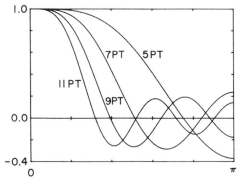

Figure 3.4 Frequency response of smoothing operators determined from a least-squares fit of 7 to 11 points to a quadratic polynomial. Operators are as follows (from Hamming, 1983): 5 point = $(\frac{1}{35})$ (−3, 12, 17, 12, −3); 7 point = $(\frac{1}{21})$ (−2, 3, 6, 7, 6, 3, −2); 9 point = $(\frac{1}{231})$ (−21, 14, 39, 54, 59, 54, 39, 14, −21); 11 point = $(\frac{1}{429})$ (−36, 9, 44, 69, 84, 89, 84, 69, 44, 9, −36).

Frequency Response of Some Simple LSI Systems

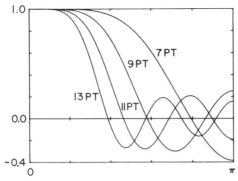

Figure 3.5 Frequency response of smoothing operators determined from a least-squares fit of 7 to 13 points to a 4th-degree polynomial. All smoothing operators derived from polynomial fitting show poor high-frequency behavior. Operators are as follows (from Hamming, 1983): 7 point = $(\frac{1}{231})$ (5, −30, 75, 131, 75, −30, 5); 9 point = $(\frac{1}{429})$ (15, −55, 30, 135, 179, 135, 30, −55, 15); 11 point = $(\frac{1}{429})$ (18, −45, −10, 60, 120, 143, 120, 60, −10, −45, 18); 13 point = $(\frac{1}{2431})$ (110, −198, −135, 110, 390, 600, 677, 600, 390, 110, −135, −198, 110).

satisfying results in the frequency domain. It turns out that this conclusion is indeed reasonable: we saw in Chapter 1 that if aliasing is to be avoided, digitizing must be confined to signals that are band limited. Experience suggests that all real signals are band limited, justifying the sampling process if care is taken in selecting the digitizing rate. On the other hand, as will become clear in Chapter 6, polynomials are not band limited; data modeling by polynomials flies in the face of an assumption in digital signal processing.

Polynomial fits also arise in other applications: for example, approximate numerical solutions to partial differential equations are represented by polynomials in a numerical scheme called the finite element method. In these two cases, data smoothing by polynomial fitting and finite element theory, we are claiming that the data can be represented by some underlying polynomial. Such a claim is called a *data model*, a concept that we will continue to explore in later chapters. In our frequency study of LSI operators, the data model is that of unaliased equally spaced samples with their associated periodic and continuous spectra. The inconsistency between this model and the polynomial model lies at the root of the problem we have experienced in attempting to obtain smooth frequency filters by polynomial fitting.

In Chapter 8, Digital Filter Design, we will pursue, in a direct manner, the problem of finding operators that have a given, specified frequency response. Our purpose here is to introduce the concept of operator frequency response and to give examples of the insight it affords. To that purpose, we next study the frequency response of digital versions of two fundamental mathematical operations.

Frequency Response of Digital Differential and Integral Operators

Integration and differentiation are, of course, quite common in the reduction and interpretation of data. A spectral analysis of the behavior of several digital implementation schemes will again allow new light to fall on old ideas. For both integration and differentiation, we know the desired frequency response exactly. In the first case, a sinusoidal signal

$$x = e^{i\omega t}$$

should yield

$$y = \frac{e^{i\omega t}}{i\omega}$$

at the output of a perfect integrator; or, in frequency domain parlence, the spectral response should be inverse to the frequency with a 90° lagging phase shift. One of the most well-known numerical integration schemes is the trapezoid rule expressed by the LSI operation

$$y_t = y_{t-1} + \tfrac{1}{2}(x_t + x_{t-1}) \tag{3.6}$$

We would naturally like to evaluate this procedure by comparing this operator's frequency response to the ideal $-1/i\omega$. Since the trapezoid rule is recursive, the scheme given by Eq. (3.6) does not give its infinite impulse response directly; we cannot proceed, as before, by evaluating Eq. (3.1) directly. Instead, by taking the Z transform of Eq. (3.6), which gives

$$Y(Z) = ZY(Z) + \tfrac{1}{2}[X(Z) + ZX(Z)]$$

we are able to solve for the ratio of the output to the input

$$\frac{Y(Z)}{X(Z)} = H(Z) = \frac{1}{2}\frac{(1+Z)}{(1-Z)} \tag{3.7}$$

Since this recursive scheme is an IIR LSI operator, we know that a sinusoidal input produces a sinusoidal output. Their ratio, $H(Z)$ above, then represents the relative amplitude and phase of the output when Z is evaluated on the unit circle, that is, $H(\omega)$ is the desired frequency response. The ratio $H(Z)$, not necessarily restricted to the unit circle, is called the *transfer function* of the system (or operator). You recognize, from our discussion in the last chapter, that the $(1-Z)$ factor in the denominator reflects the IIR nature of this operator. In fact, we have here one of the special cases mentioned in Chapter 2: the pole is right on the unit circle, causing the geometric series expansion to diverge. This makes sense, though, for an integrator; an impulsive spike at its input will pro-

Frequency Response of Digital Differential and Integral Operators 51

duce an output that is constant for all time. By our definition in Chapter 1, this is an unstable system. As a resulting limitation, the integrator cannot be implemented in the MA mode.

Now, let's lurch on to put the trapezoid rule to the acid test in the frequency domain. We evaluate $H(Z)$ from Eq. (3.7) on the unit circle:

$$H(\omega) = \frac{1}{2}\frac{(1 + e^{-i\omega})}{(1 - e^{-i\omega})} = \frac{1}{2}\frac{(e^{i\omega/2} + e^{-i\omega/2})}{(e^{i\omega/2} - e^{-i\omega/2})}$$

$$H(\omega) = \frac{\cos(\omega/2)}{2i\sin(\omega/2)} \qquad (3.8)$$

For comparison, we take the ratio

$$\frac{H_{\text{trapezoid}}}{H_{\text{desired}}} = \frac{\omega \cos(\omega/2)}{2 \sin(\omega/2)} = \frac{\omega(1 - \omega^2/2 + \cdots)}{2(\omega - \omega^3/6 + \cdots)}$$

$$= 1 - \omega^2/12 + \omega^4/720 + \cdots$$

showing that the trapezoid rule is a valid low-frequency approximation. This ratio, trapezoid to desired frequency response, is plotted in Fig. 3.6 along with the results for two other integration rules:

Simpson's: $y_t = y_{t-2} + (\frac{1}{3})(x_t + 4x_{t-1} + x_{t-2})$

Tick's: $y_t = y_{t-2} + 0.3584x_t + 1.2832x_{t-1} + 0.3584x_{t-2}$

Not surprisingly, Simpson's rule, being a longer operator, does a better job than the trapezoid rule. However, its error at high frequencies is to amplify rather than attenuate as the trapezoid rule does. Frequently, signals contain high-frequency noise that is best attenuated rather than accentuated. Tick's rule is the best of the three, and is, in fact, the result of optimizing a 3-point integration rule for the lower half of the Nyquist interval.

Looking at these integration rules in the frequency domain immediately gives an additional insight into a fundamental property of digital signal processing—aliasing. If the integration rules discussed here are to be used, the application of the simple Nyquist criterion given in Chapter 1 will lead to far too low of a sampling rate. However, if the data is intentionally oversampled, by, for example, five times the Nyquist criterion, there will be no frequencies above 0.1 on the frequency scale of Fig. 3.6. This vast improvement shows that in the integration process, the effect of sampling is always present—the only question is to what degree. This difficulty can be traced to the fact that integration operators should have a pole exactly on the unit circle.

Integration is important in digital signal processing because of its intimate relationship with many physical phenomena and mathematical ideas.

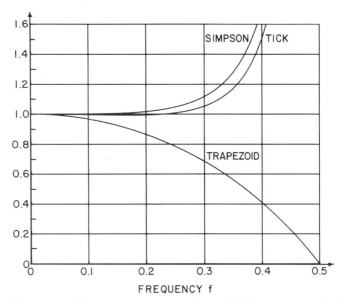

Figure 3.6 The ratio of the frequency response of three digital integrators to the frequency response of the ideal integrator, $1/i\omega$. Only the magnitude is shown; the phase is exactly correct for integration throughout the Nyquist interval. (After Hamming, 1983.)

This holds for differentiation and, likewise, digital operators attempting to duplicate the concept of an analog derivative suffer from similar limitations. In Chapter 1, we saw how different operators can be used to convert differential equations to difference equations for subsequent inclusion into a variety of digital signal processing schemes. Now, we are able to weigh the relative efficiency of several types of these difference operators by comparing their frequency response to the proper response for a derivative.

We start by observing that the spectral response from a sinusoidal input to a perfect differentiator would be $i\omega$. That is to say,

$$\frac{d}{dt}e^{i\omega} = i\omega e^{i\omega}$$

The desired frequency response is just the inverse of the integrator's: the perfect differentiator is linear in ω with a 90° leading phase shift. In Chapter 1, without much thinking, we used the central difference operator

$$y_t = \tfrac{1}{2}(x_{t+1} - x_{t-1})$$

which symmetrically spans two time intervals; clearly, only one unit would produce a better approximation. But, a span of only one unit leads to two

The Bilinear Transform

possibilities and necessarily suffers from a loss of symmetry. The cases are

Forward difference: $\quad y_t = \Delta x = x_{t+1} - x_t$

Backward difference: $\quad y_t = \Delta x = x_t - x_{t-1}$

Repeated application of these rules also can be used to generate higher differences. For example, the differences through third order are

$$\Delta x_t = x_{t+1} - x_t$$
$$\Delta^2 x_t = x_{t-1} - 2x_t + x_{t+1}$$
$$\Delta^3 x_t = \tfrac{1}{2}(x_{t-2} - 2x_{t-1} + 2x_{t+1} - x_{t+2})$$

All of these difference schemes can be compared, in the frequency domain, to the ideal. Leaving most of these comparisons for problems at the end of the chapter, we will content ourselves here with computing the frequency response of the first forward difference operator:

$$y_t = \Delta x_t = x_{t+1} - x_t$$
$$Y(Z) = (Z^{-1} - 1)X(Z)$$
$$H(Z) = \frac{Y(Z)}{X(Z)} = Z^{-1} - 1$$
$$H(\omega) = e^{i\omega} - 1 = i\omega - \omega^2/2 + \cdots$$

Like the integration operators, we see that this first forward difference operator has the correct behavior, both in magnitude and phase, but only at low frequency. A comparison of this first difference operator and another, called the bilinear transform is plotted in Fig. 3.7 relative to the ideal derivative response. Over most of the spectrum, the first difference has a superior magnitude response, but it has a linear phase shift (not shown) due to the half-step time shift from the calculated point. The bilinear transform, discussed in the next section, overcomes the objection of asymmetry and double span to make it an operator of prime significance in digital signal processing. It has a superior magnitude response at extremely low frequencies and the exact phase behavior at all frequencies at the expense of a poor, high-frequency response.

The Bilinear Transform and Its Application to Differential Equations

The problems encountered in the preceding section of converting the continuous operations of integration and differentiation to corresponding digital operators can be traced to the relation $Z = e^{-i\omega}$ or, equivalently,

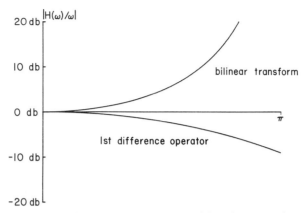

Figure 3.7 A comparison of the magnitude response of the bilinear transform and the first difference operator. The magnitude of the ratio of the actual to the desired response is plotted on the vertical scale in db. Dramatic deviation from the ideal response shows that the normal Nyquist sampling rule is far too generous for applications requiring digital approximations to derivatives. Compared to the bilinear transform, the first difference operator has a superior magnitude response, but it is obtained at the expense of linear phase delay, which reaches 90° at $\omega = \pi$. In contrast, the bilinear transform has exactly the correct phase behavior over the whole Nyquist interval.

$i\omega = -\ln Z$. The difficulty arises because digital operators are formed from rational fractions of polynomials in Z, while $i\omega$ is related to Z via the exponential function. Clearly we must tolerate approximations of some kind.

One approach, which turns out to be quite useful in digital signal processing, is to use a well-known trick leading to a rapidly converging ratio of expansions of the exponential function

$$Z = e^{-i\omega} = \frac{e^{-i\omega/2}}{e^{i\omega/2}} = \frac{1 - i\omega/2 + (i\omega/2)^2/2! \cdots}{1 + i\omega/2 + (i\omega/2)^2/2! \cdots}$$

and likewise,

$$-i\omega = \ln(e^{-i\omega}) = \ln Z = 2\left(\frac{Z-1}{Z+1}\right) + \frac{1}{3}\left(\frac{Z-1}{Z+1}\right)^3 + \cdots$$

Truncating both of the expansions leads to the powerful relations

$$Z = \frac{1 - i\omega/2}{1 + i\omega/2} \quad \text{and} \quad i\omega = 2\left(\frac{1-Z}{1+Z}\right) \tag{3.9}$$

This relation between ω and Z is called the *bilinear transform*. It can be viewed two ways: as an approximation derived by truncating the series

The Bilinear Transform

expansions, or as an exact relationship between Z and a newly defined frequency variable. If the latter view is taken, the difference between $2(1-Z)/(1+Z)$ and the true value of $i\omega$ is sometimes called *frequency warping*. Further interpretation of the bilinear transform is afforded by the frequency analysis of integral and differential operators. Equations (3.9) give the desired rational fraction to be used for $i\omega$ in applying the operations of integration and differentiation. On the one hand, it is gratifying to note that this expression for $i\omega$ given by the bilinear transform is the same as that used for the trapezoid integration rule in Eq. (3.7)—the trapezoid rule is simply an application of the bilinear transform. On the other hand, our naive guesses for differentiation operators did not lead to using $2(1-Z)/(1+Z)$ as suggested by the bilinear transform. We would expect that the infinite impulse response, indicated by the $(1+Z)$ in the denominator, would yield results superior to our previous 2-term operators. Indeed, of the three difference operators that we have considered so far, all gave us an uneasy feeling because we would have really liked to have considered unavailable data values at $t+\frac{1}{2}$ and $t-\frac{1}{2}$ in order to center the difference at the time t. (The central difference operator achieves this symmetry at the expense of using points twice as far apart—obviously a sacrifice in estimating the derivative.) To see that the bilinear approximation amounts to a superior approach to approximating a derivative by a difference, imagine that we have a ghost sequence $\hat{x}$ imbedded within x that has values at half-integer time

$$\ldots x_{t-1}, \hat{x}_{t-(1/2)}, x_t, \hat{x}_{t+(1/2)}, x_{t+1}, \ldots$$

Naturally for the best approximation to the derivative at t, we would now write

$$y_t = \hat{x}_{t+(1/2)} - \hat{x}_{t-(1/2)}$$

$$Y(Z) = (Z^{-1/2} - Z^{1/2})\hat{X}(Z)$$

Also, we would estimate

$$x_t = \frac{\hat{x}_{t+1/2} + \hat{x}_{t-1/2}}{2}$$

$$X(Z) = \left(\frac{Z^{-1/2} + Z^{1/2}}{2}\right)\hat{X}(Z)$$

Eliminating the ghost sequence $\hat{x}$ gives

$$Y(Z) = 2\left(\frac{Z^{-1/2} - Z^{1/2}}{Z^{-1/2} + Z^{1/2}}\right)X(Z)$$

or

$$Y(Z) = 2\left(\frac{1-Z}{1+Z}\right)X(Z)$$

as our approximation to $i\omega$—the bilinear transform. Another way of looking at the approximation is to observe that

$$i\omega = 2\left(\frac{1-Z}{1+Z}\right) = 2i\frac{\sin(\omega/2)}{\cos(\omega/2)}$$

or

$$i\omega \cong 2i\tan\left(\frac{\omega}{2}\right) = i\omega + \frac{i\omega^3}{12} + \cdots$$

is a much better low-frequency approximation than, for example, the 1st backward difference operator, which sets

$$i\omega = 1 - Z = 1 - e^{-i\omega}$$

or

$$i\omega \cong i\omega + \frac{\omega^2}{2} - \frac{i\omega^3}{6} + \cdots$$

Furthermore, the relation

$$Z = e^{-i\omega} = e^{-i(\operatorname{Re}\omega + i\operatorname{Im}\omega)}$$

$$Z = e^{\operatorname{Im}\omega}e^{-i\operatorname{Re}\omega}$$

maps the exterior of the unit circle in the Z plane into the upper half ω plane (ω must be complex for points off of the unit circle in the Z plane). Likewise, the interior of the unit circle in the Z plane is mapped into the lower half of the ω plane. By inspecting the definition of the bilinear transform, you can see that it has the same mapping property between the Z and ω planes, or alternately the plane of Laplace transforms (see Problem 3.15). This is extremely fortuitous because it means that the minimum phase condition is preserved under the bilinear transform. For example, we can be assured that a stable differential equation converted to a difference equation by using the bilinear transform as an approximation to the derivative will have stable digital solutions. Such is not always true when the forward and backward difference operators are used. As a simple example, we consider the following first-order differential equation:

$$\frac{dy}{dt} + ay = x(t) \tag{3.10}$$

The Bilinear Transform

One practical application of this equation is the description of current flow in an electrical circuit consisting of a source $v(t)$, an inductor L, and a resistor R connected in series. For this case, the equation would read

$$L\frac{di}{dt} + iR = v(t) \tag{3.11}$$

Whatever the application, using the forward difference operator, the backward difference operator, and the bilinear transform converts Eq. (3.10) into the following three difference equations:

Forward operator: $\quad y_t = (1-a)y_{t-1} + x_{t-1}$ (3.12a)

Backward operator: $\quad y_t = (y_{t-1} + x_t)/(1+a)$ (3.12b)

Bilinear transform: $\quad y_t = \dfrac{1}{2+a}[(2-a)y_{t-1} + x_t + x_{t-1}]$ (3.12c)

It is easy to show (Problem 3.12) that the system functions for these three computational schemes have poles at $Z_0 = 1/(1-a)$; $Z_0 = 1+a$; and $Z_0 = (2+a)/(2-a)$, respectively. The regions where these poles are outside of the unit circle determine the following regions for stable behavior of Eq. (3.10):

Forward operator: $\quad 0 < a < 2$

Backward operator: $\quad a < -2 \quad$ and $\quad 0 < a$

Bilinear transform: $\quad 0 < a$

In Fig. 3.8, we show the results of using Eqs. 3.12 for a step function excitation. Four different values of the constant a have been used, one in each of the above stability regions. Only the bilinear transform preserves the stability behavior of the original differential equation for all values of the constant a. In Fig. 3.8(a) the forward difference operator (top result) has produced an unstable output from a stable differential equation. In Fig. 3.8(d), the backward difference operator (middle result) has produced a stable output from an unstable differential equation with $a = -2.25$, corresponding to a negative damping factor. One is usually on the alert for numerical instabilities of the first kind [Fig. 3.8(a), top]. But, in the second case [Fig. 3.8(d), middle], numerical difficulties have produced an alluringly well-behaved result, fully capable of escaping suspicion in complex situations.

The introduction to Fourier analysis presented in this chapter arose naturally from considering sinusoidal inputs to digital LSI operators. The collection of eigenvalues, called the frequency spectra of the system, has

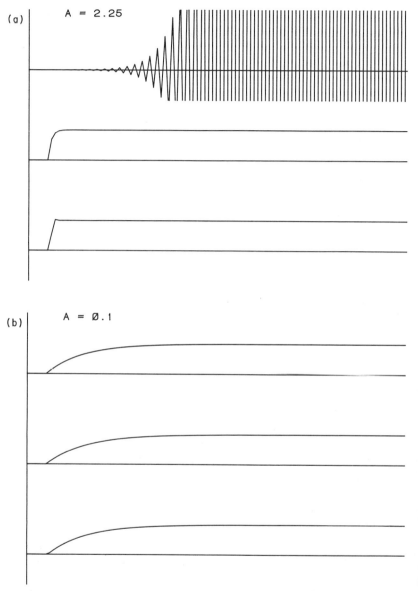

Figure 3.8 The response for a step function input of a 1st-order differential equation, Eq. 3.10, for three different digital approximations to the derivative: the forward difference operator (top); the backward difference operator (middle); and the bilinear transform (bottom). (a) The forward difference operator (top) converts a stable differential equation into an unstable difference equation for $A = 2.25$, relatively undamped. (b) All three operators convert a stable differential equation into a stable difference equation for $A = +0.1$, relatively high damping. (c) All three operators convert an unstable differential equation into an unstable difference equation for $A = -0.1$. (d) The backward difference operator (middle) converts an unstable differential equation with $A = -2.25$ into a stable differential equation. Only the bilinear transform has the correct behavior for all four cases.

The Bilinear Transform

(c) A = -0.1

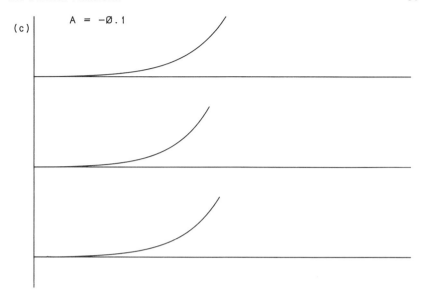

(d) A = -2.25

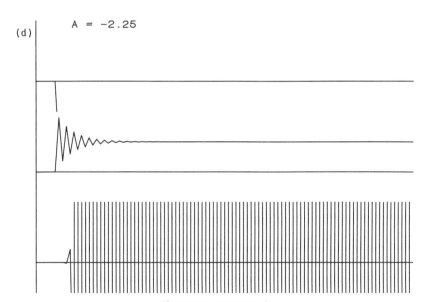

Figure 3.8 (*Continued*)

two properties quite worthy of emphasis: $H(\omega)$ is a continuous and a periodic function of frequency. As we shall see in subsequent chapters, the Fourier transform that we introduced here [Eq. (3.1.)] is one of four possible types, three of which play a central role in our thinking about processing data in a digital format. (Ironically, the Fourier series, introduced first to most students by books on continuous theory, will be the least significant in our work.)

It is possible to treat essentially the whole field of time series analysis—filtering, deconvolution, prediction, and system simulation—in the time domain without ever introducing concepts from the frequency domain (see, for example, Box and Jenkins, 1976). But, as we have seen from just a brief foraging into smoothing, integration, and differentiation, the frequency domain yields a rich harvest of insight to digital processes. More fruits can be expected.

Problems

3.1   The central point of this chapter has been that sinusoids are eigenfunctions of all LSI systems, and therefore, it is convenient and instructive to study the response of LSI operators with sine wave inputs. First, verify that, indeed, sinusoids are eigenfunctions of LSI systems by writing a computer program that convolves a given MA operator with a sine wave signal of a given frequency and unit amplitude. Examine the output of your program using several of the 5-point to 11-point MA smoothing operators from Fig. 3.4. Verify that the output has zero phase shift from the input and that the amplitudes agree with those of Fig. 3.4. Second, change some of the coefficients in your operator to produce an asymmetrical arbitrary LSI operator and verify that the output is still sinusoidal, but phase shifted from the input. Explore input frequencies above Nyquist to verify the folding nature of the spectrum.

3.2   Follow the idea of Problem 3.1 for recursive LSI operators. Use the trapezoid rule and Simpson's rule for integration as two operators for study. Observe the deviation from correct integration spectral response at high frequencies, yet the output remains sinusoidal.

3.3   The frequency response of any LSI operator could be determined using the method of Problems 3.1 and 3.2. A less mundane approach is to evaluate Eq. (3.1) directly. Show that for symmetric LSI operators

$$b_{-N} \ldots b_{-2}, b_{-1}, b_0, b_1, b_2 \ldots b_N$$

Eq. (3.1) can be written

$$H(\omega) = b_0 + 2 \sum_{n=1}^{N} b_n \cos(n\omega)$$

Use this result to write a computer problem that plots the spectrum of any symmetric LSI operator. Reproduce the plots of Figs. 3.1 through 3.5.

3.4 Derive the expressions given in the section "Frequency Response of Digital Differential and Integral Operators" in this chapter for the second- and third-order forward difference operators using repeated differences.

3.5 Find and discuss the frequency response of the operator

$$\tfrac{1}{16}(-1, 4, 10, 4, -1)$$

3.6 Derive an algebraic expression for the frequency response $H(\omega)$ of the central difference operator

$$\Delta x_t = y_t = \tfrac{1}{2}(x_{t+1} - x_{t-1})$$

3.7 If you are familiar with Fourier series, invert Eq. (3.1) to find h_k in terms of $H(\omega)$ by using the appropriate analogous procedure with the roles of time and frequency interchanged.

3.8 Suppose that, for some reason, one data point is missing from a sampled data stream and that we would like to estimate its value from neighboring points. Straight-line interpolation is one method. An extension of this idea would be to fit a polynomial to adjacent points. If we do assume that the data behave locally like a polynomial of odd degree, then the next higher even-order difference operator would give a null result when it acts on the complete data sequence, including the yet to be determined missing value. Setting the even-order difference operator to zero and solving for the value at the central point gives the interpolated value. As an example of this approach, calculate the fourth-order difference operator. Set it equal to zero and solve for the value at the midpoint. This is the interpolated value in terms of its four neighbors. Calculate the spectrum of this symmetrical operator and comment on the significance of its frequency domain behavior.

3.9 Any function can be separated into its even and odd parts. Write the impulse response function in Eq. (3.1) this way and derive separate equations for the real and imaginary parts of $H(\omega)$. Show that if h_k is even, $H(\omega)$ is real and even; if h_k is odd, $H(\omega)$ is imaginary and odd. Do these considerations lend any significance to negative frequencies?

3.10 From the Z transform, compute and plot the frequency response of the digital operator discussed in Problem 2.9. Comment on the significance of both its magnitude spectrum and its phase spectrum.

3.11 Use the forward difference operator, the backward difference operator, and the bilinear transform to generate the difference equations in Eqs. (3.12) from the differential equation given in Eq. (3.10).

3.12 Locate the poles of the system response functions implied by Eqs. (3.12). For all three system response functions, determine the values of the constant a that lead to a stable system. By comparing Eq. (3.10) to Eq. (3.11), we see that the constant a is to be interpreted as the damping factor R/L for a series resistor inductor circuit, which could never obtain negative values. Give an example of a different physical system described by Eq. (3.10) that can have negative values of the constant a.

3.13 Write a computer program to solve Eqs. (3.12) with step function and impulse function inputs.

3.14 In Problems 2.11 and 2.12, you digitized an analog RC circuit by two different methods. In the frequency domain, the system response function of this circuit is

$$H(\omega) = \frac{1}{1 + i\omega\tau} \qquad \tau = RC$$

Now use the bilinear transform to form a third discrete version of the system's response function $H(Z)$ and compare it to the first two results. For what range of τ is $H(Z)$ stable?

3.15 In Fourier theory, functions are represented by a superposition of sinusoids $\exp(-i\omega t)$, while in Laplace transform theory they are represented by a superposition of damped sinusoids $\exp(-st)$. The complex variable $s = \sigma + i\omega$ can be thought of as a generalization of frequency into the complex plane; hence, the Fourier transform is a special case of the Laplace transform. In Laplace transform theory, a continuous system is stable only if its system response function $H(s)$ has no poles in the right half of the s plane. By generalizing $i\omega$ to $\sigma + i\omega$ in Eq. (3.9), show that the bilinear transform maps the region inside of the unit circle in the Z plane into the right half of the s plane and the region outside of the unit circle into the left half of the s plane. Hence, by using the bilinear transform to digitize the impulse response of a continuous system, invariance of its stability properties is guaranteed.

3.16 In ac circuit theory, the average power dissipated in a reactive circuit

Problems

can be written $\bar{P} = \frac{1}{2}\text{Re}[V^*(\omega)I(\omega)]$. In the discussion of Fig. 2.4, we concluded that the impedance Z and the admittance Y should be minimum phase functions if they are to have the same stability and causality properties of the original analog circuit. We can pursue the classification scheme of Fig. 2.3 even further by realizing that for a passive circuit the admittance and the impedance have additional restrictions. Show that $Y(\omega)$ and $Z(\omega)$ must both have positive real parts for this case.

3.17 The circuit shown below could be called a leaky, RC highpass filter. Using the bilinear transform, convert the given analog system transfer function $H(\omega)$ to a digitized version $H(Z)$. Comment on the invertibility of both the analog and digital systems. Include a discussion of what happens as $R \to \infty$.

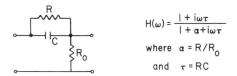

$$H(\omega) = \frac{1 + i\omega\tau}{1 + a + i\omega\tau}$$

where $a = R/R_0$
and $\tau = RC$

3.18 Using the bilinear transform, convert the transfer function $H(\omega)$ for the circuit below to a digital $H(Z)$. Comment on the invertibility of both the analog and the digital systems. In addition, make a comparison with the circuits of Problems 3.14 and 3.17 on the question of invertibility.

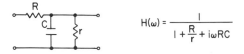

$$H(\omega) = \frac{1}{1 + \frac{R}{r} + i\omega RC}$$

4

Couplets and Elementary Filters

In Chapter 2, where we introduced the Z transform, we concluded that the most general filter that could possibly be used in a practical computing scheme would be a rational fraction of polynomials in Z. In turn, each polynomial in Z can be factored into couplets identifying the couplet as the basic building block of digital signal processing.

Next, in Chapter 3, the recognition of sinusoids as eigenfunctions of LSI operators and their associated frequency response as eigenvalues led to a whole new way of looking at digital LSI operators: the frequency domain—the spectrum. It seems natural to apply this frequency domain analysis to the basic building blocks, the couplets, in an effort to gain maximum insight into their behavior.

In this chapter then, we will study in some detail the frequency domain behavior of the simple couplet, its inverse, and the ratio of two couplets. These three elementary filters will then form a basis for the classification of most systems according to the location of their poles and zeros. Since the most general digital filter of interest can be written as a rational fraction of polynomials in Z,

$$S = \frac{(Z - a_1)(Z - a_2)(Z - a_3) \cdots (Z - a_n)}{(Z - b_1)(Z - b_2)(Z - b_3) \cdots (Z - b_m)} \quad (4.1)$$

it follows that once we are familiar with the frequency behavior of the single-zero and single-pole filter, the frequency behavior of a complex system can be readily computed; the magnitude response is the product of the single-zero couplet magnitude responses divided by the magnitude response of the single-pole couplets. For the phase response, we add the phases of the numerator couplets and subtract those of the denominator couplets. A deeper understanding of all LSI systems is certain to follow from examining the simplest ones first.

The Single-Zero Couplet

The simplest digital filter is the couplet $Z - Z_0$ with a zero located at Z_0. A quick-look evaluation of this couplet's frequency behavior is readily afforded by thinking of evaluating the vector

$$H(Z) = Z - Z_0$$

in the complex plane with Z on the unit circle to give $H(\omega)$. Figure 4.1 shows the vector relationships: the vector $Z - Z_0$ extends from the location of the zero to the unit circle. As Z moves clockwise around the unit circle from $\omega = -\pi$ through $\omega = 0$ to $\omega = \pi$, we can qualitatively sketch the magnitude and phase behavior of $H(\omega)$; it is just the magnitude and phase of the vector $Z - Z_0$. Figure 4.2(a) shows the details of the phase angle of Z itself. Because its phase angle is a negative function of ω, that is $\phi = -\omega$, it is sometimes convenient to introduce the opposite sign for phase angles, called the phase-lag angle; it is obtained by reversing the sign of all angles as shown in Fig. 4.2(b). For any operator, its phase-lag spectrum is the negative of its phase spectrum, resulting in another occupational hazard. Thus, in carefully computing the sign of the phase, we must also take care to state which angle is under consideration, the phase or phase-lag angle.

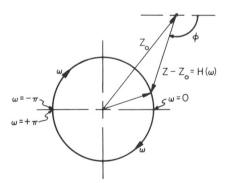

Figure 4.1 The evaluation of the couplet $(Z - Z_0)$ on the unit circle shown as the difference between the vector Z and the zero located at Z_0. The frequency response, $H(\omega)$, can be mentally estimated, both in magnitude and phase, by thinking of Z moving around the unit circle.

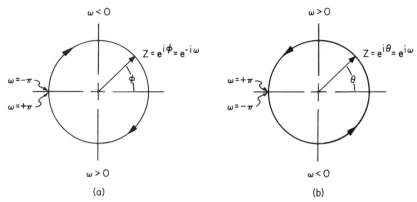

Figure 4.2 Showing two definitions of phase angles. (a) The phase angle $\phi = -\omega$; ω is positive in the lower half-plane. (b) The phase-lag angle $\theta = \omega$; ω is positive in the upper half-plane.

At any rate, an example of using this complex vector concept for estimating magnitude and phase spectra is sketched in Fig. 4.3 for two different zero locations: inside and outside the unit circle. Now, we can finally see, as promised in Chapter 2, why these two cases are called maximum phase and minimum phase couplets; the maximum phase couplet, with its

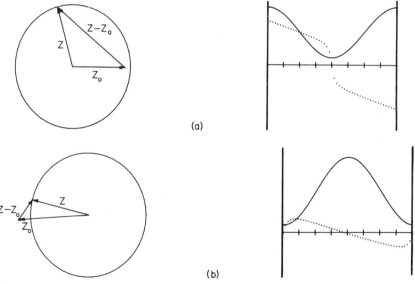

Figure 4.3 Sketch of the magnitude spectrum (solid line) and the phase spectrum (dotted line) for the single-zero filter. (a) The zero is inside of the unit circle, the maximum phase couplet. (b) The zero is outside the unit circle, the minimum phase couplet.

The Single-Zero Couplet

zero located inside the unit circle, always has its phase augmented by 2π as ω goes around the unit circle. In contrast, the minimum phase couplet, with its zero outside the unit circle, has its phase return to its initial value when ω goes around the unit circle. That is, for the minimum phase couplet, $\phi(\omega) = \phi(\omega + 2\pi)$; its phase has the minimum excursion in the Nyquist interval.

In addition to these phase characteristics, Fig. 4.1 allows us to make several useful observations on the frequency behavior of the couplet in terms of the zero location. First, we observe that $H(\omega)$ is, of course, periodic with period 2π. For the single couplet, $H(\omega)$ is necessarily complex. As the zero becomes located closer to the unit circle, the magnitude response will drop closer to zero at the frequency that places $Z - Z_0$ closest to the zero. If the zero is located right on the unit circle, $H(\omega) = 0$ at that point. Another special location for the zero is at the origin. If the zero is located there, then clearly the magnitude spectrum is constant in ω; this is called an allpass filter. The only function of an allpass filter is to delay the output. The phase varies linearly from $+\pi$ to $-\pi$ as ω goes from $-\pi$ to π. The filter becomes just Z, our familiar unit delay operator.

The foregoing readily gives us a qualitative insight into the frequency behavior of the single-zero couplet in terms of the location of the zero relative to the unit circle. In addition, it is simple enough to compute the spectrum of the single-zero couplet directly.

For this purpose, it is convenient to write the couplet as

$$H(Z) = a_1 + a_2 Z,$$

which differs from our previous form only by an overall scale factor. Evaluating H on the unit circle to get $H(\omega)$ gives

$$H(\omega) = a_1 + a_2 e^{-i\omega}$$

since we equate $Z = e^{-i\omega}$ there. Computing the magnitude spectrum gives

$$|H(\omega)|^2 = (a_1 + a_2 e^{-i\omega})(a_1^* + a_2^* e^{i\omega})$$

or

$$|H(\omega)| = \sqrt{a_1^2 + a_2^2 + 2\operatorname{Re}(a_1 a_2^* e^{i\omega})} \tag{4.2}$$

The phase spectrum can be calculated from

$$\phi(\omega) = \tan^{-1}\left[\frac{\operatorname{Im} H(\omega)}{\operatorname{Re} H(\omega)}\right] \tag{4.3}$$

A property of the magnitude spectrum that is of basic importance to spectral analysis is that Eq. (4.2) is invariant under a complex conjugate interchange of a_1 and a_2. That is to say, the two couplets

$$H_1 = a_1 + a_2 Z$$

and

$$H_2 = a_2^* + a_1^* Z$$

have exactly the same magnitude spectrum according to Eq. (4.2). The phase spectrum is not invariant under this exchange; it takes a minimum phase couplet into a maximum phase and vice versa.

This exchange property is sufficiently important to warrant an example. Consider the two couplets:

$$H_1(Z) = 0.5 + Z, \quad \text{maximum phase}$$

and

$$H_2(Z) = 1 + 0.5Z, \quad \text{minimum phase}$$

Using Eq. (4.2), we get the magnitude spectrum

$$|H(\omega)| = \sqrt{(\tfrac{5}{4}) + \cos \omega}$$

for both couplets. Using Eq. (4.3), we get for the two phase spectra

$$\phi_1(\omega) = -\tan^{-1}\left(\frac{\sin \omega}{0.5 + \cos \omega}\right)$$

and

$$\phi_2(\omega) = -\tan^{-1}\left(\frac{\sin \omega}{2 + \cos \omega}\right)$$

Figure 4.4 shows the magnitude spectra (squared) and the phase spectra of these two couplets. Many of the properties shown in these curves can be

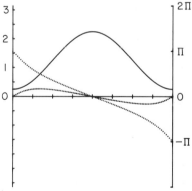

Figure 4.4 Plot of the power spectra (solid line) and the phase spectra (dotted line) of the two couplets (0.5 + Z) and (1 + 0.5Z). The power spectrum is the same for both, while the phase spectra differ; the phase spectrum of the minimum phase couplet has $\phi(\pi) = \phi(-\pi)$, while the maximum phase couplet has $\phi(\pi) = \phi(-\pi) + 2\pi$.

corroborated by using a sketch showing the zeros in the Z plane: the magnitude spectra are seen to be equal, at least at selected points, such as $\omega = 0$ and $\omega = \pi$, and are maximum where the zeros are farthest from Z, at $\omega = 0$. The minimum phase couplet has the same phase value at $\omega = -\pi$ and π, while the phase of the maximum phase couplet increases by 2π over the same interval.

The Single-Pole Couplet

The single-pole couplet is just the inverse of the single-zero couplet. As such, the same qualitative idea of the single-pole couplet's frequency response can be obtained by thinking of the complex vector $Z - Z_0$ as Z moves around the unit circle. The required mental gymnastics are slightly greater than for the single-zero couplet: we must think of the inverse magnitude and the negative phase of $Z - Z_0$ to get the spectrum of $1/(Z - Z_0)$. For example, if the pole is very close to the unit circle, $|Z - Z_0|$ becomes very small at the point on the unit circle closest to Z_0; $H(\omega)$ becomes very large there. In general, poles close to the unit circle lead to characteristic sharp, narrow peaks in the spectra that are useful in many filter applications, such as narrow bandpass filters and sharp cutoff high-pass or lowpass filters.

To explore this fundamental property of single-pole filters, we calculate the squared magnitude spectrum. This is sometimes called the *power spectrum* because it is proportional to the square of the signal as is the power dissipated in an electrical load in terms of its voltage across the load or its current through the load. (The concept of the power in a digital signal will be developed further in Chapters 6 and 12; for now we can take the term power spectrum to be a slightly more efficient way of saying squared magnitude spectrum.) Proceeding then, we get the power spectrum of $H(Z)$ by squaring it and then evaluating it on the unit circle:

$$|H(Z)|^2 = \frac{1}{(Z-Z_0)(Z^* - Z_0^*)} = \frac{1}{Z_0^2 + Z^2 - 2\operatorname{Re}(Z_0 Z^*)}$$

Using polar coordinates to specify points in the complex planes gives

$$Z = e^{-i\omega} \quad \text{and} \quad Z_0 = \rho e^{i\phi} = \rho e^{-i\omega_0}$$

where we have used the same phase for Z_0 and Z as shown in Fig. 4.2(a). The power spectrum becomes

$$|H(\omega)|^2 = \frac{1}{1 + \rho^2 - 2\rho \cos(\omega - \omega_0)} \tag{4.4}$$

To simplify matters, we restrict our study of this spectrum to values of ω near ω_0 so that we can use the first two terms in the power series expansion of $\cos(\omega - \omega_0)$:

$$\cos(\omega - \omega_0) = 1 - (\omega - \omega_0)^2/2 + \cdots$$

giving

$$|H(\omega)|^2 = \frac{1}{(1-\rho)^2 + \rho(\omega - \omega_0)^2}$$

Our case of interest is a pole very near the unit circle. We select a stable, causal filter by placing the pole a small distance, ε, outside of the unit circle:

$$\rho = 1 + \varepsilon$$

Thus, the power spectrum becomes for ω near ω_0

$$|H(\omega)|^2 = \frac{1}{\varepsilon^2 + (\omega - \omega_0)^2} \tag{4.5}$$

A sketch of this equation in Fig. 4.5 shows a narrow band centered on ω_0 with width 2ε at the $\frac{1}{2}$ power points; poles very near the unit circle produce sharp spectral peaks.

Narrow-band filters are used as one method of searching for signals of known frequency buried in broadband noise. The narrow passband admits the signal centered on ω_0, while rejecting the remaining frequencies, representing noise components. Of course, the desired signal must be likewise narrow banded so that it passes through the filter undistorted. In the limit $\varepsilon \to 0$, the signal would contain only one frequency; it would be a pure sinusoidal wave, carrying no information except, perhaps, its presence. An example of such a narrow-band filter is the radio receiving system used in continuous-wave radiotelegraphy. The signal is an on–off sinusoidal carrier wave that scarcely occupies a bandwidth of 10 to 20 Hz.

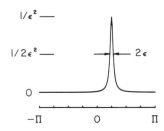

Figure 4.5 Power spectrum of the narrow-band, single-pole couplet. The pole is located at $\rho = 1.1$ and $\omega = \pi/4$.

The receiver contains a similarly narrow-banded filter that greatly suppresses all noise outside of this bandwidth. In past years, this filter has been implemented from electrical analog components, but it could be (for that matter the entire receiver could be) implemented from digital components.

The phase-locked detector is another example of a narrow-banded approach to weak signal detection. In this case, an oscillator is phase locked onto the incoming signal and the noise-free oscillator is taken as the system output. The effective bandwidth of the system can be made extremely small if the phase-locked loop is permitted a long search time at the known frequency of the signal. The equivalent situation for the digital filter has a pole extremely close to the unit circle; then the recursive filter will likewise take a long time to converge.

When the phase-locked loop has the ability to track the signal to a new frequency ω_0, the spectrum of the narrow-band filter has changed; it is a time-dependent filter that cannot be implemented by LSI systems. Digital forms of time-dependent operators can be designed in many ways, some sophisticated, others straightforward. The simplest would change the filter coefficients every so many time steps, according to a predetermined schedule.

Both the single-zero and the single-pole filters that we have just discussed are too elemental to be of practical use. For example, both have complex outputs, a feature that clearly makes them of limited practical use. However, the outputs can be made real by simply cascading the filters with another filter with its zero, or pole, at the complex conjugate of the first. In the case of the single-pole filter, we would then generate the two-pole filter

$$F(Z) = \frac{1}{(Z - Z_0)} \times \frac{1}{(Z - Z_0^*)} = \frac{1}{Z_D^2 + Z^2 - 2Z \operatorname{Re}(Z_0)}$$

which has a spectrum similar to Fig. 4.5 but with an additional peak located at $\omega = -\omega_0$. A corollary to this discussion, direct from the theory of polynomials, is that any real filter (or sequence) has its poles and zeros occurring in complex conjugate pairs.

The Single-Zero, Single-Pole, Allpass Filter

Apart from adding complex conjugate poles and zeros, the next most complicated filter that can be formed from our building-block couplets is the single-zero, single-pole filter. A most interesting arrangement of the zero and pole is that which produces some phase spectrum but a flat magnitude spectrum—the allpass filter.

Allpass filters can be divided into four different types; only the single-zero, single-pole filter deserves any detailed attention. The other three types are (1) the trivial constant-filter, $H(Z)$ = constant, with no effect on the magnitude or phase spectrum; (2) the pure delay with multiple zeros at the origin, $H(Z) = Z^n$, with a linear phase spectrum; and (3) a mathematical curiosity, called the impure delay, which we will not discuss.

The trick in making a single-zero, single-pole, allpass filter is the very special placing of the zero relative to the pole; to exactly cancel the magnitude spectra, we simply place the zero at the pole's inverse complex-conjugate location:

$$H(Z) = \frac{Z - 1/Z_0^*}{1 - Z/Z_0} \tag{4.6}$$

We next want to show that this filter has a constant magnitude spectrum or, equivalently, a constant power spectrum. The power spectrum is $H(\omega)H(\omega)^*$, but any sequence can be expressed in terms of ω or in terms of Z. We define the power spectrum in the Z domain to be

$$P(Z) = H(Z)H^*(1/Z) \tag{4.7}$$

where $H^*(1/Z)$ means to invert Z wherever it occurs and to take the complex conjugate of all coefficients. Clearly, on the unit circle this definition becomes $H(\omega)H^*(\omega)$ because there

$$Z = \rho e^{i\phi} = e^{-i\omega} \quad \text{and} \quad 1/Z = \frac{e^{-i\phi}}{\rho} = e^{i\omega}$$

Using this formulation of Eq. (4.7), our allpass filter's power spectrum becomes

$$H(Z)H^*(1/Z) = \left[\frac{Z - 1/Z_0^*}{1 - Z/Z_0}\right]\left[\frac{1/Z - 1/Z_0}{1 - 1/ZZ_0^*}\right] = 1$$

showing that the filter has a unit power spectrum; the amplitude of any sinusoidal input will be unchanged, only its phase can be altered.

To make our allpass filter of Eq. (4.6) stable and causal, we select the pole to be outside of the unit circle. Its zero is then inside and the filter is of mixed phase. Because of the zero's location, the phase of the filter will be augmented by 2π as ω goes from $-\pi$ to π. Furthermore, we will show that the filter's phase is monotonic in ω so that there is a phase delay at every frequency. To discuss this monotonic behavior, we introduce the derivative of interest by defining the group delay,

$$\tau_g = -\frac{d\phi}{d\omega}$$

The Single-Zero, Single-Pole, Allpass Filter

a concept closely related to the group velocity of a wave packet. The phase-lag, $\theta = -\phi(\omega)$, is similarly related to the phase velocity in wave propagation.

The distinction between phase velocity and group velocity is an important aspect of dispersive wave propagation. Dispersion occurs when the group delay is a function of frequency; the shape of the waveform changes during propagation. All significant causal digital systems therefore display dispersion. The case of a purely linear phase (i.e., constant group delay) and a flat magnitude spectrum is trivially achieved in digital signal processing. It is simply a time shift of the data. In sharp contrast, such behavior is very difficult to produce in analog devices where it might be highly desirable, as in amplifiers, transducers, and recorders.

Several other aspects of dispersive propagation are worth mentioning. The individual spectral components of a wave packet travel at the phase velocity while energy in the wave package is transported at the group velocity. The phase velocity of electrical signals in some materials exceeds c, the velocity of light in a vacuum, while the group velocity can never exceed c. All communication, such as the transfer of data between regions of a computer or among different computers, is limited by the group velocity of the signals in their conductors. Thus, in many applications, both digital and analog, the group delay is of more significance than the phase delay.

In Problem 4.9, we discuss how to show that the group delay of our single-pole, single-zero, allpass filter is given by

$$\tau_g = \frac{(1 - 1/Z_0 Z_0^*)}{(1 - 1/ZZ_0^*)(1 - Z/Z_0)} \tag{4.8}$$

Clearly, this is a positive function of frequency; the numerator is positive because the pole is outside the unit circle and the denominator is positive because it is in the form of a power spectrum in Z.

Further significance of a minimum phase filter can now be seen by cascading one with our allpass filter:

$$F = F_{\min}(Z) F_{\text{allpass}}(Z)$$

Since the phase of F is the sum of the phase of $F_{\min}$ and F_{allpass}, the group delays add also. Thus, F has a greater group delay at every frequency than does the minimum phase filter. It follows that the minimum phase filter has the least group delay possible at every frequency among the class of filters with the same magnitude spectrum. We emphasize that the same statement cannot be claimed for the phase delay; the minimum phase filter does not necessarily have the least phase delay at every frequency among the identical magnitude spectrum class of filters. See, for example, Problem 4.12.

Of course, complicated phase delays and group delays may be formed by cascading allpass filters of the single-zero, single-pole type considered here. Taking the pole near the unit circle causes most of the delay to occur near one frequency; moving the pole farther out spreads the delay over more of the spectrum.

Elementary Filters Classified by Their Poles and Zeros

The gross frequency behavior of ARMA filters is controlled by the relative location of their poles and zeros, relative to one another and relative to the unit circle. We would do well, then, to study in some detail the simplest ARMA filter, our single-pole, single-zero filter. In the preceding sections, we have calculated the frequency response in a complete fashion only for the single-zero filter and for the single-pole filter; for the combined case, we have limited the discussion to the particular relative placement of the zero and pole that produced the allpass filter. We will now expand the discussion with the aid of Fig. 4.6 and Fig. 4.7 to include eight possible relative positions of one pole and one zero. Cases with the pole inside of, or on, the unit circle are not considered because they represent unstable causal filters. We start with the zero at the origin and discuss the frequency response as the zero moves out along the radial line connecting the origin and the pole.

When the zero is at the origin, the numerator is Z, a pure delay; the denominator produces the single-pole response computed earlier (in the section on the single-pole couplet). The strong contribution of the linear phase arising from the pure delay is apparent in Fig. 4.6(a). As the zero moves off of the origin toward the pole, its magnitude spectrum starts to cancel that of the pole as seen in Fig. 4.6(b). When the zero reaches the allpass filter position relative to the pole, as in Fig. 4.6(c), its magnitude spectrum exactly cancels the magnitude spectrum of the pole leaving only the phase spectrum. As the zero leaves the allpass position and moves close to the unit circle, the magnitude spectrum of the zero falls to near zero at ω_0, as shown in Fig. 4.6(d), making a narrow-band notch filter. If the zero is right on the unit circle, the frequency response of the notch filter is exactly zero at ω_0, as shown in Fig. 4.7(a). Then, when the zero moves outward just off the unit circle, as in Fig. 4.7(b), a minimum phase notch filter results with a magnitude not greatly different from the case with the zero just inside the unit circle.

These notch filters are particularly useful for rejecting narrow-band interference, such as 60 Hz line noise. A more subtle application, which

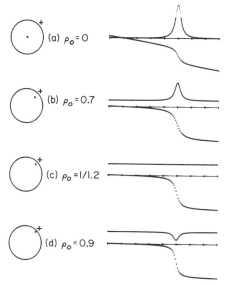

Figure 4.6 Computer sketches of four different single-pole, single-zero filters, each with a different zero position. The power spectrum (solid line) and the phase spectrum (dotted line) are plotted over the interval $-\pi$ to π. The phase spectra all have the same vertical scale, but different vertical scales have been used for the power spectra. In all plots, the pole is located at $\rho_+ = 1.2$ and $\omega_0 = \pi/4$, while the zero is on the same radial line located at the distances shown. (a) Pure delay numerator; (b) the zero starts to cancel the pole; (c) the zero is located at the allpass position; and (d) the zero dominates the pole to produce a notch filter.

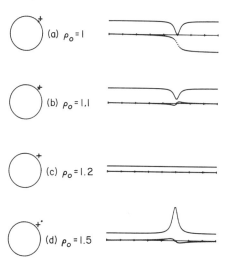

Figure 4.7 See Fig. 4.6 for the discussion. (a) The zero located on the unit circle produces a strong notch filter; (b) the zero is partially canceled by the pole to produce a weaker notch filter; (c) the pole and zero exactly cancel to produce a constant filter; and (d) the zero outside the pole's position produces the pole-on-pedestal filter.

will only be fully understood in later chapters, is to use the notch filter to subtract out nonharmonic frequencies in spectral estimation problems.

Next, a special situation occurs in Fig. 4.7(c) when the zero reaches the same position on the radius as the pole; they exactly cancel leaving only the trivial constant filter.

After the zero has moved beyond the pole, it has a diminishing effect on the frequency response. When the zero is far from the pole, it contributes only weakly to the magnitude and phase spectrum of the filter. If the pole is close to the unit circle, the familiar narrow-band filter will result, except an added constant from the distant zero shifts the magnitude response upward as depicted in Fig. 4.7(d); it is called a pole-on-pedestal filter. This pole-on-pedestal filter is similar to the single-pole, narrow-band filter except for the asymptotic behavior away from ω_0. The pole-on-pedestal has a flat response away from ω_0 while the single-pole, narrow-band filter decays away from ω_0. The pole-on-pedestal is more convenient for designing complicated filters by cascading multiple poles because its magnitude response has little interaction with those of the other poles away from ω_0.

The approach taken in plotting the single-zero, single-pole filter response of Figs. 4.6 and 4.7 can be readily expanded to include any number of zeros and poles. But, as explored in Problems 4.4 and 4.7, the poles that are useful in providing sharp transitions in the spectra cause long decay transients and wild phase behavior. The problem inverse to the present one under discussion is to specify a given magnitude and phase response and ask for the time domain operator that gives that response. This is the problem of digital filter design, discussed in Chapter 8.

In this chapter, we have studied the frequency response of very simple filters that were represented by two-term sequences—couplets. In the preceding chapter, where we first introduced the concept of frequency response via the sinusoidal eigenfunctions and their associated eigenvalues, we calculated the spectral response of longer but still quite simple sequences. So far, our computational methods have been analytic. Clearly, to proceed further—to conveniently compute the spectrum of a long sequence of numbers representing a complicated filter or a large data stream—we need a digital computational scheme for evaluating the spectrum. We address that interesting and important problem in the next chapter.

Problems

4.1 Give a qualitative discussion explaining the broad passband of a single-zero filter compared to the narrow passband of a single-pole filter; for example, compare Fig. 4.3(b) with Fig. 4.5.

Problems 77

4.2 Design a two-pole, two-zero, 2-Hz wide notch filter to reject 60-Hz line noise from data sampled at 600 Hz.

4.3 Loran is a long-range radio navigation system that employs a pulsed narrow-band signal at 100 kHz. Design a 3-term, narrow-band recursive filter with a 100 Hz bandwidth for use in improving the signal-to-noise ratio of Loran data sampled at 1000 kHz.

4.4 The penalty for employing narrow-band filters is their long decay time. Estimate the decay time of the filter used in Problem 4.3. Write a computer program to operate on synthetic data with this filter and observe the decay of its transient response by feeding it a 100 kHz sine wave sampled at 1000 kHz.

4.5 Equation (4.7) defines the power spectrum in the complex Z plane. Calculate the power spectrum for a couplet with a zero at $\rho_0 \exp(i\omega_0)$, display its real and imaginary parts, and show that the imaginary part is identically zero only for Z on the unit circle.

4.6 Multiple-zero, multiple-pole filters can have their phase vary rapidly through many values of 2π during the Nyquist interval. For example, an n-zero maximum phase filter will have its phase augmented by $2n\pi$ over the Nyquist interval. Most methods of computing this phase will yield the result modulo 2π, yet we know the phase of such a polynomial is continuous. Devise an algorithm to produce this continuous phase. This is called phase unwrapping.

4.7 Write a computer program to reproduce the spectral plots in Figs. 4.6 and 4.7, using your phase unwrapping ideas from Problem 4.6. Include the ability to specify arbitrary multiple poles and multiple zeros. Study single-zero, single-pole filters with the zero and pole located at different frequencies. Study multiple-pole, multiple-zero filters including those with poles and zeros occurring in conjugate pairs.

4.8 Show that as an operator relation, operating on any function of Z on the unit circle, one can write

$$\frac{d}{d\omega} = -iZ\frac{d}{dZ}$$

4.9 Using the results of Problem 4.8, show that the group delay of the allpass filter of Eq. (4.6) is indeed given by Eq. (4.8). Recall that the complex logarithm is given by

$$\ln y(Z) = \ln|y(Z)| + i\phi$$

so that for the special case of the allpass filter,

$$\phi = \frac{1}{i}\ln[H(Z)]$$

since $\ln|H(Z)| = 0$.

4.10 Using the above definition of the complex logarithm, show that for any complex spectrum $H(\omega)$, sampled at points ω and $\omega + \Delta\omega$, a reasonable approximation for the group delay is

$$\tau_g = \frac{2}{\Delta\omega} \text{Im}\left[\frac{H(\omega + \Delta\omega) - H(\omega)}{H(\omega + \Delta\omega) + H(\omega)}\right]$$

4.11 Integrate τ_g of the single-zero, single-pole allpass filter over the Nyquist interval and find the filter's average group delay when it operates on 1000 Hz data. Interpret your results.

4.12 Show that the two couplets

$$F_1 = (Z - \sqrt{2}e^{-i3\pi/4})$$

and

$$F_2 = \sqrt{2}(Z - e^{-i3\pi/4}/\sqrt{2})$$

have the same magnitude spectrum, that F_1 is minimum phase, and that F_2 is maximum phase. Plot the zeros in the Z plane and sketch the phase lags to show that the minimum phase couplet does not have less phase delay at all frequencies.

4.13 What is the nature of the pole and zero locations of a causal digital filter?

4.14 Discuss the inverse of the allpass filter.

4.15 Show that $\ln y(Z) = \ln|y(Z)| + i\phi(Z)$.

5

The Discrete Fourier Transform

We have spent the last two chapters studying the frequency response of various operators. In each case, we performed an analytic calculation that resulted in a closed-form expression for $H(\omega)$. Traditionally, closed-form solutions—for any mathematical problem—have been prized for their intrinsic elegance and beauty. Indeed, they are a satisfying outcome for any problem. But frequently in using such solutions, we like to make graphs. In making a graph, we evaluate some function at selected points, either by hand or with the aid of a computer. If a computer is used, commonly we would evaluate the function at equally spaced points sufficiently close together to give a complete picture of the solution.

In this chapter, we apply this idea of evaluating, or sampling, to the frequency response of LSI systems and operators. This frequency sampling of $H(\omega)$ will lead us to the discrete Fourier transform, a powerful mathematical tool in digital signal processing.

Sampling the System Response in the Frequency Domain

Because sinusoids are eigenfunctions of discrete LSI systems, we have found their eigenvalues, the spectrum, extremely useful in thinking about the behavior of digital operators. In Chapter 3, we showed that the spectral response of any LSI operator can be computed from its impulse response function, h_κ, from

$$H(\omega) = \sum_{\kappa=0}^{N-1} h_\kappa e^{-i\omega\kappa} \tag{5.1}$$

Two important basic properties of $H(\omega)$ are (1) that it is a continuous function of ω, and (2) that it is periodic in ω with a period of 2π. We want to sample $H(\omega)$ at equally spaced points in frequency:

$$\omega = 2\pi\nu/M \qquad \nu = 0, 1, 2, \ldots, M-1$$

where we have limited the maximum frequency to $\omega = 2\pi$ because no new information is gained above that point because of the periodicity of $H(\omega)$. This sampling gives the M values of H:

$$H_\nu = \sum_{\kappa=0}^{N-1} h_\kappa e^{-i2\pi\nu\kappa/M} \qquad \nu = 0, 1, 2, \ldots, M-1$$

representing a transformation of N numbers, h_κ, to M other numbers, H_ν. Naturally, we would expect that a more powerful picture would emerge if this transformation could be easily inverted; that is if the h_κ could be found from the H_ν. With this objective in mind, we limit the number of frequency points M to equal N, the number of points in time. Although this restriction is not absolutely necessary, we do expect it to lead to a convenient inversion of the resulting N equations and N unknowns. Indeed, the inverse turns out to be quite simple: taking

$$H_\nu = \sum_{\kappa=0}^{N-1} h_\kappa e^{-i2\pi\nu\kappa/N} \qquad \nu = 0, 1, 2, \ldots, N-1$$

and multiplying both sides by $\exp(i2\pi\nu\kappa'/N)$ and summing on ν gives

$$\sum_{\nu=0}^{N-1} H_\nu e^{i2\pi\nu\kappa'/N} = \sum_{\kappa=0}^{N-1} \sum_{\nu=0}^{N-1} h_\kappa e^{i2\pi\nu(\kappa'-\kappa)/N} \qquad (5.2)$$

Next, we introduce a fundamental relation for discrete exponential functions:

$$\sum_{\nu=0}^{N-1} e^{i2\pi\nu(\kappa'-\kappa)/N} = N\delta_{\kappa'\kappa} \qquad (5.3)$$

where δ is the Kronecker delta symbol, equal to one if $\kappa' = \kappa$ and zero otherwise. This *orthogonality relation*, as Eq. (5.3) is sometimes called, is easily verified by thinking of the summation as vector addition in the complex plane: for all cases where $\kappa' \neq \kappa$, the vector diagram is a closed polygon whose resultant is zero. If $\kappa = \kappa'$, the segments of the polygon form a straight line on the real axis composed of N vectors of unit length.

Returning to the evaluation of Eq. (5.2), this orthogonality property of discrete exponentials gives us

$$\sum_{\nu=0}^{N-1} H_\nu e^{i2\pi\nu\kappa'/N} = \sum_{\kappa=0}^{N-1} h_\kappa N\delta_{\kappa'\kappa} = Nh_{\kappa'}$$

Sampling the System Response

or, dropping the prime on κ,

$$h_\kappa = \frac{1}{N} \sum_{\nu=0}^{N-1} H_\nu e^{i2\pi\nu\kappa/N}$$

which is the desired inverse transformation from the H_ν to the h_κ. This pair of equations, which we rewrite for reference,

$$H_\nu = \sum_{\kappa=0}^{N-1} h_\kappa e^{-i2\pi\nu\kappa/N} \qquad (5.4a)$$

$$h_\kappa = \frac{1}{N} \sum_{\nu=0}^{N-1} H_\nu e^{i2\pi\nu\kappa/N} \qquad (5.4b)$$

is collectively called the *discrete Fourier transform*. Sometimes, one is more specific and calls Eq. (5.4a) the discrete Fourier transform (the DFT), and Eq. (5.4b) is called the inverse discrete Fourier transform (the IDFT). Our discussion was motivated by thinking of the h_κ as the real impulse response of a digital LSI system and then the H_ν are sampled complex values of its frequency response. As a mathematical entity, the DFT can be thought of as a more general mapping of N complex numbers into N other complex numbers. We hasten to issue an alert: DFT definitions are not completely standard and you may find some that differ from Eqs. (5.4). In particular, the factor of N can appear in either equation (or even be symmetrically disposed by a factor $\sqrt{N}$), and the forward and inverse definitions can be reversed. This lack of standard definitions stems from an important property of the DFT—the duality of Eqs. (5.4)—where time and frequency play indistinguishable mathematical roles. Any property connecting time to frequency will have a corresponding statement connecting frequency to time.

The importance of the DFT to digital signal processing stems from two properties: first, it has the obvious relationship to the frequency domain; and second, the transformation has a high degree of symmetry, permitting a fast computing scheme called the *fast Fourier transform* or FFT. There are many ways of designing FFT algorithms, all with different advantages and limitations. Details can be found in many references, such as Elliott (1988), Press *et al.* (1986), IEEE (1979), and Rabiner and Gold (1975). But, you need not write your own FFT program. They are readily available in most scientific computing subroutine packages, they are even available in firmware on some microcomputers, and there are FFTs available on chips. Many FFT algorithms require data whose length is a power of two, but we will see that this is hardly ever a limitation. Problems 5.14, 5.15, and 5.16 discuss the computation of DFTs, both by direct calculation and by an FFT algorithm. To take advantage of these fast computing algorithms, we do need to fully understand the properties of the DFT and its use in performing LSI operations in the frequency domain.

Properties of the DFT

The DFT is a restricted version of the Z transform; it is the Z transform sampled at equally spaced points around the unit circle. Consequently, most of the important properties of the DFT will be familiar to us from our Z transform experience.

Perhaps the most basic property of the DFT is its periodicity. The periodicity of the spectrum of sampled time domain data has been a central point in our discussion of digital signal processing. But now, sampling in the frequency domain by the DFT has forced periodicity in the time domain:

$$h_{\kappa+N} = \frac{1}{N} \sum H_\nu e^{i2\pi\nu(\kappa+N)/N} = \frac{1}{N} \sum H_\nu e^{i2\pi\nu\kappa/N}$$

$$h_{\kappa+N} = h_\kappa \quad \text{because} \quad e^{i2\pi\nu} = 1$$

This periodicity in both domains can be represented in two ways, as shown in Fig. 5.1. Thinking of the DFT points as lying on a circle is frequently helpful when thinking of the properties of the transformation.

But what about this, perhaps unexpected, perhaps even unwanted, periodicity imposed on the time domain data? There are some applications, occurring relatively rarely in digital signal processing problems, in which we do indeed have repetitive time domain data. However, in the majority of applications, the time domain data are not periodic, leading us to an intriguing dilemma that strikes at the very heart of digital signal processing: in order to justify digitizing data without aliasing, its spectrum must be band limited. If the data is band limited in frequency, the data must be either periodic or infinitely long; in practice, neither is ever exactly true. In subsequent chapters, we will explore the consequences of this inconsistency between reality and Fourier theory in more detail. But beware, it will forever haunt us. For now we will present less subtle properties of the DFT.

For starters, we note that because of the periodicity of the DFT, the summation in Eqs. (5.4) can start at any point on the circle and continue through one period. For example, it is common to write these equa-

Properties of the DFT

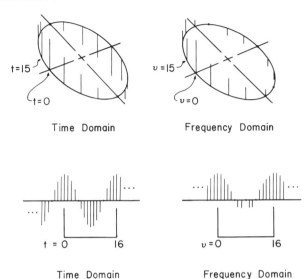

Figure 5.1 Showing the repetitive nature of the DFT data in both domains. At the top, the periodicity is represented by arranging the data on circles. At the bottom, the periodicity is shown by implying infinitely long periodic sequences. Two sets of figures are actually required in each domain: one for the real part of the data and one for the imaginary part.

tions as

$$H_\nu = \sum_{-N/2}^{N/2-1} h_\kappa e^{-i2\pi\nu\kappa/N} \tag{5.5a}$$

$$h_\kappa = \frac{1}{N} \sum_{-N/2}^{N/2-1} H_\nu e^{i2\pi\nu\kappa/N} \tag{5.5b}$$

so that the indices run from $-N/2$ through zero to $N/2 - 1$. For lack of better terminology, we will call Eqs. (5.4) the one-sided format and Eqs. (5.5) the centered format. The only difference is where one starts counting elements on the DFT circle. For example, the one-sided format gives

$$\text{DFT}(1, 0, 0, 1) = (2, 1 + i, 0, 1 - i) \tag{5.6}$$

where the centered format of the same sequence would read

$$\text{DFT}(0, 1, 1, 0) = (0, 1 - i, 2, 1 + i) \tag{5.7}$$

Since the one-sided format uses only positive indices, it is frequently more convenient for computer implementation, whereas the centered version is more useful for discussing the symmetry properties that we take up next.

One may think it strange that the DFT transforms N real numbers h_κ into $2N$ numbers, resulting from the real and imaginary parts of the N terms of H_ν. This situation is resolved by the symmetry of the transformation that results when h_κ is real. In the more general case, h_κ can be complex; the DFT then transforms N complex h_κ into N complex H_ν. To investigate the symmetry properties of the DFT, we take the complex conjugate of Eq. (5.5a) to get

$$H_\nu^* = \sum_{-N/2}^{N/2-1} h_\kappa^* e^{i2\pi\nu\kappa/N}$$

and then relabel ν to $-\nu$:

$$H_{-\nu}^* = \sum_{-N/2}^{N/2-1} h_\kappa^* e^{-i2\pi\nu\kappa/N}$$

The previous two equations show that

$$H_{-\nu}^* \leftrightarrow h_\kappa^* \tag{5.8a}$$

$$H_\nu^* \leftrightarrow h_{-\kappa}^* \tag{5.8b}$$

Also, it is easy to see that

$$H_{-\nu} \leftrightarrow h_{-\kappa} \tag{5.8c}$$

where the double arrow indicates taking the DFT when going from κ, in time, to ν, in frequency, or taking the IDFT in the opposite direction. For the special case when h_κ is real, Eqs. (5.8) say that

$$H_{-\nu}^* = H_\nu \tag{5.9}$$

a property of H that is sometimes called *Hermitian*; it means that H has a symmetric real part and an antisymmetric imaginary part. Thus, H only has N independent components. If h is purely imaginary, Eqs. (5.8) say that

$$H_{-\nu}^* = -H_\nu \tag{5.10}$$

a property of H that is sometimes called *anti-Hermitian*; it means that H has an antisymmetric real part and a symmetric imaginary part.

Further special symmetries result if h_κ or H_ν is an even or odd function of their index. For example, if h_κ is real and even about $\kappa = 0$, in the case of a zero-phase real operator, then Eq. (5.8b) says that

$$H_\nu^* \leftrightarrow h_{-\kappa}^* = h_{-\kappa} = h_\kappa$$

or

$$H_\nu^* = H_\nu$$

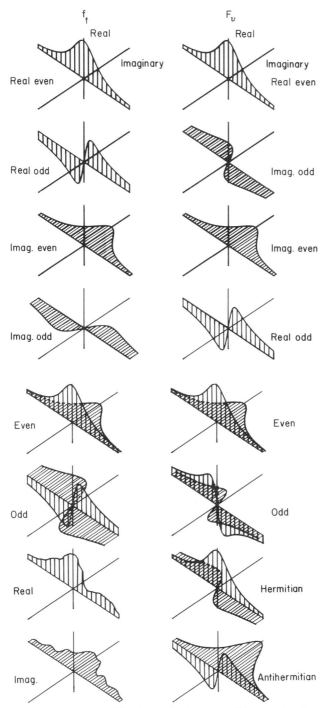

Figure 5.2 Symmetry properties of the DFT using the centered format whereby the indices t and ν run from $-N/2$ to $N/2 - 1$. For even N, the center of symmetry occurs at t and $\nu = 0$.

and

$$H_\nu = H_{-\nu}$$

That is, H is purely real and even about $\nu = 0$. A useful summary of all such symmetry properties, which are likewise easily verified, is shown in Fig. 5.2.

Special Values of the DFT

Next, we consider the transformation equations, Eqs. (5.5), for some special values of their indices. For $\nu = 0$, Eq. (5.5a) reads

$$H_0 = \sum_{-N/2}^{N/2-1} h_\kappa \tag{5.11}$$

The zero-frequency value of the spectrum is the sum of the time terms. When divided by N, this sum becomes the average value of the time sequence—the dc value in electrical circuit terms. Similarly, for $\kappa = 0$, Eq. (5.5b) reads

$$h_0 = (1/N) \sum_{-N/2}^{N/2-1} H_\nu \tag{5.12}$$

sometimes simply called the first value of the time sequence.

Another special value occurs one-half way around the DFT circle:

$$H_{N/2} = \sum h_\kappa e^{-i2\pi N\kappa/2N} = \sum e^{-i\pi\kappa} h_\kappa$$

or

$$H_{N/2} = \sum_{-N/2}^{N/2-1} (-1)^\kappa h_\kappa \tag{5.13}$$

which is called the Nyquist value of H. Likewise, for h,

$$h_{N/2} = \frac{1}{N} \sum H_\nu e^{i2\pi\nu N/2N} = \frac{1}{N} \sum e^{i\pi\nu} H_\nu$$

$$h_{N/2} = \frac{1}{N} \sum_{-N/2}^{N/2-1} (-1)^\nu H_\nu \tag{5.14}$$

Another important relation, called *Parseval's* theorem, readily follows from the definition of the DFT:

$$N \sum_{-N/2}^{N/2-1} |h_\kappa|^2 = \sum_{-N/2}^{N/2-1} |H_\nu|^2 \tag{5.15}$$

The sum of the squares of the amplitude in a signal is sometimes called the total energy of the signal in analogy with an electrical potential across a load.

These symmetry properties and special values that we have discussed in Eqs. (5.8) through Eq. (5.15) are of great practical value in checking and using DFT calculations. Note that many of these properties can be easily demonstrated for the DFT shown in Eq. (5.7).

The Phase-Shift Theorem

The value of the frequency domain is due in large part to providing an alternate viewpoint to the time domain. The usefulness of this alternate view hinges critically on developing connecting relationships between the two domains. We have just discussed relationships between numbers calculated in each domain. Next, we wish to explore relationships between operations in both domains. These operations, intimately associated with the DFT's granddaddy—the Z transform—will play a central role in using the DFT in digital signal processing.

The first of these operations is directly associated with the delay operator Z^τ. To delay a sequence τ units in time, its Z transform is multiplied by this delay operator. The DFT of a delayed sequence is similarly related to the DFT of the original sequence by

$$\mathrm{DFT}(h_{\kappa-\tau}) = e^{-i2\pi\nu\tau/N}\mathrm{DFT}(h_\kappa) \qquad (5.16)$$

which is a relation, easily verified by direct substitution, called the *phase shift theorem*; a shift in time is a phase factor in the frequency domain. As an example, we consider the 4-long real symmetric sequence $(0, 0, 1, 0)$; its magnitude spectrum is all one's $(1, 1, 1, 1)$ and its phase spectrum is zero. We can write phase of the DFT $(0, 0, 1, 0) = (0, 0, 0, 0)$. If the one is shifted through the sequence by progressive delays, the spectrum is multiplied by

$$e^{-i\pi\tau/2}$$

which leaves the magnitude spectrum unchanged and introduces a linear phase with a negative slope of 90° per unit of time shift. We can write

phase of the DFT $(0, 0, 0, 1) = (180°, 90°, 0, -90°)$
phase of the DFT $(1, 0, 0, 0) = (360°, 180°, 0, -180°)$
phase of the DFT $(0, 1, 0, 0) = (540°, 270°, 0, -270°)$

This phase-shift theorem has many important applications. For example, the one shifting through the 4-long sequence might represent wave propagation to the right in time. During propagation, the wave's magnitude

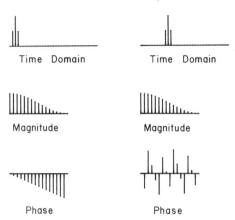

Figure 5.3 Two examples of the phase-shift theorem. At the top left, a time domain signal is shifted one unit from the symmetric position. Its phase spectrum, at the bottom left, shows a phase that varies linearly from 0 to $-\pi$ over one-half the Nyquist interval. At the top right, the same signal is shown shifted 9 units from the symmetrical position. The magnitude spectrum is unchanged; its phase spectrum varies so rapidly that unwrapping, to show the linear phase dependency, is difficult. Only one-half of the frequency spectrum of the 32-point DFT is shown because the real time domain signal produces a symmetric magnitude spectrum and an antisymmetric phase spectrum.

spectrum remains unchanged while its phase spectrum undergoes clockwise rotation. The calculation of DFT phases can run into two difficulties, particularly on natural data, that frequently make the phase spectrum look like noisy nonsense. First, the phase may be the computed modulo 2π, or π, so that phases that are changing rapidly are difficult to unwrap. Second, the phase is usually calculated from the arc tangent of the ratio of the imaginary part of the spectrum to the real part. If the real part meanders about zero, wild fluctuations result in the phase angle. Examples are shown in Fig. 5.3.

The Convolution Theorem

Like the phase-shift theorem, the convolution theorem reflects a fundamental property of the Z transform. Convolution in time corresponds to multiplying Z transforms, and therefore, it also corresponds to multiplying DFT spectra. We know that the most general operation an LSI system can perform on a time sequence is a convolution; clearly, the convolution theorem promises to be a major link between the two domains.

We prove the DFT convolution theorem by multiplying two spectra and transforming back to time. Capital letters will be used for frequency

The Convolution Theorem

domain data and lowercase letters for time domain data:

$$F_\nu = \sum_\kappa f_\kappa e^{-i2\pi\kappa\nu/N}$$

$$G_\nu = \sum_{\kappa'} g_{\kappa'} e^{-i2\pi\kappa'\nu/N}$$

The product of the spectra becomes

$$F_\nu G_\nu = \sum_\kappa \sum_{\kappa'} f_\kappa g_{\kappa'} e^{-i2\pi\nu(\kappa'+\kappa)/N}$$

and taking the IDFT of the product gives

$$\text{IDFT}(F_\nu G_\nu)_\tau = \frac{1}{N} \sum_\nu e^{i2\pi\nu\tau/N} \sum_\kappa \sum_{\kappa'} f_\kappa g_{\kappa'} e^{-i2\pi\nu(\kappa'+\kappa)/N}$$

$$= \frac{1}{N} \sum_\kappa \sum_{\kappa'} f_\kappa g_{\kappa'} \sum_\nu e^{-i2\pi\nu(\kappa'+\kappa-\tau)/N}$$

$$= \frac{1}{N} \sum_\kappa \sum_{\kappa'} f_\kappa g_{\kappa'} N \delta_{\kappa', \tau-\kappa}$$

where we have used the orthogonality property of Eq. (5.3) to evaluate the sum over ν. Finally, the δ symbol only contributes the $\tau-\kappa$ term from the sum over τ':

$$\text{IDFT}(F_\nu G_\nu)_\tau = \sum_\kappa f_\kappa g_{\tau-\kappa} \quad (5.17a)$$

Likewise, if we started by multiplying two time sequences and transformed them into the frequency domain, we would see that

$$\text{DFT}(f_\tau g_\tau)_n = \frac{1}{N} \sum_\nu F_\nu G_{n-\nu} \quad (5.17b)$$

so that convolving in one domain corresponds to multiplication in the other domain.

A good reason for taking time to prove the DFT convolution theorem is that inspection of the proof reveals that all the summations used are around the DFT circle; the convolution sums in Eqs. (5.17) are called *circular convolution*.

The distinction between circular convolution and *linear convolution*, the type that we have considered until now, is an important detail of DFT data processing. Perhaps the best way to elaborate on the difference is by means of an example. We select $(1, 3, 0, 2) * (1, 0, 2, 2)$. Under circular convolution,

$$(1, 3, 0, 2) * (1, 0, 2, 2) = (7, 6, 10, 7)$$

while under linear convolution,

$$(1, 3, 0, 2) * (1, 0, 2, 2) = (1, 3, 2, 10, 6, 4, 4)$$

The results are, of course, completely different—they even have different lengths. Hand calculations of short circular convolutions are best performed by repeating one factor and then shifting from the zero lag position so as to maintain complete overlap; schematically, the above case would look like

$$\begin{array}{c} (\ 2\ 0\ 3\ 1\ ) \rightarrow \\ \cdots\ 1\ 0\ 2\ 2\ 1\ 0\ 2\ \cdots \end{array}$$

Linear convolution can be included within the circular convolution framework by appending zeros to eliminate the overlap between cycles. A schematic representing this looks like

$$\begin{array}{c} (\ 0\ 0\ 0\ 0\ 2\ 0\ 3\ 1\ ) \rightarrow \\ \cdots\ 1\ 0\ 2\ 2\ 0\ 0\ 0\ 0\ 1\ 0\ 2\ 2\ 0\ 0\ 0\ 0\ \cdots \end{array}$$

It follows that the advantage of high-speed FFT algorithms can be used for linear convolution: the two factors are appended with zeros so as to be of equal length, at least twice as long as the longer of the two. The product of the DFTs of each sequence is then transformed back into the time domain.

Symmetry Properties

Time	$\longleftrightarrow$	Frequency
f_t^*		$F_{-\nu}^*$
f_{-t}^*		F_ν^*
f_{-t}		$F_{-\nu}$

Convolution Relations

Time	$\longleftrightarrow$	Frequency
$f * g$		$F_\nu G_\nu$
$f * g_{-t}$		$F_\nu G_{-\nu}$
$f_{-t} * g_{-t}$		$F_{-\nu} G_{-\nu}$
$f_t * g_{-t}^*$		$F_\nu G_\nu^*$
$f_t * g_t^*$		$F_\nu G_{-\nu}^*$
$f_{-t} * g_{-t}^*$		$F_{-\nu} G_\nu^*$
$f_{-t} * g_t^*$		$F_{-\nu} G_{-\nu}^*$
$f_{-t}^* * g_{-t}$		$F_\nu^* G_\nu^*$
$f_{-t}^* * g_t$		$F_\nu^* G_{-\nu}^*$
$f_t^* * g_t^*$		$F_{-\nu}^* G_{-\nu}^*$

Figure 5.4 Symmetry properties of the DFT and their application to 10 forms of the convolution theorem.

For sufficiently long sequences, the speed of the FFT more than compensates for what seems, at first sight, like extra calculations.

Clearly, the convolution theorem will play a central role in the remainder of our discussions. It is not surprising then, that application of the symmetry properties of the DFT discussed earlier in the chapter leads to some equally interesting forms of the convolution theorem. Figure 5.4 reviews these symmetry properties and lists 10 forms of the convolution theorem resulting from them. In the next section, we study one particular form.

Cross-Correlation and Autocorrelation

A version of the fourth form of the convolution theorem listed in Fig. 5.4, obtained by a simple relabeling, is

$$\text{DFT}(f^*(-) * g) = F^* G \quad (5.18)$$

It has particular significance. First, let us write out the convolution indicated on the left-hand side; the τth element is

$$f^*(-) * g = \sum_t f^*_{-(\tau-t)} g_t = \sum_t f^*_{(t-\tau)} g_t$$

We define the particular convolution sum to be the complex *cross-correlation* of f^* with g and use the notation

$$f^* \otimes g = \sum_t f^*_{t-\tau} g_t \quad (5.19a)$$

By a simple change of variables

$$t' = t - \tau$$

we can see, after dropping the prime notation, that

$$f^* \otimes g = \sum_t f^*_t g_{t+\tau} \quad (5.19b)$$

Clearly, the cross-correlation is similar to convolution; the difference is that the order is not reversed in the cross-correlation computation as it is for convolution. For example, to compute the cross-correlation between the two 4-long sequences of the preceding section,

$$(1, 3, 0, 2) \otimes (1, 0, 2, 2)$$

the computational schematic looks like

$$(\; 1 \; 3 \; 0 \; 2 \;)$$
$$\cdots \; 1 \; 0 \; 2 \; 2 \; 1 \; 0 \; 2 \; 2 \; \cdots$$

where neither sequence is reversed. The result is

$$(1,3,0,2) \otimes (1,0,2,2) = (5,8,8,9)$$

Also

$$(1,0,2,2) \otimes (1,3,0,2) = (5,9,8,8)$$

Comparison of these two results show that factors of the cross-correlation do not commute, but they are time reversed versions of one another. (Remember that these results are on the DFT circle; time reversal implies a cyclic progression in the opposite sense.) That is, if

$$f \otimes g = R_t$$

then

$$g \otimes f = R_{-t}$$

An important special case of cross-correlation occurs when $g = f$. Then, Eq. (5.19a) becomes the *autocorrelation* of f, which reads

$$f^* \otimes f = \sum_t f^*_{t-\tau} f_t \qquad (5.20a)$$

and

$$f^* \otimes f = \sum_t f^*_t f_{t+\tau} \qquad (5.20b)$$

showing that the autocorrelation is symmetric in its lag variable τ. Furthermore, Eq. 5.18 now says

$$\text{DFT}(f^* \otimes f) = F^*F = |F|^2 \qquad (5.21)$$

or, in words, the DFT of the autocorrelation is the DFT power spectrum. It is possible to prove that the autocorrelation is a maximum at zero lag. For this reason, sometimes it is normalized to unity at $\tau = 0$ by writing

$$f^* \otimes f = \sum f^*_t f_{t+\tau} / \sum f^*_t f_t$$

The DFT power spectrum of a sequence and its autocorrelation contain the same information. In the frequency domain, it is clear that the sequence's phase information is lost. All sequences with the same power spectrum have the same autocorrelation, even though the sequences differ. For example, $(4, 0, -1)$ and $(2, 3, -2)$ have the same autocorrelation. Autocorrelation will play an important role in later chapters in our discussions of inverse filtering, spectral factorization, and power spectrum estimation.

Both cross-correlation and autocorrelation are easily described in the Z plane. Since cross-correlation differs from convolution in its time reversal, it is clear that the complex cross-correlation between two sequences, $A(Z)$ and $B(Z)$, is given by

$$A^* \otimes B = A^*(1/Z)B(Z)$$

and the autocorrelation of a sequence A is

$$A^* \otimes A = A^*(1/Z)A(Z) \tag{5.22}$$

which is the power spectrum in the Z plane introduced in Eq. (4.7).

In this chapter, we have found that the concept of the DFT, a discrete time, discrete frequency Fourier transform, arises quite naturally from the desire to sample the continuous spectrum of a discrete time sequence. We have only presented the fundamental properties of the DFT, the mechanics we might say. Before we can profitably turn the wheels of this DFT machinery and apply it to practical problems, we need a deeper insight into the subtleties required to relate DFT data to the real world. This relationship can only be brought to light within the conceptual framework of continuous time and continuous frequency. In the next chapter, we will develop the appropriate Fourier transform for this case and then use it to illustrate the role of the DFT in data processing.

Problems

5.1 Clearly, the DFT of Eq. (5.4a) is not a shift invariant operation on the h_κ. Show, however, that for $N = 4$ Eq. (5.4a) can be written as the matrix operation

$$\begin{pmatrix} H \end{pmatrix} = \begin{pmatrix} 1 & 1 & 1 & 1 \\ 1 & W & W^2 & W^3 \\ 1 & W^2 & W^4 & W^6 \\ 1 & W^3 & W^6 & W^9 \end{pmatrix} \begin{pmatrix} h \end{pmatrix}$$

where $W = \exp(-i2\pi/N)$. A matrix whose elements satisfy $A_{\nu\kappa} = A^{(\nu-1)(\kappa-1)}$ is called a Van Der Monde matrix. Show that the above matrix satisfies this definition. Show that the inverse to the above matrix just has each element inverted, that is the inverse equation is

$$\begin{pmatrix} h \end{pmatrix} = \frac{1}{N} \begin{pmatrix} 1 & 1 & 1 & 1 \\ 1 & W^{-1} & W^{-2} & W^{-3} \\ 1 & W^{-2} & W^{-4} & W^{-6} \\ 1 & W^{-3} & W^{-6} & W^{-9} \end{pmatrix} \begin{pmatrix} H \end{pmatrix}$$

5.2 It is sometimes useful to have a stock list of DFTs for reference. Show that the following are DFT transforms (we are using the centered format):

$$\text{DFT}(0,0,1,0) = (1,1,1,1)$$
$$\text{DFT}(1,1,1,1) = (0,0,4,0)$$
$$\text{DFT}(0,0,1,1) = (0,1+i,2,1-i)$$
$$\text{DFT}(0,-1,1,1) = (1,1+2i,1,1-2i)$$
$$\text{DFT}(0,1,1,0) = (0,1-i,2,1+i)$$
$$\text{DFT}(0,-1,0,1) = (0,2i,0,-2i)$$
$$\text{DFT}(0,-i,0,i) = (0,-2,0,2)$$

5.3 Identify the dc and the Nyquist values of the DFTs in Problem 5.2. Show that they satisfy Eqs. (5.12) and (5.13). Show that the DFTs satisfy the symmetry properties for even and odd functions.

5.4 The third and fifth sequences listed in Problem 5.2 differ by a shift of one unit. Convert the DFT spectra shown into magnitude and phase, showing that these spectra differ only by an overall phase factor.

5.5 Multiply the spectra of examples three and five from Problem 5.2. Compute the IDFT of the resulting product and show that it is equal to $(0,0,1,1) * (0,1,1,0)$ as an example of the convolution theorem.

5.6 Prove Parseval's theorem, Eq. (5.15), and verify that the examples of Problem 5.2 satisfy it.

5.7 Our definition of the cross-correlation started with the convolution $f^*(-)*g$. Show that if you start with $f*g^*(-)$ and proceed in a similar fashion, the resulting cross-correlation is the time-reversed version of our definition.

5.8 Compute both the linear and the circular autocorrelation of the following sequences: $(2,5,2)$, $(4,4,1)$, and $(1,4,4)$.

5.9 Prove the phase-shift theorem of Eq. (5.16) by direct substitution into the DFT equations.

5.10 In the paragraph following Eqs. (5.4), we spoke of the duality of the DFT equations. Find the statement that is dual to the phase-shift theorem stated in Eq. (5.16).

5.11 Figure 5.4 shows 10 forms of the convolution theorem. Use these relationships to write down at least 5 similar expressions that relate autocorrelation in the time domain to multiplication in the frequency domain.

5.12 In Chapter 1, we noted that convolutions can be written as matrix operations. Show that for circular convolution the matrix must be a *circulant*, that is, a special Toeplitz structure where each row is a circular shift of the one above. As a consequence of this circulant structure, show that the eigenvalues of the matrix are given by the DFT of the first row. That is,

$$\mathbf{C}\mathbf{V}_i = \lambda_i \mathbf{V}_i$$

where **C** is the circulant matrix, **V** are the eigenvectors, and λ are the eigenvalues. How are the DFTs of the other rows related to the eigenvalues?

5.13 Discuss the time reversed version of a real minimum phase sequence within the DFT formalism.

5.14 It is important to realize that the DFT is a simple computation that may be done directly when the extra speed of the fast Fourier transform is not required. Use the direct calculation coded below to verify the transforms given in Problem 5.2. In this program, the complex input array is $X + iY$ and the output is $R + iI$. Of course, there is no limitation on their length N.

```
      REAL X(N), Y(N), R(N), I(N)
      DO 11 K = 1, N
      W = 2*PI*(K − 1)/N
      XO = X(1)    YO = Y(1)
      DO 9 J = 2, N
      P = W*(J − 1)
      C = COS(P)   S = SIN(P)
      XO = XO + C*X(J) + S*Y(J)
    9 YO = YO + C*Y(J) − S*X(J)
      R(K) = XO
   11 I(K) = YO
```

5.15 The above program requires $2N^2$ evaluations of the sine and cosine functions while only N different values are required. A table lookup approach will solve this problem, and will be even more efficient when multiple DFT computations of the same length are needed because the table need only be computed once. Modify the above program to operate by table lookup.

5.16 The neat and efficient fast Fourier transform algorithm listed below (after Claerbout, 1976) uses complex arithmetic. Rewrite the program to use only real numbers for use on computers not permitting complex arithmetic. The complex array X is both the input and the output of length N, which must be a power of two. The flag F should be 1 for the forward transform and -1 for the inverse transform. Check the operation of your program against the transforms of Problem 5.2 and the results of the program in Problem 5.14.

```
      SUBROUTINE FFT (X, N, F)
      COMPLEX X(N), A, T, W
      J = 1
      S = SQRT(N**(F - 1))
      DO 14 I = 1, N
      IF (I.GT.J) GOTO 9
      T = S * X(J)
      X(J) = S * X(I)
      X(I) = T
   9  M = N/2
  10  IF (J.LE.M) GOTO 14
      J = J - M
      M = M/2
      IF (M.GE.1) GOTO 10
  14  J = J + M
      L = 1
  16  D = 2 * L
      DO 23 M = 1, L
      A = -(0.0, 1.0) * PI * F * (M-1)/L
      W = CEXP(A)
      DO 23 I = M, N, D
      T = W * X(I + L)
      X(I + L) = X(I) - T
  23  X(I) = X(I) + T
      DO 23 I = M, N, D
      T = W * X(I + L)
      X(I + L) = X(I) - T
  23  X(I) = X(I) + T
      L = D
      IF (L.LT.N) GOTO 16
      RETURN
      END
```

6

The Continuous Fourier Integral Transform

Digital signal processing has become extremely important in recent years because of digital electronics. In treating the processing of analog signals, which require any reasonable amount of computation, it is usually beneficial to digitize the signals and to use digital computers for their subsequent processing. The advantage results from both the extremely high computation speeds of modern digital computers and the flexibility afforded by computer programs that can be stored in software or firmware or hardwired into the design. Low-cost, large-scale integrated circuits, and more recently VLSI circuits, have made this approach beneficial even for devices limited to special-purpose computing applications or restricted by throwaway economics. This computational asset has been a major impetus for thinking of signals as discrete time sequences. An additional advantage of representing signals in discrete time has been their pleasant mathematical simplicity; continuous-time theory requires far more advanced mathematics than the algebra of polynomials, some simple trigonometric function theory, and the behavior of the geometrical series that we have employed. The digital revolution seduces us into viewing every situation in its terms; still, we are haunted by the concept of underlying continuous relationships. We need to know the effect of digitizing continuous-time signals and the essential difference between these digitized signals and those signals that are inherently digital from the start. Are continuous-time and discrete-time versions simply alternate models of the real physical world from which we are free to chose? Some say yes; yet, there are essential differences.

For example, meteorological data, such as the barometric pressure at a given location, would certainly seem to be a continuous signal that conceptually extends infinitely far from the past into the future. Physical considerations force us to conclude that its spectrum is band limited. The pressure wave from a nuclear blast, on the other hand, has a definite beginning and extends with decaying amplitude infinitely long thereafter. We shall see shortly that such a signal must have a frequency spectrum that is not band limited and therefore cannot be digitized without aliasing. Still other signals seem to be inherently digital: the price of a stock issue is determined only by its trading, an event that occurs only at discrete-time intervals. There can be no underlying analog signal. Furthermore, no business lasts forever; its stock trading has a definite beginning and ending. Like the finite-length operators that we have studied so far, the stock quote's spectrum must be repetitive as well as inherently broadbanded.

The spectra of the signals in these three examples are quite different. Clearly, to apply digital signal processing in an intelligent manner, we need to know more about continuous-time functions. We can no longer ignore these concepts—we need to bite the bullet, develop a continuous time theory of signals, and then use it to gain insight into its relationship with digital signal processing.

The Fourier Integral Transform Developed from the DFT

Our development of mathematical machinery is following a natural course, motivated only by the study of LSI digital systems and operators. The concept of the spectrum arose because sinusoids are eigenfunctions of LSI systems. The convenience of sampling the spectrum at equally spaced intervals gave us the DFT. The DFT, with its equally spaced points in both time and frequency, places us in a position to easily take the mathematical limit $N \to \infty$ to pass over to continuous time and frequency. We start with the inverse transform

$$f_t = \frac{1}{N} \sum_{-N/2}^{N/2-1} F_\nu e^{i2\pi\nu t/N} \quad (6.1)$$

and recognize that this sum may be evaluated for any t; it may be considered a continuous function of time. [Just like the similar sum for the spectral response $H(\omega)$ of an LSI operator, Eq. (6.1) can be evaluated at any time.] The frequency interval used in this summation is

$$\Delta\omega = 2\pi/N$$

The Fourier Integral Transform Developed from the DFT

Therefore, the frequency (in radians/unit time) is

$$\omega = 2\pi\nu/N \qquad (6.2)$$

and as N becomes infinite,

$$1/N = \Delta\omega/2\pi \to 0$$

giving for the limit of the sum in Eq. (6.1)

$$f_t = \lim_{\substack{N\to\infty \\ \Delta\omega\to 0}} \sum_{N/2}^{N/2-1} F_\nu e^{i\omega t}\, \Delta\omega/2\pi$$

In the limit $N \to \infty$, this sum becomes an integration—a continuously infinite number of terms separated by the infinitesimal frequency interval $\Delta\omega \to d\omega$:

$$f(t) = \frac{1}{2\pi} \int_{-\infty}^{\infty} F(\omega) e^{i\omega t}\, d\omega \qquad (6.3)$$

Now, both $F(\omega)$ and $f(t)$ are continuous functions. To invert this equation, the orthogonality relation,

$$\frac{1}{N} \sum_{-N/2}^{N/2-1} e^{i2\pi\nu(t'-t)/N} = \delta_{t't}$$

must likewise be converted to continuous time and frequency. The same limiting process, $N \to \infty$ and $1/N = \Delta\omega/2\pi \to 0$, gives the continuous version:

$$\frac{1}{2\pi} \int_{-\infty}^{\infty} e^{i\omega(t'-t)}\, d\omega = \delta(t'-t) \qquad (6.4)$$

Before continuing, we need to elaborate a little on this result: the Kronecker δ has gone over in the limit into a continuous function called the Dirac δ function. Strictly speaking, it is not a function in the mathematical sense at all; nonetheless, it has been put on a firm mathematical basis by Lighthill (1958). For our purposes, the δ function can be thought of as the limiting form of a narrow symmetrical pulse located at t' whose width goes to zero and height goes to infinity such that its area is constant and normalized to unity:

$$\int_{-\infty}^{\infty} \delta(t-t')\, dt = 1$$

Figure 6.1 shows an example of this limiting concept along with the development of the sampling property of the δ function:

$$\int f(t)\delta(t-t')\, dt = f(t')$$

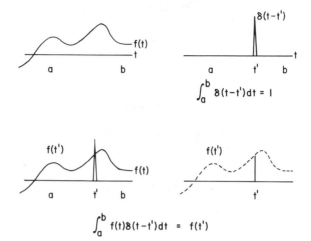

Figure 6.1 The sampling property of the δ function under integration. The area under the δ function is normalized to unity; an integrand composed of the product $f(t)\delta(t-t')$ just renormalizes the δ function's area to $f(t')$.

where $f(t)$ is any continuous function of time and $f(t')$ is its sampled value at t'. This sampling property of the δ function provides an important connection between continuous-time theory and discrete-time theory. In addition, the orthogonality of complex sinusoids expressed by Eq. (6.4) clearly possesses a companion relation, obtained by simply relabeling variables:

$$\frac{1}{2\pi}\int_{-\infty}^{\infty} e^{i(\omega-\omega')t}\,dt = \delta(\omega-\omega') \qquad (6.5)$$

Now, we can return to find the inverse of Eq. (6.3) by using Eq. (6.5). Multiplying both sides of Eq. (6.3) by $\exp(-i\omega't)$ and integrating over time gives

$$\int_{-\infty}^{\infty} e^{-i\omega't}f(t)\,dt = \frac{1}{2\pi}\int_{-\infty}^{\infty} e^{-i\omega't}\int_{-\infty}^{\infty} F(\omega)e^{i\omega t}\,d\omega\,dt$$

$$= \frac{1}{2\pi}\int_{-\infty}^{\infty} F(\omega)\int_{-\infty}^{\infty} e^{i(\omega-\omega')t}\,dt\,d\omega$$

The last integral on the right yields the δ function from Eq. (6.5). Thus,

$$\int_{-\infty}^{\infty} e^{-i\omega't}f(t)\,dt = \int_{-\infty}^{\infty} F(\omega)\delta(\omega-\omega')\,d\omega = F(\omega')$$

The Fourier Integral Transform Developed from the DFT

which is the desired relation giving $F(\omega)$ in terms of $f(t)$. For reference, we rewrite this result and Eq. (6.3) as a transform pair:

$$F(\omega) = \int_{-\infty}^{\infty} f(t)e^{-i\omega t}\,dt \qquad (6.6a)$$

$$f(t) = \frac{1}{2\pi}\int_{-\infty}^{\infty} F(\omega)e^{i\omega t}\,d\omega \qquad (6.6b)$$

This pair of equations affords a continuous-time and continuous-frequency Fourier transformation, which are collectively simply called the *Fourier transform*. Sometimes one is more specific and calls Eq. (6.6a) the forward Fourier transform of $f(t)$, and then Eq. (6.6b) is called the inverse Fourier transform of $F(\omega)$. Note the logical resemblance of these equations to Eqs. (5.4) for the DFT. Again, as in the DFT case, there is an obvious duality of the Fourier transform that results from an interchange of time and frequency by a simple relabeling, or redefinition, of variables. One consequence of this duality is the lack of a standard definition for the Fourier transform; sometimes the forward and inverse versions of the transform in Eqs. (6.6) are interchanged. Different placements of the factor of 2π provides further possibilities for definitions of the Fourier transform.

As we shall see, the similarities between the Fourier transform and the DFT will allow us to exploit the computational advantages of the DFT. But, the differences between the Fourier transform and the DFT, though perhaps few in number, are profound in character. We have approached the Fourier transform from a desire to represent both time and frequency as continuous variables. The resulting transformation equations contain integrals over all values of these variables from minus infinity to plus infinity. This property is a double-edged sword. On the one hand, it does let us represent signals that the DFT does not allow, such as a one-sided transient that decays infinitely far into the future. But, on the other hand, to exactly compute the frequency response of such a signal using a numerical scheme, we would need a continuously infinite number of data points. We will see how to deal with, but not completely solve, this problem later. Another concern, which is nonexistent for the DFT, arises because of the Fourier transform's integrals; we need to know something of their convergence properties.

The convergence of Fourier integrals is a fascinating subject of Fourier theory, explored by many famous mathematicians, such as Plancherel, Titchmarsh, and Wiener. Various conditions have been found that prove the convergence of Fourier integrals for functions displaying rather strange behavior compared to our view of naturally occurring signals. Because our interest is limited to realistic signals and systems, we can afford to start our discussion with an overrestrictive (sufficient but not necessary) convergence condition: the Fourier integral transform of $f(t)$ exists if $f(t)$ is absolutely integrable over the open interval, that is,

$$\int_{-\infty}^{\infty} |f(t)| \, dt < \infty$$

Under these conditions, the repeated integral

$$\hat{f}(t) = \frac{1}{2\pi} \int_{-\infty}^{\infty} e^{i\omega t} \int_{-\infty}^{\infty} f(t')e^{-i\omega t'} \, dt' \, d\omega \qquad (6.7)$$

called the *Fourier integral representation* of f, converges to the average value of $f(t)$ at a discontinuity. That is,

$$\hat{f}(t_0) = \tfrac{1}{2}[f(t_0^+) + f(t_0^-)]$$

when there is a discontinuity at t_0.

Some functions, such as step functions, impulses, and sinusoids, never really occur in nature; nonetheless, they are very convenient for thinking about signals and systems. The absolutely integrable limitation immediately disqualifies many of these favorite functions; clearly, any periodic function, including sinusoids themselves, are excluded from functions possessing Fourier transform pairs if we accept this condition. However, a sufficiently rich class of functions possessing Fourier integral transforms will result if we allow the Dirac δ function to be included. Lighthill (1958) has shown with mathematical rigor how to include δ functions in Fourier integral theory. We simply note that Eq. (6.5) is, indeed, a Fourier transform of a complex sinusoid. This equation shows that the spectrum of $\exp(-i\omega't)$ is eminently reasonable; it contains exactly one pure frequency at ω'.

Furthermore, after our discussion in the next section on the convolution theorem, we will show how Wiener was able to include signals, such as periodic functions and random noise, in frequency analysis. Even though these signals do not possess a Fourier integral transform, they may have a power spectrum.

Properties of the Fourier Integral Transform

Symmetry properties and a few simple theorems played a central role in our thinking about the DFT. These same properties carry over to the Fourier integral transform as the DFT sums go over into Fourier integrals. Since these properties are generally quite easy to prove, and we are already

Properties of the Fourier Integral Transform   103

familiar with them from our discussion of the DFT, we will simply list them in this section, leaving proofs to the problems at the end of the chapter.

The symmetry properties are, of course, crucial for understanding and manipulating Fourier transforms. They can be summarized by

$$\text{FT } f^*(t) = F^*(-\omega) \tag{6.8a}$$

and

$$\text{FT } f(-t) = F(-\omega) \tag{6.8b}$$

The applications of these basic symmetry properties leads to

$$\text{FT } f^*(-t) = F^*(\omega)$$

and for two special cases of interest we can show that

if $f(t)$ is real, $F(\omega)$ is Hermitian, $F^*(-\omega) = F(\omega)$;
if $f(t)$ is imaginary, $F(\omega)$ is anti-Hermitian, $F^*(-\omega) = -F(\omega)$.

These symmetry properties can be reviewed by applying Fig. 5.2 to the integral Fourier transform.

The similarity theorem results from a simple change of variable in the Fourier integral:

$$\text{FT } f(at) = \frac{1}{|a|} F(\omega/a) \tag{6.9}$$

and likewise for the familiar shift theorem,

$$\text{FT } f(t - \tau) = e^{-i\tau\omega} F(\omega) \tag{6.10}$$

The power theorem, which states that

$$2\pi \int_{-\infty}^{\infty} f(t) g^*(t) \, dt = \int_{-\infty}^{\infty} F(\omega) G^*(\omega) \, d\omega \tag{6.11}$$

can be specialized to Rayleigh's theorem by setting $g = f$:

$$2\pi \int_{-\infty}^{\infty} |f(t)|^2 \, dt = \int_{-\infty}^{\infty} |F(\omega)|^2 \, d\omega \tag{6.12}$$

In more mathematical works, Rayleigh's theorem is sometimes called Plancherel's theorem.

Of course, the important and powerful convolution theorem—meaning linear convolution—is valid in continuum theory also:

$$\text{FT} \int_{-\infty}^{\infty} f(\tau) g(t - \tau) \, d\tau = F(\omega) G(\omega) \tag{6.13}$$

The many variations of the convolution theorem arising from the symmetry properties of the Fourier transform apply as well. Review the lower portion of Fig. 5.4 for some of these forms of the convolution theorem. The fourth entry in that figure, when written in continuous theory, gives us the important autocorrelation theorem:

$$\text{FT}\int_{-\infty}^{\infty} f(\tau)g^*(t+\tau)\,d\tau = F(\omega)G^*(\omega) \qquad (6.14)$$

The function FG^* is called the cross-power spectrum. When we set $g = f$, this equation states the important result that the Fourier transform of the autocorrelation of a function is its power spectrum:

$$\text{FT}\int_{-\infty}^{\infty} f(\tau)f^*(t+\tau)\,d\tau = |F(\omega)|^2 \qquad (6.15)$$

The formal similarity between these continuous-theory properties and those of the DFT makes them easy to remember and to visualize. But, there are essential differences between the two. The DFT with its finite sum has no convergence questions. The Fourier integral transform, on the other hand, has demanded the attention of some of our greatest mathematicians to elucidate its convergence properties. As we saw, the absolute integrable condition is only a start; it can be relaxed—quite easily at the heuristic level—to include the sine wave/δ function pair. The sine wave's ill behavior is characteristic of a wide class of functions of interest in digital signal processing that do not decay sufficiently fast at infinity for them to possess a Fourier transform in the normal sense.

The Wiener–Khintchine Theorem

An important method of treating such signals that do not decay at infinity, due independently to Wiener in 1930 and Khintchine in 1934, starts from the form of the convolution theorem in Eq. (6.15). The inverse Fourier transform of this equation is

$$\int_{-\infty}^{\infty} f(\tau)f^*(t+\tau)\,d\tau = \frac{1}{2\pi}\int_{-\infty}^{\infty} |F(\omega)|^2 e^{i\omega t}\,d\omega \qquad (6.16)$$

For many signals of interests, such as sinusoids, step functions, and random noise of fixed statistical properties, the autocorrelation integral on the left does not converge. But, if we define a truncated version of $f(t)$ by

$$f_T(t) = \begin{cases} f(t) & -T < t < T \\ 0 & \text{otherwise} \end{cases}$$

The Wiener–Khintchine Theorem

then we can write

$$\int_{-\infty}^{\infty} f_T(\tau) f_T^*(t+\tau)\, d\tau = \frac{1}{2\pi} \int_{-\infty}^{\infty} |F_T(\omega)|^2 e^{i\omega t}\, d\omega \qquad (6.17)$$

where $F_T(\omega)$ is the Fourier transform of $f_T(t)$. Dividing by the time interval $2T$ and taking the limit, Eq. (6.17) becomes

$$\lim_{T \to \infty} \frac{1}{2T} \int_{-T}^{T} f(\tau) f^*(t+\tau)\, d\tau = \frac{1}{2\pi} \int_{-\infty}^{\infty} \lim_{T \to \infty} \frac{|F_T(\omega)|^2}{2T} e^{i\omega t}\, d\omega$$

Wiener (1949) was able to show that, under the condition that the limit on the left exists, the limit inside of the right-hand integral converges to a function:

$$P(\omega) = \lim_{T \to \infty} \frac{|F_T(\omega)|^2}{2T} \qquad (6.18)$$

which we call the power spectrum density of f. Using these revised definitions of autocorrelation,

$$\phi(t) = \lim_{T \to \infty} \frac{1}{2T} \int_{-T}^{T} f(\tau) f(t+\tau)\, d\tau \qquad (6.19)$$

and power spectrum, our result now reads

$$\phi(t) = \frac{1}{2\pi} \int_{-\infty}^{\infty} P(\omega) e^{i\omega t}\, d\omega \qquad (6.20)$$

which is called the *Wiener–Khintchine theorem*; the autocorrelation is the inverse Fourier transform of the power spectrum. This is a very significant result, not a simple restatement of our starting point, Eq. (6.16). In Eq. (6.16), both $f(t)$ and $F(\omega)$ must be square integrable, that is, they must contain finite energy over all time and frequency. In Eq. (6.20), $f(t)$ must only be sufficiently well behaved so that

$$\lim_{T \to \infty} \frac{1}{2T} \int_{-T}^{T} |f(t)|^2\, dt < \infty$$

That is to say, $f(t)$ is only required to have finite power (signal squared per time) to have a power spectrum, but $f(t)$ must have finite energy (i.e., be square integrable or, with additional restrictions, be only absolutely integrable) to possess a Fourier transform. Two classes of functions of interest, periodic functions and some types of random noise, satisfy the first condition but not the second.

We will exploit the Wiener–Khintchine theorem further in Chapter 11, Power Spectral Estimation. Here, we have introduced it to show how

Fourier integral theory can be generalized to include functions with infinite energy but finite power. Having presented this vignette of the theory of *generalized Fourier integrals*, we now feel free to abandon further convergence questions in our heuristic discussions of Fourier transforms pairs.

The Time-Limited Band-Limited Theorem

We have seen that the behavior of signals and their Fourier transforms at infinity can be a major concern. Indeed, in any practical digital computing scheme, signals and the Fourier transforms have to be of limited length. Naturally then, any general statement that can be made about the length of Fourier transform pairs will have considerable bearing on digital signal processing, both in theory and in practice.

Our discussion addresses time-limited and band-limited signals. A time-limited signal is one that is confined to a finite length of time and is zero outside that interval. For a time-limited signal, its total energy is contained in an interval $2a$:

$$A = \int_{-a}^{a} |f(t)|^2 \, dt$$

Likewise, a band-limited signal has its spectrum completely confined to a finite frequency interval, so that its total energy is

$$B = \int_{-b}^{b} |F(\omega)|^2 \, d\omega$$

The time-limited/band-limited theorem says that no signal can be both time-limited and band-limited, except for the trivial case where $f(t)$ is identically equal to zero. We prove this important theorem by assuming $f(t)$ is both time-limited and band-limited; then, we show that $f(t)$ necessarily must be the null function. First, observe that not only is this signal

$$f(t) = \frac{1}{2\pi} \int_{-b}^{b} F(\omega) e^{i\omega t} \, d\omega = 0 \quad \text{for} \quad |t| \geq a$$

but all of its derivatives must also be zero at $|t| \geq a$. Therefore differentiation with respect to time under the integral n times gives

$$\int_{-b}^{b} F(\omega) e^{i\omega a} \omega^n \, d\omega = 0 \quad \text{for} \quad n = 0, 1, 2, \ldots$$

Next, we write the inverse Fourier transform of a band-limited signal in a special form. For such a signal, we can write

$$f(t) = \frac{1}{2\pi} \int_{-b}^{b} F(\omega) e^{i\omega t} d\omega = \frac{1}{2\pi} \int_{-b}^{b} F(\omega) e^{i\omega(t-a)} e^{i\omega a} d\omega$$

Then, using the power series expansion for the exponential function allows term-by-term integration to give

$$f(t) = \frac{1}{2\pi} \sum_{n=0}^{\infty} \frac{[i(t-a)]^n}{n!} \int_{-b}^{b} F(\omega) e^{i\omega a} \omega^n d\omega$$

as an alternate representation of a band-limited signal in terms of its spectrum. But, we have shown that if the signal is also time limited, each of the integrals in this sum is identically zero. Hence, $f(t) \equiv 0$ is the only function that can be both time limited and band limited.

This theorem immediately raises a specter of a fundamental nature for digital signal processing because it says that every signal must be infinitely long in either the time domain or in the frequency domain, or both. We will see the consequences of this requirement in the next chapter where we develop the relationship between continuous signals and their sampled counterparts.

So far, our discussion of the continuous Fourier integral transform has been on a general level, giving powerful theorems and properties applicable to a wide variety of continuous functions. Next, to exemplify these theorems and also to form a basis for further discussion, we introduce a repertoire of Fourier transforms particularly important to digital signal processing.

A Repertoire of Transforms and Their Importance

There are tremendous advantages to thinking about signals and systems in the frequency domain. Generally speaking, deeper understanding can be gained when a subject is viewed from more than one angle. To better understand how the frequency domain relates to the time domain and the implication (and limitation) of this relationship for digital signal processing, we need a repertoire of transforms available at our fingertips. This repertoire together with the properties of the Fourier transform discussed in the preceding sections will be indispensable for understanding and applying Fourier transforms in practical applications.

Most of the transforms of this repertoire are easy to compute. Therefore, we will concentrate on their implications for our subject, leaving their

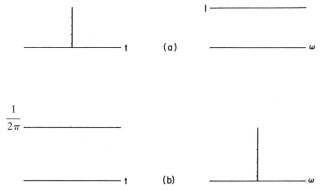

Figure 6.2 The Fourier transform of the δ function and its dual.

derivation, for the most part, to the problems. Furthermore, to simplify the repertoire, the time domain signal will be either real-symmetric or real-antisymmetric so the transform will be either real-symmetric or odd-imaginary, simplifying the drawings and discussions.

The delta function is perhaps the easiest function to integrate. Substitution of $f(t) = \delta(t)$ in Eq. (6.6a) immediately tells us that the Fourier transform of a spike at the origin is a constant of unit height as shown in Fig. 6.2(a). Physical reasoning tells us that the δ function could not exist in any real analog system; its spectrum substantiates this by demanding an infinite bandwidth. Nor can this spike exist in a digital system—it is infinitely large. Some of the Fourier transform properties of the previous section can be applied to the δ function. Others defy ordinary mathematics—try Rayleigh's theorem. The δ function is like the girl with the little curl on her forehead: when she was good, she was very, very good; when she was bad, she was horrid. Mathematically, the δ function is indeed somewhat horrid. On the other hand, it is quite good—really indispensable—for relating continuous theory to discrete signals and systems.

The complementary transform is equally easy to compute. Substituting $f(t) = $ constant in Eq. (6.6a) and using Eq. (6.5) tells us that the Fourier transform of a constant is a δ function as shown in Fig. 6.2(b). Thus, the Fourier transform concentrates the dc component (the average value) of a signal into the zero frequency contribution of the spectrum. Conversely, from Fig. 6.2(a), a signal concentrated at one point in time is spread out over all frequencies. The infinitely sharp time domain signal has an infinitely broad spectrum; the infinitely broad time domain signal has an infinitely sharp spectrum. It turns out that this is a manifestation of a general property of the Fourier transform.

The Gaussian function is one of the few waveforms that possesses the same functional form in both domains; the Fourier transform of a Gaussian

A Repertoire of Transforms and Their Importance

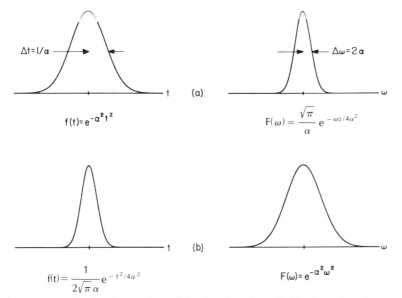

Figure 6.3 (a) The Fourier transform of the Gaussian. (b) Its dual is shown for reference.

is a Gaussian. These functions, and their half-width located $1/e$ down from the maximum, are shown in Fig. 6.3. Using these half-widths, the spreading property observed for the δ function can be easily quantified for the Gaussian pair:

$$\Delta t \, \Delta \omega = (1/\alpha)(2\alpha) = 2$$

The product of the widths is a constant for all Gaussians. If we imagine the Gaussian time function getting progressively narrower, then its spectrum will get increasingly broader, and vice versa. This reciprocal time–frequency bandwidth relationship is an inherent property of the Fourier transform and is best put on a firm mathematical basis by defining the widths in a somewhat more complicated fashion than we have done here. With this more complicated definition of width, it can be shown (see Bracewell, 1978) that

$$(\text{width in } t)(\text{width in } \omega) \geq \tfrac{1}{2} \quad (6.21)$$

for all Fourier transform pairs. With this definition of width, the Gaussian pair satisfies the equality, a fairly unusual situation for a transform pair. This relationship between widths in the two domains is called the *uncertainty principle* in quantum mechanics, but it was well known to mathematicians and electrical engineers well before the foundations of quantum theory were developed.

110   6/ The Continuous Fourier Integral Transform

Usually it is sufficient to think of the $\Delta t \, \Delta\omega$ product as being roughly unity. For example, we can then observe that if a certain amplifier is designed to pass a 1 μsec wide pulse, then its bandwidth must be on the order of 1 MHz. Of course, the actual design requirements of such an amplifier depend on its purpose. Any practical bandwidth will necessarily distort the pulsed signal. The tolerance for such distortion may depend on signal-to-noise considerations versus fidelity requirements.

As we introduce this repertoire of Fourier transform pairs, it will be instructive to apply some of the fundamental properties of the preceding section to them. The convolution theorem, for example, immediately tells us that the convolution of a Gaussian is another, wider Gaussian. We see this easily by thinking of the operation in the other domain: there multiplication of two Gaussians clearly produces another, narrower Gaussian.

The boxcar/sinc transform pair gives us a greater insight into both Fourier transforms and digital signal processing than perhaps any other single transform pair. The Fourier transform of the boxcar function, shown in Fig. 6.4(a), is of the form $\sin(x)/x$. This type of function is commonly called a *sinc* function. Think of it as a sine wave that decays as $1/x$ for large x, and note that $\sin(x)/x$ converges to 1 at $x = 0$.

Taking the Fourier transform of the boxcar is a straightforward integration that you can easily do. The reverse, the inverse Fourier transform of the sinc function, turns out to require advanced integration techniques that are tangential to our discussion. Perhaps it is not surprising that this inverse

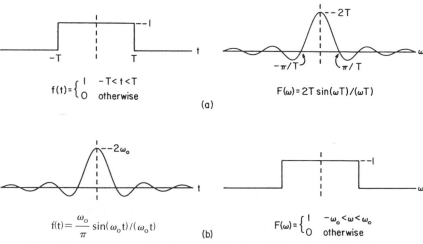

Figure 6.4 (a) The boxcar function and its Fourier transform. (b) Its dual is shown.

Fourier transform,

$$\text{boxcar} = \frac{1}{2\pi} \int_{-\infty}^{\infty} \frac{\sin(T\omega)}{T\omega} e^{i\omega t} d\omega$$

requires some tricky integration because it has a rather curious behavior: its value is completely independent of t for t between $-T$ and T. Then, when t exceeds this interval, the value drops abruptly to zero for all other t. This situation, where an ordinary continuous function is related to a discontinuous one, is a common occurrence among Fourier transform pairs. On the other hand, the forward Fourier transform,

$$\text{sinc} = \int_{-T}^{T} e^{-i\omega t} dt$$

is an elementary definite integral relating two continuous functions.

Note, again, the relationship between the boxcar's duration and its frequency bandwidth: if we define the sinc's bandwidth by the first zero crossings, $\Delta t \, \Delta \omega = 4\pi$. Narrow pulses have wide bandwidths and vice versa.

For another example of the utility of the convolution theorem, we discuss a theorem that originally arose in probability theory and was given the name *central limit theorem* by G. Polya in 1920. The connection with probability occurs because the probability of drawing several numbers from a distribution that add up to a predetermined sum can be written as a convolution. The theorem, stated in terms of these convolutions, says that if a large number of functions are convolved together, the result tends toward the Gaussian functions (normal distribution in probability theory).

Let us see if the boxcar/sinc pair satisfies this theorem. In Fig. 6.5, we show successive convolutions of the boxcar and the associated successive multiplications of the sinc in the frequency domain. The boxcar does appear to be approaching the shape of a Gaussian. The frequency domain version likewise appears to be approaching a Gaussian shape; indeed, it must, since the transform of a Gaussian is another Gaussian.

Now that we have demonstrated that the boxcar appears to satisfy the central limit theorem, it is natural to ask if the sinc function does also. Convolving successive sinc functions can easily be done mentally by thinking of multiplying boxcars in the other domain. Clearly, the product of any two boxcars is another boxcar. Therefore, successive convolutions of sinc functions with other sinc functions always produces sinc functions; the Gaussian is never approached.

The central limit theorem applies to convolutions among a wide class of functions, but as we have seen, it is not true for all functions. However, it does seem to apply for many naturally occurring functions. In these cases,

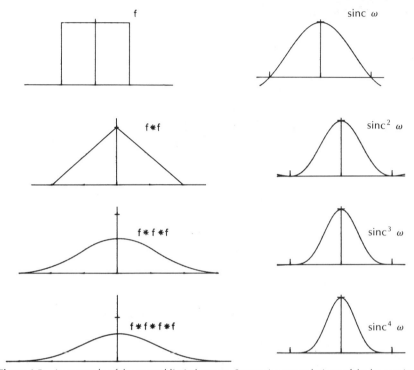

Figure 6.5 An example of the central limit theorem. Successive convolutions of the boxcar in time and the corresponding successive multiplications of its sinc function in the frequency domain approach a Gaussian looking shape.

the effect of the convolution operation is to produce a smeared out result resembling the Gaussian shape, and frequently this approach to this bell-shaped curve is surprisingly fast. See Problem 6.11, for example.

On the other hand, many specially designed convolution operators of interest in signal processing do not obey the central limit theorem. Differential and inverse operators, for example, serve to sharpen a signal rather than smooth it. Some required properties of functions satisfying the central limit theorem can be found by using the fundamental properties of the Fourier transform, as we suggest in Problems 6.12 and 6.13.

Another very important application of the boxcar/sinc pair arises when we think of the boxcar as truncating a data stream. Imagine, for example, that a radio astronomer wishes to analyze a signal received from a distant source for certain periodicities. The researcher can only record a small portion of this signal for spectral analysis. We can say that the data are recorded over a window in time, represented by multiplication of the natural signal by a boxcar of width $2T$. This truncation of the data stream in

A Repertoire of Transforms and Their Importance 113

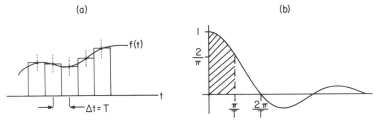

Figure 6.6 (a) Sampling intervals of time $\Delta t = T$ in effect assumes a constant value of the function between samples. The sampling process convolves the samples with this boxcar and hence multiplies its spectrum by the sinc function shown. (b) The resulting falloff of the spectrum is called aperture error.

the time domain manifests itself as a convolution of the data's spectrum with the sinc function of width $2\pi/T$ (width between first zero crossings). The effect of this convolution is to smear out the frequency spectrum of the data, thereby reducing the resolution of the spectrum analysis. This loss of resolution due to a finite data record is unavoidable in every experimental situation. If the data record is very long, the sinc function is very narrow. In the limit of an infinitely long data record, the sinc function associated with truncation will be infinitesimally wide. Convolution with this spike is the same as convolution with a δ function; it is the identity operation and no loss of frequency resolution occurs. If, on the other hand, the data window is very short, the associated sinc function will be very wide, resulting in a severe loss of frequency resolution because of convolution with a very broad sinc function. We will see in Chapter 7 important ramifications of this data truncation when we discuss the relationship of the DFT to the Fourier transform.

One effect of sampling analog data can be seen by looking at the boxcar/sinc pair shown in Fig. 6.6. Here, the sampling time of an analog-to-digital converter is represented by a boxcar of width $\Delta t = T$ (called the sampling aperture). In this worst-case A/D conversion, the signal is averaged over one whole sample period, thereby losing some high frequencies. This sampling is represented by a convolution of the ideal samples with the boxcar. In the frequency domain, the signal's spectrum is multiplied by the sinc function, shown in Fig. 6.6(b), which has a lowpass filter effect, producing a roll-off that is down $2/\pi$ at Nyquist. This unavoidable distortion of the signal's spectrum is called *aperture error*.

For our last example, extolling the importance of the boxcar/sinc pair, we look to an optical diffraction experiment. Consider monochromatic light waves impinging on the aperture shown in Fig. 6.7(a). Let the wave's amplitude vary across the aperture according to $A(x)$. The disturbance emanating from an element of the aperture dx is, using the complex

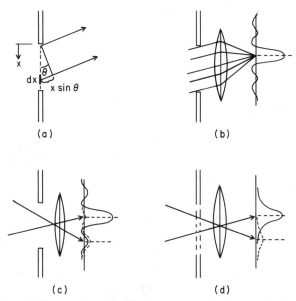

Figure 6.7 Optical diffraction from the objective aperture of a telescope: (a) phase shift caused by travel path difference; (b) a lens focuses parallel rays, forming a diffraction pattern in its focal plane; (c) overlapping sinc diffraction patterns of two stars, one much weaker than the other; and (d) apodization of the objective aperture broadens but smoothes the diffraction patterns, allowing better identification of two objects.

exponential form for sinusoids,

$$A(x)e^{-i\omega t}\,dx$$

On a screen, sufficiently far enough away so as to be considered at infinity, these contributions produce an image resulting from interference due to phase shifts caused by different travel paths. As the figure shows, the phase shift for the ray pictured is

$$\delta = \frac{2\pi x \sin\theta}{\lambda} \simeq \frac{2\pi x \theta}{\lambda}$$

for small θ. The disturbance arriving at the screen from dx is then

$$A(x)e^{-i\omega t - i2\pi x\theta/\lambda}\,dx$$

The total illumination, $T(\theta)$, reaching the screen from the entire aperture is just the superposition of all these contributions:

$$T(\theta) = \int A(x)_{\text{aperture}} e^{-i\omega t - i2\pi x\theta/\lambda}\,dx$$

$$T(\theta) = e^{-i\omega t} \int A(x)_{\text{aperture}} e^{-i2\pi x \theta/\lambda} dx$$

The complex exponential in front of the integral expresses the sinusoidal time dependence of the signal at every point on the screen. The integral is the amplitude of this signal versus θ at each point on the screen; we recognize it as the Fourier transform of the amplitude across the aperture. Waves of any kind propagate to infinity via the Fourier transform. To avoid screens at infinity, a lens can be used between the aperture and the screen to focus rays on the focal plane of the lens as shown in Fig. 6.7(b). Now, the aperture, lens, screen, and a plane-wave light source form a simple and practical Fourier transformer. If a semitransparent film with its transmission varying as $A(x)$ is placed over the aperture, the Fourier transform of $A(x)$ is cast onto the screen. Before the days of the modern digital computer, clever data processing engineers and scientists used equipment of this type to perform Fourier transforms, convolutions, cross-correlations, and filtering.

The boxcar/sinc pair arises in the case where $A(x)$ equals a constant across the aperture. For example, an astronomical telescope focused on Sirius, a binary star system in the constellation Canis Major, would project two sinc-type images onto its focal plane, one for each member of the system, as shown in Fig. 6.7(c). We know from our discussion of Fourier transforms that the width of these sinc functions varies inversely with the aperture size. Furthermore, additional optics cannot increase the separation between these sinc patterns; they can only magnify the entire picture. In the case of Canis Major, it turns out that one star is considerably fainter than the other and in most telescopes it gets lost in the side lobes of the sinc function of the brighter star. In this case, resolution is not the problem—it is the side lobes produced by the boxcar-like aperture. But, a knowledge of Fourier transforms saves the day and allows the weaker star to be detected: by coating the objective lens (the aperture) so as to reduce its transmission gradually away from the center, the shape of the original sinc function can be altered to a new pattern having smaller side lobes. For example, if this coating reduced transmission in a Gaussian fashion, each star's image would be Gaussian. Although the Gaussian does have a wider central peak than the sinc function, it has the virtue of lacking side lobes. The superposition of two Gaussians of greatly differing strengths will always reveal the location of the weaker peak, given sufficiently sensitive detection methods. This technique of tapering the transmission characteristics of optical instruments, called *apodization*, has been successfully applied to telescopes. It has an important counterpart in signal processing as well. In coming chapters, we shall see several applications of tapering data streams, as opposed to truncating them boxcar fashion.

Before leaving the discussion of the boxcar/sinc pair, we mention its Fourier dual pair, the sinc/boxcar in Fig. 6.4(b), which is, of course, easily obtained from the first pair by a relabeling of variables. One immediate significance of this sinc/boxcar pair for signal processing is that the boxcar represents an ideal lowpass filter in the frequency domain. We see that it requires an infinitely long time domain convolution operator to achieve this lowpass filter—a physical impossibility. Designing substitute time-domain operators of tolerable lengths is the topic of Chapter 8.

The Fourier transform of sines and cosines plays a central role in the study of linear shift-invariant systems for three quite related reasons: (1) they are eigenfunctions of LSI systems, (2) the spectrum consists of the eigenvalues associated with these eigenfunctions, and (3) the weighting factor (sometimes called the kernel) in the Fourier transform is a complex sinusoid. Clearly, we want to include sinusoids in our Fourier transform repertoire.

We have already seen from Eq. (6.5) that the Fourier transform of the complex exponential is a δ function located at the frequency of the complex sinusoid. That is, rewriting Eq. (6.5) slightly gives

$$\int_{-\infty}^{\infty} e^{i\omega_0 t} e^{-i\omega t} \, dt = 2\pi\delta(\omega - \omega_0)$$

and letting $\omega_0 \to -\omega_0$

$$\int_{-\infty}^{\infty} e^{-i\omega_0 t} e^{-i\omega t} \, dt = 2\pi\delta(\omega + \omega_0)$$

The Fourier transform of the cosine and the sine follows directly from these two equations. Adding the two equations gives

$$\int_{-\infty}^{\infty} \cos(\omega_0 t) e^{-i\omega t} \, dt = \pi[\delta(\omega + \omega_0) + \delta(\omega - \omega_0)] \qquad (6.22)$$

while subtracting them gives

$$\int_{-\infty}^{\infty} \sin(\omega_0 t) e^{-i\omega t} \, dt = \pi i[\delta(\omega + \omega_0) - \delta(\omega - \omega_0)] \qquad (6.23)$$

These transforms are just the combinations of δ functions that are required by the basic symmetry properties of the Fourier transform; because the cosine is a real and even function, its transform is real and symmetric. The sine is real and odd; therefore, it has an odd imaginary transform.

Alternately, we could derive the sine and cosine transforms by using the phase-shift theorem; shifting a function along the axis in one domain introduces a complex sinusoid in the other domain. For example, if we want to generate the dual pairs to Eq. (6.22) and Eq. (6.23), we apply the

A Repertoire of Transforms and Their Importance

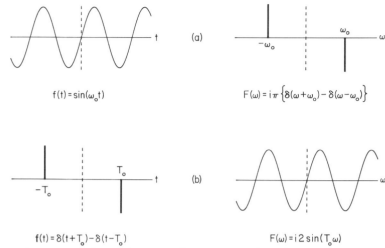

Figure 6.8 (a)The sinc function and its Fourier transform. (b) Its dual is shown.

phase shift theorem of Eq. (6.10) to the δ function and write

$$\text{FT}\{\delta(t - t_0)\} = e^{-it_0\omega}$$

and

$$\text{FT}\{\delta(t + t_0)\} = e^{it_0\omega}$$

Adding and subtracting these two equations gives

$$\text{FT}\{\delta(t - t_0) + \pi\delta(t + t_0)\} = 2\cos(\omega t_0) \qquad (6.24)$$

and

$$\text{FT}\{\delta(t + t_0) - \pi\delta(t - t_0)\} = i2\sin(\omega t_0) \qquad (6.25)$$

The sine and cosine transforms and their duals are shown in Figs. (6.8) and (6.9).

Thus, the Fourier transforms of sines and cosines can be viewed as resulting from forcing certain symmetries into the δ function transform after it is shifted along the axis: shifting the δ function off the origin in the frequency domain and then requiring a symmetric spectrum results in Eq. (6.22). An antisymmetric spectrum leads to Eq. (6.23). Analogous statements apply to Eqs. (6.24) and (6.25) for δ function shifts along the time axis.

The δ funtions in Eqs. (6.22) through Eq. (6.25) make for easy convolutions leading to an important observation. For example, convolving

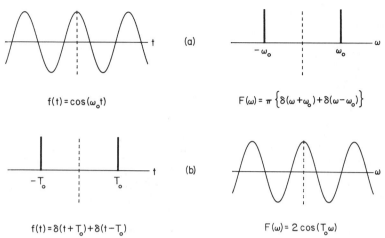

Figure 6.9 (a) The cosine function and its Fourier transform. (b) Its dual is shown.

Eq. (6.22) in the frequency domain with a given function $F(\omega)$ and appropriately multiplying by $f(t)$ in the time domain gives

$$\text{FT}\{f(t)\cos(\omega_0 t)\} = \pi[F(\omega + \omega_0) + F(\omega - \omega_0)] \quad (6.26)$$

This equation and its sine wave counterpart are sometimes referred to as the modulation theorem. They show that modulating the amplitude of a function by a sinusoid shifts the function's spectrum to higher frequencies by an amount equal to the modulating frequency. This effect is well understood in television and radio work where a sinusoidal carrier is modulated by the program signal; see Problem 6.15. Obtaining transforms of sines, cosines, and related functions has been a good example of exploiting the basic Fourier transform properties. Our next example is not so obliging.

The Fourier transform of $1/t$ is essentially a step function. We shall see that convolution with $1/t$ leads to an important new integral transform. Specifically, the Fourier transform of $(-1/\pi t)$ is $i \cdot \text{sgn}(\omega)$. This pair and its dual are shown in Fig. 6.10. Since $1/t$ has a pole at the origin, its Fourier integral diverges there and $1/t$ has a transform only in a limiting sense. (It can be evaluated by using contour integration.) Likewise, the Fourier integral of $\text{sgn}(x)$, in computing either the forward or inverse transform, does not exist in the conventional sense because $\text{sgn}(x)$ is not absolutely integrable. The transform of $\text{sgn}(x)$ can be defined as outlined in Problem 6.10 by considering a sequence of transformable functions that approaches $\text{sgn}(x)$ in the limit. We do not let these mathematical inconveniences deter us any more than they did in our discussion of δ functions and sinusoids, for the $(1/t)/\text{sgn}$ pair has some intriguing properties.

A Repertoire of Transforms and Their Importance 119

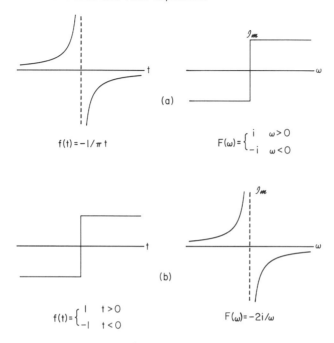

Figure 6.10 (a) The function $-1/\pi t$ and its Fourier transform. (b) Its dual is shown.

Since $(1/t)$ is real and odd, its transform is odd and pure-imaginary. But more interestingly, its magnitude spectrum is obviously constant. (Could there be a δ function lurking nearby? See Problem 6.20.) The interest in this transform pair comes from convolving a function with $(-1/\pi t)$ in the time domain. This convolution integral, called the *Hilbert transform*,

$$\mathcal{H}[f(t)] = -\frac{1}{\pi} \int_{-\infty}^{\infty} \frac{f(t')\,dt'}{t - t'} \qquad (6.27)$$

arises in many applications of mathematical physics. The Cauchy principal value, which is a kind of average value familiar to those knowledgeable in contour integration, is to be used over singularities of the integrand. This mathematical inconvenience is avoided in the frequency domain where we can easily visualize the effect of the Hilbert transform.

Multiplication by $i \cdot \text{sgn}(\omega)$ in the frequency domain produces a 90° phase shift at all frequencies. The phase of $F(\omega)$ is advanced a constant 90° for all positive frequencies and retarded a constant 90° for all negative frequencies. The magnitude spectrum of $F(\omega)$ is unchanged since the spectrum of $i \cdot \text{sgn}(\omega)$ is flat. This Hilbert transform operation in the frequency domain is summarized in Fig. 6.11. The Fourier transform of the given function has its phase shifted 90°, in opposite directions for positive and negative

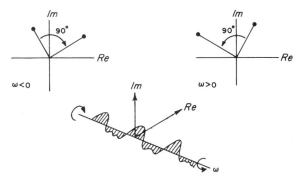

Figure 6.11 A summary of the Hilbert transform operation in the frequency domain. The spectrum undergoes a 90° phase retardation for $\omega < 0$ and 90° phase advance for $\omega > 0$. The magnitude spectrum is unchanged.

frequencies; then, the inverse Fourier transform produces the time domain result.

This exact 90° phase shift [including the sgn(ω) dependence] occurs in several different instances in wave propagation: the reflection of electromagnetic waves from metals at a certain angle of incidence involves an exact 90° phase shift independent of frequency; special teleseismic ray paths produce the same type of phase shifts; and the propagation of point sources for all types of waves includes a 90° phase shift for all frequencies in the far field.

For a more elementary example of the Hilbert transform type of phase shift, we turn to the sine and cosine functions that are frequently said to be 90° phase shifts of each other. However, the actual phase shift required to transform between sines and cosines is not this simple 90° phase shift; rather it is, again, the somewhat curious 90° phase shift of the Hilbert transform where positive and negative frequencies are treated oppositely. Figure 6.12(a) shows that the $i \cdot \text{sgn}(\omega)$ phase shift applied to the sine spectrum produces the cosine spectrum. Likewise Fig. 6.12(b) shows that the Hilbert transform phase shift applied to the cosine spectrum produces the negative of the sine spectrum. Thus, sines and cosines are Hilbert transform pairs:

$$\cos(\omega t) = -\frac{1}{\pi} \int_{-\infty}^{\infty} \frac{\sin(\omega' t')}{(t - t')} \, dt' \qquad (6.28a)$$

$$\sin(\omega t) = +\frac{1}{\pi} \int_{-\infty}^{\infty} \frac{\cos(\omega t')}{(t - t')} \, dt' \qquad (6.28b)$$

A Repertoire of Transforms and Their Importance   121

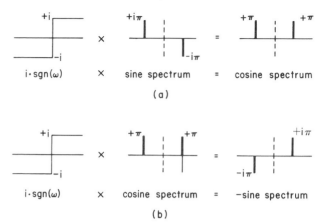

Figure 6.12 The sine (a) and cosine (b) functions are Hilbert transforms of one another.

This manipulation of the line spectra of sines and cosines can be done in yet another way, which again relates to the Hilbert transform. In Fig. 6.13, we show that the cosine spectrum plus i times the sine spectrum is equal to the complex exponential's spectrum. That is, we observe that

$$e^{i\omega t} = \cos \omega t + i \sin \omega t$$

Additional significance of this familiar result can now be seen. While both the sine and cosine have two-sided spectra, the complex exponential only contains one positive frequency. We see now why the complex exponential

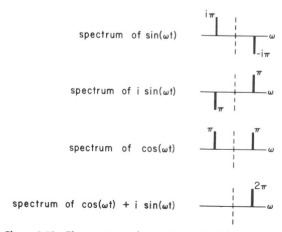

Figure 6.13 The spectrum of $\cos(\omega t) + i \sin(\omega t)$ is one sided.

form is so suitable in describing rotors in ac circuit theory; it represents a rotor moving counterclockwise with only one frequency. On the other hand, a sine or cosine represents a superposition of two rotors moving in opposite directions, making phases impossible to track.

Using the fact that the sine is the negative Hilbert transform of the cosine, we can write the complex exponential as

$$e^{i\omega t} = \cos(\omega t) - i\mathcal{H}[\cos \omega t]$$

This generation of a complex function from the real function $\cos(\omega t)$ can be similarly applied to any function:

$$g(t) = f(t) - i\mathcal{H}[f(t)] \quad (6.29)$$

and then $g(t)$ is called the *analytic signal* belonging to $f(t)$. Sometimes $H[f(t)]$ is called the *quadrature function* or, equivalently, the *allied function* of $f(t)$. This analytic signal, like the exponential, is one sided in the frequency domain—it only contains positive frequencies. This is easily seen, for the Fourier transform of Eq. (6.29) is

$$G(\omega) = F(\omega) - i[i\,\text{sgn}(\omega)]F(\omega)$$
$$= F(\omega)[1 + \text{sgn}(\omega)]$$
$$G(\omega) = \begin{cases} 2F(\omega) & \omega > 0 \\ 0 & \omega < 0 \end{cases} \quad (6.30)$$

If we turn this idea around by generating analytic signals in the frequency domain, they will be one-sided, that is, causal, in the time domain. Thus, Hilbert transforms are intimately related to causality. We explore that relationship in Chapter 10.

Our last application of the Hilbert transform is the *envelope* function given by

$$E(t) = |g(t)| = \sqrt{f^2(t) + (\mathcal{H}f)^2} \quad (6.31)$$

This nonlinear function of $f(t)$ has several interesting properties. First, the envelope is tangent to $f(t)$ at points where $E(t)$ and $f(t)$ meet; see Problem 6.24. Since the envelope is clearly greater than $f(t)$ everywhere, the envelope circumscribes $f(t)$, giving the envelope its name. An example of this property is shown in Fig. 6.14.

As a consequence of this tangency property, the envelope can arise as a so-called singular solution to certain nonlinear, first-order differential equations. A family of solutions $f(t)$ is generated by using different integration constants. But the envelope, although not a member of the family, still satisfies the differential equation because it has the same derivative at each point that members of the family have.

A Repertoire of Transforms and Their Importance

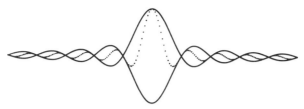

Figure 6.14 The envelope of the sinc function, with sin(ωt)/t shown by dotted line. All functions formed by performing a constant phase rotation ϕ sgn (ω) on the spectrum of the sinc function will be inscribed within this envelope.

The envelope becomes particularly interesting to signal processing when we identify the family of functions $f(t)$ that form the envelope in terms of the frequency domain. We can show (see Problem 6.23) that a constant frequency phase rotation ϕ sgn(ω) of the spectrum of $f(t)$, a generalized Hilbert transform, leaves the envelope of $f(t)$ unchanged. Thus, the envelope outlines all the possible functions that are obtainable from a given $f(t)$ by rotating its phase spectrum through an arbitrary frequency-independent angle. This characteristic of the envelope is useful in describing the effects that filters of unknown phase spectra may have had on data.

Examples come from instrumentation transducers and seismic prospecting. It may be difficult or impossible to measure the transfer characteristic of a transducer because the measurement of the input may require yet another transducer. In the seismic case, the source signal is generally impossible to measure. Frequently, however, the relevant magnitude spectrum can be determined, leaving only the phase spectrum in question. Then, the effects of the transducer's (or seismic source's) magnitude spectrum can be removed by division in the frequency domain. A display of the resulting envelope then provides a result that is independent of the unknown phase shifts. The assumption is made that the unknown phase shift is constant across the bandwidth of the signal, a reasonable assumption for our examples where it is known that $\phi(\omega)$ is at least slowly varying over a narrow bandwidth.

This chapter is but a brief introduction into continuous Fourier transform theory with examples devoted to signal processing. Many texts are available for more general and more comprehensive treatments. See, for example, Bracewell (1978) or Champeney (1973) which both include a pictorial atlas of Fourier transforms. More often than not, digital signals are derived from continuous ones; hence, a sound understanding of continuous theory is necessary. Fortunately, the knowledge of the basic symmetries, properties, and the five transform pairs of our repertoire will suffice for pursuing our study of digital signal processing.

Problems

6.1 By changing variables of integration in the definition of the Fourier integral transform, derive the symmetry properties stated in Eq. (6.8).

6.2 Likewise, derive the similarity theorem of Eq. (6.9).

6.3 Derive the shift theorem of Eq. (6.10) by making an appropriate change of integration variables.

6.4 Verify the power theorem of Eq. (6.11) by direct substitution of $F(\omega)$ and $G(\omega)$ into the right-hand integral.

6.5 Prove the convolution theorem by computing directly the Fourier transform of the left-hand side of Eq. (6.13).

6.6 Prove that $FT\{(f*g)(u*v)\} = (FG)*(UV)$.

6.7 Give three examples of functions that are not square integrable and hence do not possess Fourier transforms in the normal sense, but do have power spectra according to the Wiener–Khintchine theorem. Show explicitly that each of your functions contains finite power.

6.8 There are infinitely many sequences that approach the δ function in a limit (called representations of the δ function). Three are listed here:

Lorentzian: $\lim\limits_{\lambda \to 0} \dfrac{\lambda}{\pi(\lambda^2 + t^2)} \to \delta(t)$

Gaussian: $\lim\limits_{\lambda \to 0} \dfrac{1}{\lambda\sqrt{\pi}} e^{-(t/\lambda)^2} \to \delta(t)$

Sinc: $\lim\limits_{\lambda \to 0} \dfrac{1}{\pi} \dfrac{\sin(t/\lambda)}{\lambda(t/\lambda)} \to \delta(t)$

Show that when used in integration, each of these functions satisfy the following properties of the δ function:

$\delta(-t) = \delta(t)$

$t\delta(t) = 0$

$\delta(at) = \delta(t)/a \quad a > 0$

$\delta[f(t)] = \delta(t - t_0)/(df/dt) \quad f(t_0) = 0$

$\dfrac{d}{dt}\delta(t) = \delta'(t) = -\dfrac{1}{t}\delta(t)$

6.9 Find the Fourier transform of the unit step function by considering a superposition of a constant function and $\text{sgn}(t)$.

Problems

6.10 Compute the Fourier transform of sgn(t) by considering a sequence of transformable functions $\exp(-\alpha|t|)$ sgn(t). Calculate the Fourier integral and then take the limit $\alpha \to 0$.

6.11 Explore the rate at which successive convolutions approach the Gaussian form (the central limit theorem) by convolving $(1,1)$, $(1,1,1,1)$, and $(3,2,1)$ with themselves N times. For a convenient test of the Gaussian shape, plot the logarithms of the convolution results versus the time squared.

6.12 A simple way for a function to violate the central limit theorem is for it to have zero area. Show that if just one function of a set of successive convolutions has zero area, it will produce a result whose spectrum is zero at zero frequency and hence the result will have a zero area in time domain—obviously not a Gaussian form.

6.13 Show that if the Fourier transform of just one member of a set of successive convolutions has a sharp corner (a discontinuous derivative) at the origin, an approach to the Gaussian can never be made, thereby violating the central limit theorem.

6.14 If two simple lenses are placed so that their focal planes coincide, rays of an image passing through both lenses will form an inverted image. Relate this result to two successive Fourier transforms.

6.15 Transmitters employed in AM radiotelephony modulate a carrier signal of frequency ω_0 with audio program material according to $[1 + f(t)]\cos(\omega_0 t)$. Sketch the spectrum of this AM signal assuming that $f(t)$ is an audio signal that is band limited between 20–5000 Hz.

6.16 Show that the Hilbert transform of a real function is real.

6.17 Show that the Hilbert transform of the Hilbert transform of a function is the negative of that function.

6.18 Show that the Hilbert transform of an even function is odd and that the Hilbert transform of an odd function is even.

6.19 Show that $\mathcal{H}(f * g) = (\mathcal{H}f) * g = f * (\mathcal{H}g)$.

6.20 Most autocorrelation functions tend to be wider than the function that they are generated from. Show, however, that

$$(1/\pi t) \otimes (1/\pi t) = \delta(t)$$

6.21 Clearly, functions with δ-function autocorrelations must possess a flat power spectrum. Use this to devise several functions that have δ-function autocorrelations.

6.22 To appreciate why the Hilbert transform has phase shifts of opposite sense for positive and negative frequencies, assume that it does not: show that if we insist on a constant 90° phase shift in the same sense for both positive and negative frequencies, a function $f(t)$ is uninterestedly transformed into $f(t) \to \pm i f(t)$.

6.23 Show that if the spectrum $F(\omega)$ of a function is given a constant phase rotation $\phi\,\mathrm{sgn}(\omega)$, the analytic function belonging to $f(t)$ is transformed to $g(t) \to g(t)(\cos\phi - i\sin\phi)$, and that hence the envelope function is invariant under this phase rotation.

6.24 Show that the Hilbert transform of a function is zero at points where the function is equal to its envelope. As a consequence of this result, show that the function is tangent to its envelope where the two meet.

6.25 The envelope of $\sin(\omega t)/t$ is shown in Fig. 6.14. Show that it is given by

$$E(t) = (1/t)\sqrt{2[1-\cos(\omega t)]}$$

6.26 Modify the discussion in the text to show that no band-limited signal can be zero over a finite interval, except, of course, the trivial case $f(t) \equiv 0$. Similarly, show that no time-limited signal can have a spectrum that is zero over any finite frequency band.

6.27 Compute the Fourier transform of

$$f(t) = \begin{cases} e^{-\alpha t} & t > 0 \\ 0 & t < 0 \end{cases}$$

Take the limit of the transform $F(\omega)$ as $\alpha \to 0$ and compare your result with that of Problem 6.9.

7

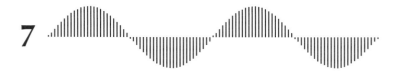

Application of the Fourier Transform to Digital Signal Processing

We have invested in the previous chapter on continuous signal theory because many of the signals that find their way into digital signal processing are thought to arise from some underlying continuous function. In this chapter, we discuss the relationship between these underlying continuous functions and the discrete signals that are produced by their sampling. Fundamental problems will be encountered. A parallel development of the sampling process, in both the time and frequency domains, will clarify these problems and show how to cope with them. However, the deeper insight provided by this development will permit us to evaluate the severity of the limitations encountered in a given circumstance. Finally, after we understand the relationship between the Fourier integral transform and the DFT, new useful ideas will emerge on interpolation, decimation, and modulation.

In the preceding chapters, we have pursued a natural course through three kinds of time–frequency transformations. First, evaluation of the frequency response of discrete LSI systems led to the discrete-time/continuous-frequency description of digital impulse responses and their spectra. Then, by computing this spectra at discrete points in frequency, we were led to the discrete-time/discrete-frequency domain of the DFT. In the last chapter, suitable limits took the DFT over into the Fourier integral transform of continuous time and frequency.

To complete our journey into time–frequency transformations, we next address the fourth possibility: continuous time and discrete frequency.

Continuous Time, Discrete Frequency: The Fourier Series

Discrete spectra are familiar from areas such as spectroscopy, where they are called line spectra. Viewed as a mathematical entity by itself, apart from continuous signals, we have seen that the DFT relates equally spaced discrete spectra to equally spaced data points in time. We now ask what are the consequences of demanding equally spaced discrete spectra from a set of continuous time domain data. That is, we require that

$$F(\omega) = 2\pi \sum_{n=-\infty}^{\infty} a_n \delta(\omega - n\omega_0) \qquad (7.1)$$

which places spectral lines of magnitude $2\pi a_n$ at multiples of the fundamental frequency ω_0. The time domain signal with such a spectrum is

$$f(t) = \frac{1}{2\pi} \int_{-\infty}^{\infty} F(\omega) e^{i\omega t} \, dt = \frac{1}{2\pi} \int 2\pi \sum a_n \delta(\omega - n\omega_0) e^{i\omega t} \, d\omega$$

or

$$f(t) = \sum_{n=-\infty}^{\infty} a_n e^{in\omega_0 t} \qquad (7.2)$$

Clearly, $f(t)$ is still a continuous function of time, but the discrete, equally spaced spectrum has produced periodicity in the time domain of period

$$T = 2\pi/\omega_0$$

since

$$f(t + 2\pi/\omega_0) = f(t)$$

Equation (7.2) is exactly equivalent to Eq. (3.1), connecting discrete, equally spaced time domain data to continuous periodic spectra, with the roles of time and frequency interchanged.

To complete the continuous-time/discrete-frequency picture, we should find the forward transform that takes $f(t)$ into $F(\omega)$. Because $f(t)$ is periodic, no new information is contained outside of the interval $-T/2$ to $T/2$. Thus motivated, we integrate Eq. (7.2) over these limits with the appropriate Fourier kernel:

$$\int_{-T/2}^{T/2} f(t) e^{-im\omega_0 t} \, dt = \sum_n a_n \int_{-T/2}^{T/2} e^{i(n-m)\omega_0 t} \, dt$$

Direct elementary integration shows that

$$\int_{-T/2}^{T/2} e^{i(n-m)\omega_0 t} \, dt = T\delta_{nm} \qquad (7.3)$$

Continuous Time, Discrete Frequency: The Fourier Series

giving

$$\int_{-T/2}^{T/2} f(t)e^{-im\omega_0 t}\, dt = Ta_m$$

This result, combined with Eq. (7.2), provides the desired pair of transformations, relating continuous-time periodic functions with their equally spaced discrete spectra:

$$a_n = \frac{1}{T}\int_{-T/2}^{T/2} f(t)e^{-in\omega_0 t}\, sdt \qquad (7.4a)$$

$$f(t) = \sum_{n=-\infty}^{\infty} a_n e^{in\omega_0 t} \qquad (7.4b)$$

Equations (7.4), called the Fourier series, play a major role in continuous theory because periodic, continuous-time signals are so commonplace in many applications. Equation (7.4a) is sometimes called the analysis equation because it separates the periodic time signal into its component line spectra, a_n. Equation (7.4b) represents the Fourier synthesis of a periodic, continuous-time signal from the superposition of complex sinusoids with frequencies that are multiples of the fundamental. The sum may or may not extend to infinity. The a_n are called Fourier coefficients. Three common examples of Fourier series, the square wave, the triangle wave, and the full wave rectified sine wave are shown in Fig. 7.1.

It is interesting to note that the Fourier series, which has dominated Fourier theory since its inception, takes on only a very minor part in

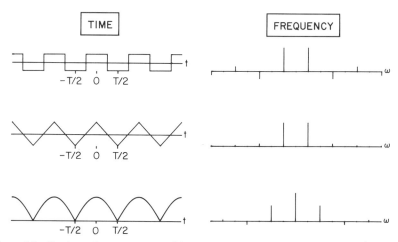

Figure 7.1 Fourier series components of three common continuous repetitive waveforms. All three have spectral components extending to infinitely high frequencies.

modern digital signal processing. In fact, we introduce it here mostly to complete the discussion of the four types of Fourier transforms arising from the four possible combinations of discrete/continuous and time/frequency. We include the traditional continuous-time, discrete-frequency Fourier series in our discussion for two purposes: to examine its least-squares convergence properties and to introduce the sampling function.

For reference, we summarize the four types of Fourier transformations in Fig. 7.2, reviewing just how they arose in our discussions. Clearly, there

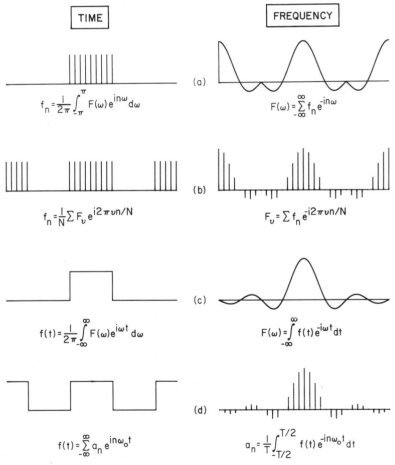

Figure 7.2 Summary of the four kinds of Fourier transforms: (a) the discrete-time, continuous-frequency Fourier series; (b) the DFT; (c) the continuous Fourier integral transform; and (d) the traditional Fourier series with its continuous time and discrete frequency.

are two types of transformations that might properly be called Fourier series: discrete-time and continuous-time Fourier series. Mathematically, these two transforms are identical, but with the roles of time and frequency interchanged. Next, we explore the interesting convergence properties of these two types of Fourier series.

The Least-Squares Convergence of the Fourier Series

The convergence property of the Fourier series is both curious and interesting in its own right; additionally, it has important implications for digital filter design. Our approach is to pretend that we do not know how to compute the Fourier coefficients as given in Eq. (7.4a). Instead, we wish to take an alternate tack to the course of the previous section and determine the a_n by curve fitting. That is, following the notion of Fourier synthesis, suppose that we wish to synthesize a given continuous periodic function $f(t)$ from a superposition of a finite number of sinusoids equally spaced in frequency. The result of such a superposition of N sinusoids is

$$\hat{f}(t) = \sum_{n=0}^{N} a_n e^{in\omega t} \qquad (7.5)$$

Generally, when this finite sum is used, $\hat{f}(t)$ will be different from the desired $f(t)$. Of all the possible criteria for minimizing this difference, we choose to minimize the total squared error over the period of $f(t)$; that is, we wish to determine the a_n that minimizes

$$\text{the sum squared error} = \int_{-T/2}^{T/2} |f(t) - \hat{f}(t)|^2 \, dt \qquad (7.6)$$

The minimum is found by differentiating with respect to a_m^* (a_m and a_m^* are independent; see Problem 7.2):

$$\frac{\partial}{\partial a_m^*} \int_{-T/2}^{T/2} \left| f(t) - \sum_{n=0}^{N} a_n e^{in\omega t} \right|^2 dt = 0$$

giving

$$\int_{-T/2}^{T/2} (f(t) - \sum a_n e^{in\omega t}) e^{-im\omega t} \, dt = 0$$

or

$$\int f(t) e^{-im\omega t} \, dt = \sum_n a_n \int_{-T/2}^{T/2} e^{i(n-m)\omega t} \, dt = T \sum_n a_n \delta_{nm}$$

Thus, the coefficient that minimizes the total squared error is

$$a_m = \frac{1}{T} \int_{-T/2}^{T/2} f(t) e^{-im\omega t} dt$$

just the normal Fourier series coefficient.

The conclusion is perhaps somewhat surprising; simple truncation of a function's Fourier series expansion approximates the function in the least-squares sense. This serendipitous result allows us to take satisfaction in the knowledge that as each term is added to the partial sum in Eq. (7.5), the best possible fit to $f(t)$ is obtained, for that number of terms, in the least-squares sense. In addition, a second surprise concerning the nature of Fourier series convergences occurs at a discontinuity.

Three examples of partial sums for a square wave synthesis are shown in Fig. 7.3. In each case, an overshoot occurs near the discontinuity, and the partial sums equal the average value of the jump at the discontinuity. The unexpected behavior of this convergence is that the amount of overshoot remains constant as more terms are added. In fact, it can be shown (see Dym and McKean, 1972) that this overshoot approaches a constant 8.9% of the jump at the discontinuity as the number of terms in the partial sum

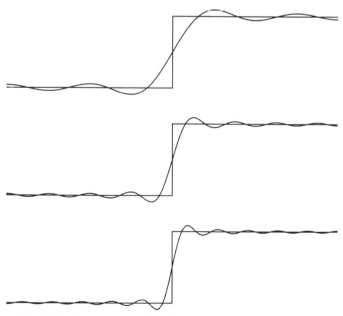

Figure 7.3 The Gibbs phenomenon resulting from the least-squares convergence property of the Fourier series at the discontinuity of a square wave for $n = 7$, 15, and 21.

tends to infinity. It might be natural to expect that this overshoot tends to zero with an increasing number of terms, but this is not the case. This unexpected convergence behavior was first described by British mathematician H. Wilbraham, but it was made more well known by a public exchange in *Nature* between American physicists A. Michelson and J. W. Gibbs. In a 1898 letter to *Nature*, Michelson explained that he was having difficulty in synthesizing a discontinuity from the summation of the first 80 terms of a Fourier series. He suspected that the device he used to determine the Fourier coefficients, called a harmonic analyzer, was defective or that the mathematicians were all wet in saying that a discontinuity could be replaced by a summation of continuous functions. In a responding letter one year later, Gibbs explained the true nature of the convergence, which is now called the Gibbs phenomenon.

As terms are added to the partial sum, the integrated squared error of Eq. (7.6) does indeed decrease uniformly: the wriggles of the Gibbs phenomenon increase in frequency and shift toward the discontinuity as seen in Fig. 7.3, decreasing the squared area between them and the desired function. The amplitude of the largest wriggle does not decrease, but the integrated squared error does decrease uniformly.

The importance of the Gibbs phenomenon to digital signal processing can be appreciated by considering the design of an ideal lowpass filter. The frequency response of such a filter is a continuous periodic function with a discontinuity at the cutoff frequency. The discrete time domain operator consists of the Fourier coefficients of this frequency response function. These coefficients are easily computed for any cutoff frequency, but the operator will be infinitely long. The situation is similar to that shown in Fig. 7.3, with the roles of time and frequency interchanged. If the infinitely long time domain operator is simply truncated to produce a workable digital filter, the Gibbs phenomenon is produced in the frequency response, making the ideal lowpass filter unobtainable with a finite length operator. The question then of how to obtain the best time domain coefficients of a given length becomes the interesting and challenging problem of digital filter design. In Chapter 8, we present several approaches to designing digital filters, given their desired frequency response.

The Sampling Function

We next exploit the continuous-time Fourier series to introduce the sampling function, a transform pair that will serve as a major structural element in building a bridge of insight and interpretation between continuous signals and the DFT. In the continuous-time Fourier series representation, $f(t)$ is periodic of period T. By selecting $f(t) = \delta(t)$, a series of delta-function spikes is formed in the time domain equally spaced by T. Such a series of spikes would be written

$$f(t) = \sum_{n=-\infty}^{\infty} \delta(t - nT)$$

The Fourier coefficients of this spike train are given by

$$a_n = \frac{1}{T} \int_{-T/2}^{T/2} \delta(t) e^{-in\omega_0 t} dt, \quad \omega_0 = \frac{2\pi}{T}$$

or

$$a_n = \frac{1}{T}$$

So the spectrum is also a series of equally spaced spikes, with frequency spacing $\omega_0 = 2\pi/T$. Rewriting the time domain function and substituting $a_n = 1/T$ in Eq. (7.1) gives for this transform pair of equally spaced spikes

$$f(t) = \sum_{-\infty}^{\infty} \delta(t - nT) \tag{7.7a}$$

$$F(\omega) = \frac{2\pi}{T} \sum_{n=-\infty}^{\infty} \delta(\omega - n\omega_0) \tag{7.7b}$$

where $\omega_0 = 2\pi/T$. Either member of this transform pair, shown in Fig. 7.4, is called a *sampling function* because multiplication (in either domain) with a continuous function just produces equally spaced samples of that function. Notice again the reciprocal relationship between time spacing T and frequency spacing ω_0 of this pair: $T\omega_0 = 2\pi$. The sampling function, like the Gaussian, is one of those rare functions that is its own transform, that is, the functional dependence is the same in both domains.

We have taken the circuitous route via the continuous-time Fourier series to introduce the sampling function because it seems like the

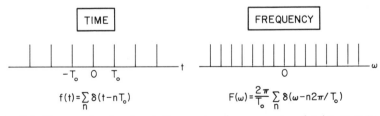

Figure 7.4 The sampling function is its own transform—a series of δ functions in both domains. In spite of its obstreperous mathematical properties, the sampling function serves well in relating the FT to the DFT.

easiest approach. Indeed, the computation of the spectrum of Eq. (7.7a) using the continuous Fourier integral formulation is impossible since all the conditions for the existence of the Fourier integral are violated. The sampling function with its infinite number of infinite discontinuities and its nonconvergent Fourier integral is the most pathological function that we will need. Yet, in the next section, it will prove most easy to manipulate and most useful in understanding the sampling of signals.

The Relationship between the FT and the DFT: Resolution and Leakage

In many applications our digital signal is the result of sampling a continuous signal. Examples abound: the voltage of an electronic signal may be sampled with an A/D converter, a meteorologist may read a barometer at the same time each day, and a navigator may determine the position of his vessel with a noontime sun shot using a sextant or he may determine his position every second with electronic aids, such as Loran or satellite navigation. In each case, it may be desired to treat the resulting equally spaced data with an LSI digital operator. For example, the satellite navigation data may need to be smoothed in some manner to better estimate the ship's position. Whatever LSI operation is performed on the data, it can be performed by either convolution in the time domain or multiplication in the frequency domain. Therefore, it will be beneficial to see the effects of sampling in both domains.

We start with the continuous signal that exists before sampling. It and its continuous magnitude spectrum are shown in Fig. 7.5(a); we omit the phase spectrum for simplicity. In many situations, the signal and its spectrum will be rather broad, as shown.

Figure 7.5(b) shows the sampling function with spacing T. Multiplication in the time domain produces the sampled signal; convolution in the frequency domain produces its resulting spectrum. The result is shown in Fig. 7.5(c). Convolution with a row of δ functions just places the original function at each δ-function location, replicating the original function. Thus sampling in the time domain has induced periodicity in the spectrum. If the spectrum is wider than π/T, an overlap superimposes portions of the true spectrum causing a distortion. This overlapping of the spectrum is *aliasing*. The amount of the overlap depends on the spectral width compared to π/T, the spacing between repeated spectra. For any given spectrum, the effect can be reduced by more time domain sampling, which spaces the replicated spectra farther apart. The Nyquist sampling theorem discussed in Chapter 1 applies only to a signal that is band limited so that the

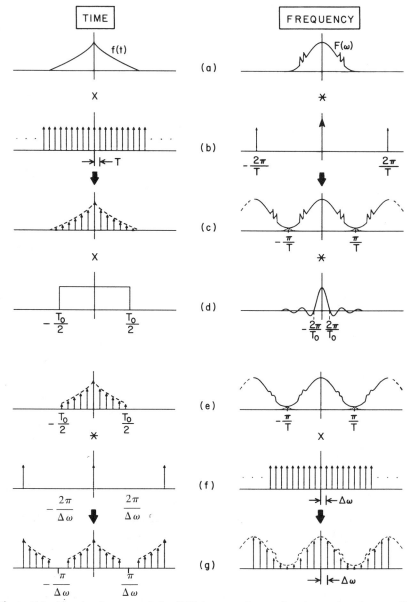

Figure 7.5 The development of the DFT from continuous functions using the sampling function. In this example, the DFT applied to a continuous broadband signal results in a spectrum with aliasing, limited resolution, and leakage.

spectrum has exactly zero components above frequency π/T. We know from the time-limited/band-limited theorem of Chapter 6 that such a band-limited signal would have to be infinitely long—an impossibility. Thus, fundamentally, there are no band-limited signals and aliasing becomes a matter of degree. In practice, however, the long tail of the spectrum may be masked by other effects, such as noise, so that a sufficiently high sample rate can separate the replicated spectra to a point where the only meaningless information exists in the vicinity of the Nyquist frequency. The result of any further increase in the sampling rate depends on the nature of the contaminating noise. For example, if the noise had large, high-frequency components, they may be folded down into the low-frequency portion of the signal's spectrum by an insufficient sampling rate. In this case, a higher sampling rate separating the noise spectrum from the signal's spectrum will improve the data's usefulness for subsequent signal processing.

Putting noise considerations aside, there is another fundamental limitation that we must impose on the digitized data shown in Fig. 7.5(c). In many situations, we can only record a portion of the actual signal. In effect, this truncation is multiplication of the data stream with the boxcar data window of length T_0, shown in Fig. 7.5(d). In the frequency domain, the spectrum of the data is convolved with the Fourier transform of the data—a sinc function of width $4\pi/T_0$. This convolution with the sinc function smears out the sharp features of the spectrum, limiting the *spectral resolution*. The two sharp lines in our example are now smeared out as shown in Fig. 7.5(e). The longer the data window T_0, the sharper is the sinc function, and hence the less loss of resolution. In the limit of an infinitely long data window, its associated sinc function becomes a δ function, resulting in the identity operation when convolved in the frequency domain; only then is there no loss of resolution. The dependence of frequency resolution on data window length is a manifestation of the uncertainty principle: it requires a long sample to obtain a precise frequency determination. Note that the frequency resolution depends only on the length and the shape of the data window. An increased sampling rate only separates the replicated spectra, reducing aliasing, but it does not increase spectral resolution.

In addition to limiting spectral resolution, this convolution of the spectrum with the sinc function has another effect. The slowly decaying side lobes of the sinc function mix spectral energy from one spectrum to its adjacent replicas, as shown in Fig. 7.5(e). This interspectral mixing is called *leakage* and is distinct from the loss of resolution that occurs within one spectral replica. Unlike resolution, leakage can be reduced by increased time domain sampling because of the increased separation between spectral replicas.

Both resolution and leakage depend on the shape of the data window. We have used the boxcar for this discussion; clearly others would produce a different effect on the spectrum. It turns out that compared to other reasonable window shapes the boxcar's sinc function has a relatively narrow central peak with comparatively large side lobes. It thus produces good spectral resolution at the expense of poorer leakage properties. Whenever we are using only a portion of a data stream, sampled over some length of time as discussed here, we can never observe the true spectrum, but we see it through this data window: we see an estimate of the spectrum, smoothed by a moving average forced on us by the convolution operation. The selection of the best shape for the data window, which in turn determines the kind of smoothing done on the spectrum, is called *windowing*, one approach to spectral estimation discussed in Chapter 11.

Up to this point, we have only discussed the acquisition of sampled data over a finite window length. These data have a continuous spectrum that is distorted by both resolution and leakage. To view this spectrum we need to calculate it from the sampled data, using a convenient computational scheme—the DFT, of course. The DFT samples the spectrum at equally spaced points in frequency. This frequency domain sampling is represented by multiplication with the sampling function, as shown in Fig. 7.5(f). In the time domain, the data is convolved with the sampling function, producing the replicated data samples shown in Fig. 7.5(g). Again, sampling in one domain forces periodicity in the other so that our data is now unavoidably periodic. In our example, we have sampled the spectrum at sufficiently close frequency intervals, $\Delta\omega$, that time domain replicas have gaps (i.e., zero values) between them. This development shows how to use the DFT to compute the spectrum at any number of frequency points—simply append zeros to the data. This is called *zero padding* and is normally a good practice in computing spectra of discrete data via the DFT. The spectrum of discrete data is a continuous function of frequency; zero padding is simply a method of computing the spectrum at a large number of points using the DFT. Figure 7.5 is an important summary of our discussion relating the DFT to continuous signals.

This relationship is sufficiently important for using the DFT in practical applications to warrant some examples. A particularly instructive case is no zero padding. Then, the data window, shown in Fig. 7.6(a), is replicated with no gaps. This condition occurs when the replicating spikes in the time domain are separated by the length of the window, T. Then, the frequency sampling points, which we will call the *DFT frequencies*, are spaced by $\Delta\omega = 2\pi/T$ up to $\pi N/T$ for an N-point DFT. The important observation is that the zeros of the data window's sinc function exactly fall on the DFT frequencies at multiples of $2\pi/T$.

The Relationship between the FT and the DFT: Resolution and Leakage   139

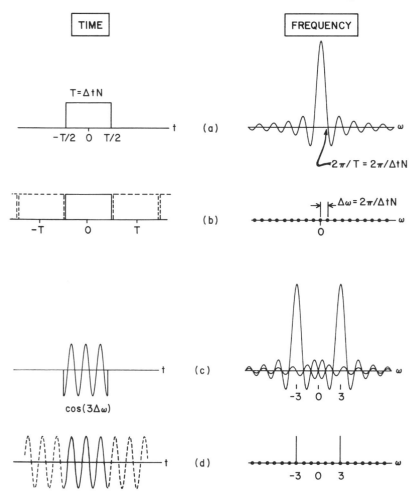

Figure 7.6 A special DFT case occurs when the data window is an exact multiple of sinusoidal components of the signal. (a) The data window and its corresponding sinc function. (b) The DFT forces periodicity in both domains. (c) A sinusoid that has exactly an integral number of cycles in the data window has its spectral lines falling exactly on DFT computation points in the frequency domain. The true spectrum of the three-cycle sinusoid burst consists of the two sinc functions, but their zero crossings lie exactly on the DFT computation frequencies, preventing the DFT from revealing the correct spectrum. (d) If the actual signal under study is indeed an infinitely long sinusoid, then the DFT calculation is correct, giving pure spectral lines, but this conclusion requires a knowledge of the signal beyond the data window.

The significance of this result can be seen by considering sinusoidal data with frequencies that are multiples of $2\pi/T$. Then, an integer number of periods, n, just fits inside the data window, as shown in Fig. 7.6(c) for $n = 3$. This continuous cosine wave has a δ-function spectrum located at $n(2\pi)/T$. Convolution with the data window's sinc function produces two superimposed sinc functions located at $\pm n(2\pi)/T$, as shown in Fig. 7.6 (c). But, because of the special frequency of the sine wave, the DFT's frequency sampling just hits the zeros of these combined sinc functions at every frequency except $\pm n(2\pi)/T$, where the sinc's central peak is sampled. The resulting DFT spectrum, shown in Fig. 7.6(d), is deceptively accurate; the time domain signal has been smoothly joined by its replicas to make an infinitely long sine wave whose spectrum is properly described by two zero-width spikes.

Have we beaten the uncertainty principle in this example, demonstrating infinite resolution and zero leakage? By no means. The sinc function associated with the boxcar data window is really lurking in the background of the DFT spectrum. Zero padding reveals it. The spectrum is continuous, and it can be computed at any frequency points that we desire by using the DFT with zero padding. Figure 7.7 shows the result of 8:1 zero padding for our example.

One may argue that the zero padding of Fig. 7.7 is an abomination of our original sine wave data. This suggestion brings us to an extremely important point for interpreting DFT results. To properly use the DFT, one needs to have more information about the signal than just that available in the data window. For example, if we know that our signal is the cosine burst of Fig. 7.7, then a lot of zero padding is the correct thing to do. It makes the repetitive time domain signal of the DFT look more like an isolated burst. Also, because we know that an isolated burst has an infinitely broad spectrum, the higher sampling rates will reduce aliasing. On the other hand, if it were known that the signal under study were periodic with a period exactly equal to the data window, then no zero padding is the correct application of the DFT. If this signal were also band limited and unaliased, the resulting DFT line spectrum would then be exact.

By using the proper procedure, it is possible to use the computational convenience of the DFT to compute all four kinds of Fourier transforms that we have discussed. In some cases, exact results are possible; in others, we must be content with approximations. A systematic discussion of six cases is left for Problem 7.6. To see the kind of thinking required to use the DFT in Fourier transform problem consider the following scenario.

Imagine a digitized sinusoid as sketched in Fig. 7.8(a). Approximately, not exactly, one-half of a period of the sinusoid occupies the data window. Therefore, energy will be spread out across the entire DFT spectrum

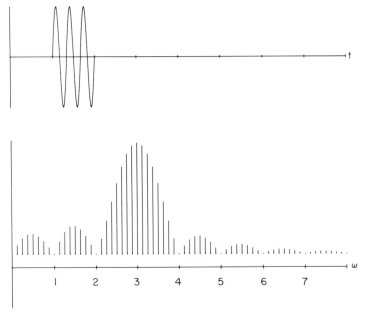

Figure 7.7 The magnitude of the hidden sinc function discussed in the example of Fig. 7.6 revealed by 8:1 DFT zero padding. With no zero padding, only the spectral values at the labeled frequencies would be computed.

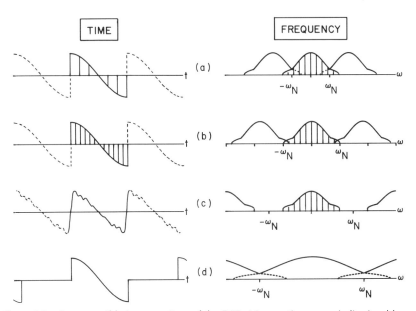

Figure 7.8 Some possible interpretations of the DFT: (a) a continuous periodic signal has a broadband spectrum with large amounts of aliasing; (b) increased sampling reduces aliasing; (c) frequency domain zero padding shows the band-limited signal that has the spectrum of (b); and (d) in the limit of large sampling and time domain zero padding, the continuous FT spectrum of a nonperiodic signal is approached, but never completely realized.

because the frequency of the sinusoid does not fall right on a DFT computation frequency. What then, is the interpretation of the DFT spectrum sketched in Fig. 7.8(a)? To answer this question, we require knowledge of the signal beyond the data window. If we assume it is periodic with a period just equal to the data, then the signal fits the DFT formulization. Is the spectrum exact then? No, because the periodic extension of the signal, suggested by the dotted line in the figure, has discontinuities that require it to be broadbanded. This infinite bandwidth signal is therefore necessarily aliased, and the DFT spectrum is inexact. A higher sampling rate, as shown in Fig. 7.8(b), will reduce the aliasing and improve the spectral estimation. This demonstrates the method of using the DFT to estimate the Fourier series coefficients of a broadbanded continuous function.

We might ask what signal does have the exact spectrum shown in Fig. 7.8(b)? Zero padding in the frequency domain, as shown in Fig. 7.8(c), gives the answer. This zero padding is saying that the DFT spectrum is exact out to Nyquist and there is no spectral energy beyond. That is, we are now saying that the signal is band limited. This band-limited signal is recovered by taking the IDFT of the zero-padded spectrum as shown in Fig. 7.8(c). This procedure is really interpolating between the discrete points in the time domain by assuming that the signal is band limited and unaliased. Any amount of interpolation is possible; it just requires more frequency domain zero padding to get more interpolated points in the time domain. In the limit, a continuous, band-limited time domain signal is approached. It is this signal that has the DFT spectrum as Fourier series coefficients.

The last possibility that we consider is that the signal is a single burst of the sinusoid. Such a signal, of course, is broadbanded, having spectral components extending to infinity. Zero padding in the time domain will make the periodic DFT representation look more like a single burst, as suggested in Fig. 7.8(d). As the sampling rate is increased to lower aliasing and zero padding is increased, the DFT spectrum approaches the broadbanded, nonperiodic, continuous spectrum of the sinusoidal burst, but never completely reaches it.

In summary, then, essentially the same data can be used to generate DFT spectra with greatly different interpretations. In two cases, the band-limited nature of a signal permits the DFT to produce an exact spectrum. The most common of these two cases is the discrete, finite-length sequence whose spectrum is naturally periodic. Time domain zero padding reveals the exact spectrum. The less common case is the continuous periodic signal that is known to be band limited. If the data window can be matched to exactly one period and a sufficiently high sample rate is used, the DFT will produce the true line spectrum.

In two other cases, the broadband nature of the signals allows only an estimate of the true spectrum via the DFT. If the signal is periodic and sampled for an integer multiple of its fundamental period, the DFT spectrum will approach the exact line spectrum as the sample rate increases. If the signal is known to be zero outside of the data window, the DFT spectrum will approach the correct continuous Fourier spectrum with increased sampling and increased zero padding.

It is a central point in DFT signal processing that one needs to know information about the data beyond the data window to determine the proper DFT treatment. This formulation of a signal's behavior beyond the observed data window is called *data modeling*. In addition to the four models discussed here, many more are possible; we will examine others in Chapter 11 on spectral estimation. Next, we will use the sampling function and the properties of zero padding that we have discussed to explore some important manipulations of data and their DFT spectra.

Interpolation, Decimation, and Multiplexing

Frequently, there is the need in digital signal processing to change the sampling rate of existing data. Increasing the number of samples per unit time, sometimes called *upsampling*, amounts to interpolation. Decreasing the number of samples per unit time, sometimes called *downsampling*, is *decimation* of the data. (The original meaning of the word decimation comes from losing one-tenth of an army through battle or from self-punishment; we apply it to data, using various reduction ratios). Of course, interpolation and decimation can occur in frequency as well as in time.

In fact, we have already encountered frequency domain interpolation; zero padding in time followed by the DFT interpolates the hidden sinc functions in the DFT spectrum as we saw in Fig. 7.7. In Fig. 7.8(c), just the opposite was done; zero padding in the frequency domain produced an interpolated time function. We will now investigate this type of upsampling, applied to interpolation of time domain data, in a little greater detail.

Consider the discrete data stream shown in Fig. 7.9(a) along with its continuous spectrum. As we now realize, this DFT spectrum has different possible interpretations, depending on our data model. For purposes of discussion, let us say that this data results from sampling a band-limited (or nearly band-limited) continuous signal. Then, in the limit of a very long data window, sampled at a sufficiently high rate, no leakage or aliasing occurs. Time domain interpolation will correctly recover the original analog signal if it does not alter the spectrum in Fig. 7.9(a). The periodicity

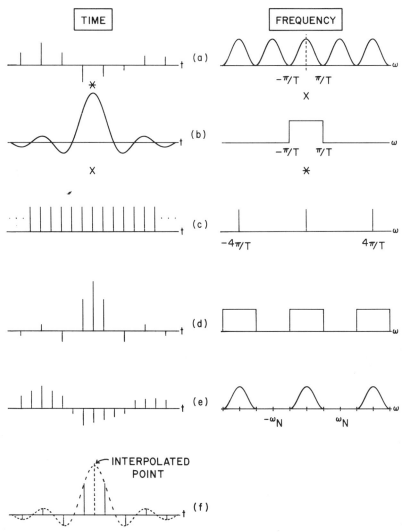

Figure 7.9 Upsampling by sinc interpolation. (a) The original discrete signal. (b) Frequency domain multiplication by the boxcar separates out one spectral replica, forcing the signal to be continuous in time by virtue of the convolution with the associated continuous sinc function. The continuous signal thus produced can be sampled at any rate. (c) For example, twice the original sampling rate can be used. (d) This sampling of the interpolating sinc function produces the midpoint interpolation operator. (e) Midpoint interpolation separates spectral replicas, leaving the upper one-half Nyquist interval empty. (f) As a discrete convolution operator, the midpoint interpolation operator acts only on the original data. Its zeros coincide with the midpoint positions.

Interpolation, Decimation, and Multiplexing 145

induced into the spectrum by the data sampling process can be eliminated by extracting just one replica. This extraction, accomplished by frequency domain multiplication with the boxcar shown in the right side of Fig. 7.9(b), convolves the discrete time domain data with the continuous sinc function to reproduce the original analog signal. It thus seems evident that a truly band-limited signal can be recovered completely from its sampled version providing that the sampling rate is sufficiently high and that the sample is sufficiently long. The statement is commonly made that a band-limited analog signal can be uniquely recovered from its sampled version provided that it is sampled at a rate greater than twice the highest frequency contained in its spectrum; this statement is called the *sampling theorem*. This theorem has been discussed by many authors and several aspects of it have been proven in mathematical detail. However, from our theorem in Chapter 6, any such band-limited signal must be infinitely long, making the exact determination of its spectrum impossible in the first place.

Thus, in practice, we must always be content with an approximate reconstruction of the original analog signal. Preferring a digital scheme for this reconstruction, we convolve the boxcar spectral window of Fig. 7.9(b) with the sampling function shown in Fig. 7.9(c). The result tells us how to exploit the DFT for the recovery of the analog signal—use zero padding in the frequency domain. In our example, we use 2:1 zero padding, which produces the *midpoint interpolation* operator shown in Fig. 7.9(d). The result of this operator acting on the original data in Fig. 7.9(a) is shown in Fig. 7.9(e). In the frequency domain, one simply appends zeros to the DFT spectrum. It is interesting to note that during the convolution process the sinc operator in the time domain appropriately has its zeros aligned with the unknown midpoints except at the point currently being interpolated: every interpolated point is a linear combination of all other original points, weighted by the sinc function; see Fig. 7.9(f). This interpolation, sometimes called *sinc interpolation*, can only be carried out in an approximation because the sinc function will have to be truncated somewhere. In the frequency domain, the result of truncating the sinc manifests itself as a convolution of the ideal lowpass filter of Fig. 7.9(d) with a narrow sinc arising from the truncation of the interpolating sinc operator. As a result, the final unsampled data has the same spectrum as the original data only to some approximation.

Decimation, or *downsampling*, is the reverse operation of the sinc interpolation. To decimate 2:1 with no loss of information from the original data, the data must be oversampled to begin with. Figure 7.10(a) shows data that is nearly oversampled by 2:1 to produce a spectrum that has very little energy in the upper half of the Nyquist interval. As is usually done,

146   7/ Application of the Fourier Transform

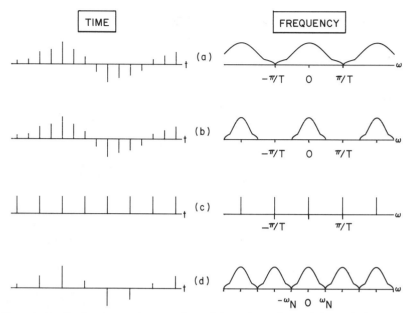

Figure 7.10 Decimation or downsampling of data. The well-sampled signal of (a) is lowpass filtered to clear out the upper half of the Nyquist interval as shown in (b). The sampling function in (c) takes every other sample, resulting in the decimated signal in (d).

we lowpass filter in preparation for decimation. In our example of Fig. 7.10(b), the upper half of the Nyquist interval has been filtered out with an appropriate filter. Then, the 2:1 decimation operation simply consists of extracting every other sample in the time domain. This operation can be perceived as multiplication in time and convolution in frequency, with the sampling function shown in Fig. 7.10(c). The decimated signal, in Fig. 7.10(d), now has a new sampling rate and Nyquist frequency—its spectrum just filled in to meet the new Nyquist criterion. The lowpass filtering has assured that no aliasing occurs in the decimated data.

The next two examples of manipulating data and their spectra employ combinations of filtering, sampling, interpolation, and decimation. Consider the spectrum shown in Fig. 7.11(a), which is divided into four separate bands. Each of these bands contains information that we wish to separate from the original spectrum. In one important case in communication applications, each frequency band contains an independent information channel. The modulation theorem, expressed in continuous form by Eq. (6.26), shows that if we modulate a given channel with a sinusoid of frequency ω_0, the spectrum is translated by $\pm\omega_0$. Thus, each of the four

Interpolation, Decimation, and Multiplexing 147

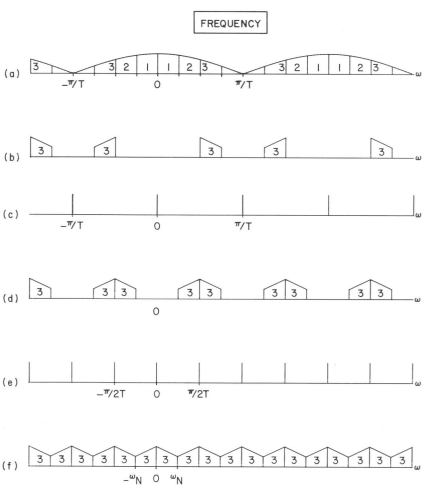

Figure 7.11 Frequency division demultiplexing: (a) the spectrum divided into four bands; (b) band three separated out by bandpass filtering; (c) and (d) decimation by 2:1; and another 2:1 decimation in (e) and (f). The final result has the whole Nyquist interval filled with band three.

frequency bands of Fig. 7.11 could represent separate channels formed by *frequency division multiplexing* (FDM), using an appropriate carrier frequency, ω_0, $2\omega_0$, $3\omega_0$, and $4\omega_0$, for each band. Analog versions of FDM have been extensively used for years in many communication applications, such as AM radio, stereo broadcasting, television, and radiotelemetry. Digital FDM is similar, except the spectrum is repetitive.

Recovering a given channel, called *demodulation* or *demultiplexing*, is accomplished by first isolating the selected channel using bandpass filtering and then decimating the result. Figure 7.11 shows channel three demultiplexed by filtering followed by a 4:1 decimation. The resulting digital data has a new sampling rate, meeting the Nyquist criterion. If the original channels are well sampled, gaps occur in between the spectral bands of Fig. 7.11(a), which are called *guard bands*.

Another application of isolating a given frequency band in this fashion occurs when we simply desire to pick off a given portion of the spectrum of a signal for more detailed examination and processing. In this case, the original spectrum of Fig. 7.11(a) belongs to just one digital signal, and the bands are portions of the spectrum of special interest. In our example then, band three has been selected for closer examination. The process has given us time domain data that require only one-fourth the original samples, an important savings in some applications where further processing on the spectrum is desired, such as in spectral estimation. When used in this fashion, this procedure is called *zoom processing* because it zooms in on the portion of the spectrum of interest.

For our second example of multiplexing, we address a situation that is complementary to FDM. In FDM, the information channels are mixed in a complicated way in the time domain because of the modulation of sinusoids, but the channels are quite separate in the frequency domain. The reverse situation has the channels easily separated in time, but mixed in frequency. For our example, we consider only two different digital information channels. An obvious way to combine them in time is to interlace the samples, with every other sample belonging to the same channel, called *time division multiplexing* (TDM). Multiplexing and demultiplexing in the time domain is then a simple matter of using every other sample. However, let us explore the frequency behavior of this process.

In Fig. 7.12(a), we show one of the two data channels, called channel A. It is oversampled by 2:1 so that its spectrum occupies only one-half of the Nyquist interval. Decimation using the sampling function of Fig. 7.12(b) yields the result shown in Fig. 7.12(c). But instead of redefining the sampling rate as in normal decimation, we put a twist into the processing by interpreting the results of Fig. 7.12(c) as having the same sampling rate as the original data. Thus, the time domain data has zeros at every other point. This *zero interlacing* produces a spectrum that is folded at one-half the Nyquist frequency as shown. To conserve energy using this interpretation, the spectrum must be renormalized to one-half the original values; see Problem 7.17. The other channel, the channel B, is similarly oversampled by 2:1, and then it is decimated by the shifted sampling function shown in Fig. 7.12(d). Again, its spectral amplitudes are reduced by a factor of one-half as a consequence of the zero interlacing. Finally, the TDM is completed by adding the results of the two channels: Figs. 7.12(c) and 7.12(e) sum to Fig. 7.12(f). As anticipated in TDM, while the time

Digital Control Systems

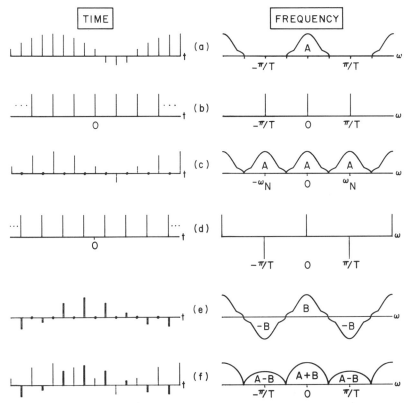

Figure 7.12 Time division multiplexing: (a) one channel oversampled 2:1; (b) the sampling function used; (c) the result of zero interlacing; (d) the $\frac{1}{2}$-shifted sampling function used on the other channel; (e) the zero interlacing of the other channel; and (f) the sum of the two channels from (c) and (e).

data are easily separated, the frequency data are mixed. Even so, note that now the Nyquist interval is filled with nonredundant information that can be used to separate the spectrum of the two channels since $A+B$ and $A-B$ are linearly independent. Clearly, TDM demultiplexing could be done in either domain.

Digital Control Systems

Because digital control systems frequently include continuous-time elements as well, they provide additional fertile ground for discussing the relationship between continuous-time and discrete-time systems. Many

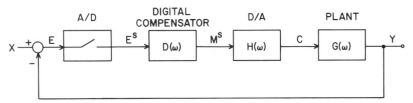

Figure 7.13 Block diagram of a simple control system showing a digital system $D(\omega)$ followed by a D/A converter. The state of the controlled device, called the plant, is compared to command signal X to generate an error signal which serves as the input to the digital system.

interconnections between these elements are possible, but all the essential features can be discussed in terms of the fairly simple configuration shown in Fig. 7.13. Here an analog device (frequently called the *plant* in control theory) is driven to a desired state as commanded by a reference signal. The current state of the plant, represented by the variable Y, is compared to the desired state in a negative feedback loop, generating an error signal E. This error signal is then used to drive the plant to the desired state. In this example, a digital algorithm $D(Z)$ (sometimes called a *compensator* in control theory) generates a control signal C which forces the plant to follow the reference signal within some acceptable tolerance. This system contains four different cascaded elements and one feedback loop. Given the transfer functions of each of these elements, we would like to compute the overall system transfer function $Y(\omega)/X(\omega)$ and analyze the result. The ideal result would have a flat spectrum out to the highest frequencies of interest, have minimal phase lag, and have no instabilities. Our first step in this analysis is to consider the details of the A/D and D/A conversions that provide the connections between the digital and analog worlds.

The most useful observation in connecting these two environments is shown in Fig. 7.14 in which a sampler is inserted in an otherwise continuous-time system. As usual, we denote sampled data with subscripted variables and continuous-time functions with the variable in parentheses. We are interested in comparing the two cases shown in Fig. 7.14(b) and 7.14(c). The output sampler in both of these cases serves only to look at the sampled version of our results (the samplers are synchronized). The comparison then is between convolving x and h and then sampling, as opposed to sampling both x and h first and then convolving. Thus we are studying the effect of inverting the orders of convolving and sampling. By picturing these processes in the frequency domain it is clear that they are different. When the sampling is applied first individually to h and x, followed by the convolution $h_t * x_t$, cross-products are produced from the aliased portions that are absent in $(h*x)_t$. Therefore $(h*x)_t \neq h_t * x_t$. See Problem 7.22.

Digital Control Systems

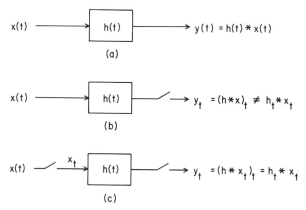

Figure 7.14 Effects of mixing continuous and discrete systems. (a) A continuous system; (b) the output is sampled after convolution; and (c) the input is sampled before convolution. All three results are different.

However, if a sampler also precedes h, the spectrum of x becomes periodic before the convolution operation. The significance of this is easily seen by working in the frequency domain. The periodic nature of the spectrum of any discrete sequence A_t can be expressed as

$$A_t(\omega) = \sum_{n=-\infty}^{\infty} A(\omega - n\omega_s)$$

where $A(\omega)$ is the spectrum of the sequence over the Nyquist interval, ω_s, is the sampling frequency, and $A_t(\omega)$ means the spectrum of the sampled version of A. Using this notation to express the sampling at the output sampler in Fig. 7.14(c) gives

$$[X_t(\omega)H(\omega)]_t = \sum_{-\infty}^{\infty} X(\omega - n\omega_s)H(\omega - n\omega_s)$$

But shifting the periodic spectrum of $X_t(\omega)$ by $n\omega_s$ effects no change in it, so $X_t(\omega - n\omega_s) = X_t(\omega)$ giving

$$[X_t(\omega)H(\omega)]_t = \sum_{-\infty}^{\infty} X_t(\omega)H(\omega - n\omega_s)$$

$$= X_t(\omega) \sum_{-\infty}^{\infty} H(\omega - n\omega_s)$$

$$[X_t(\omega)H(\omega)]_t = X_t(\omega)H_t(\omega) \qquad (7.8)$$

These results of A/D conversion are of primary importance for dealing with mixed continuous-time, discrete-time, systems.

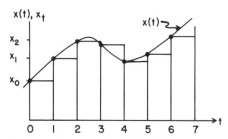

Figure 7.15 The result of first-order sample and hold D/A conversion. The values between sample times are interpolated as constant and equal to the last sampled value. For comparison, x(t) represents the results of a smoother D/A conversion, such as higher order holds or sinc interpolation.

Next we address the D/A conversion required in the control system of Fig. 7.13. To drive the analog plant, a continuous signal must be generated from the discrete output of the digital compensator. Many interpolation schemes are possible for filling in between the sample times of a discrete sequence. One example is sinc interpolation which, as we have seen, assumes that the digital data is an unaliased sampled version of a band-limited signal. Another commonly used assumption is that the analog function behaves locally as a polynomial passing through the sampled points. While this assumption is normally difficult to justify, it does allow for some particularly simple implementations of D/A converters. By far the most commonly used in digital control systems is the *zero-order hold*, which fits a zeroth-order polynomial (i.e., a constant) through the latest sampled point as shown in Fig. 7.15. Others, such as the high-order polynomials of first- and second-order holds, obtain smoother outputs at the expense of time delays, a trade-off that is not desirable from stability considerations in closed-loop control systems.

We are thus brought to the problem of determining the transfer function for the zero-order hold for incorporation into the overall system response function. Because this hold function is in continuous time we cannot use Z transforms, so we will work in the frequency domain of continuous functions. Its spectrum will be continuous and nonperiodic. We start with the spectrum of the boxcar and generate the zero-order hold output of Fig. 7.15 by using superposition and the phase-shift theorem. The spectrum of the first box in Fig. 7.15 of height x_0 is

$$X_0(\omega) = x_0 \frac{2\sin(\omega T/2)}{\omega} e^{-i\omega T/2}$$

where we used the spectrum from Fig. 6.4, relabeled the boxcar width to T, and shifted it to $T/2$ to the right. Generating all of the boxcars in Fig. 7.15

Digital Control Systems 153

gives
$$X(\omega) = \sum_{n=0} x_n \frac{2\sin(\omega T/2)}{\omega} e^{-i\omega[n+(1/2)]T}$$

Writing complex exponentials for the sine gives
$$X(\omega) = \frac{1 - e^{-i\omega T}}{i\omega} \sum_n x_n e^{-i\omega n T}$$

Of course we recognize the summation as the spectrum of the digital signal x_t, giving us
$$X(\omega) = H(\omega) \sum x_n e^{-i\omega n T}$$
where
$$H(\omega) = \frac{1 - e^{-i\omega T}}{i\omega} \tag{7.9}$$

must be the system transfer function of the zero-order hold. Not surprisingly, we see that it is a mixture of a discrete-time and a continuous-time transfer function. Its numerator is $1 - Z$, representing the step $x_t - x_{t-1}$ and its denominator represents continuous-time integration. The zero-order hold simply produces its continuous output by integrating the last increment until the next sample arrives. It is thus a lowpass filter. In fact, as can be easily seen from Eq. (7.9), its magnitude response is just $2\sin(\omega T/2)/\omega$ which is a pretty poor lowpass response, having frequencies extending to infinity. In practice, additional lowpass filtering is often used to attenuate frequencies above $\omega_s/2$.

Now we are in a position to return to the block diagram of the control system in Fig 7.13. We wish to know the overall system response Y/X. We will work in the frequency domain, denoting the spectrum of sampled signals with a superscript "s" for sampled. (Some authors on control theory use an asterisk, but we reserve that for convolution and for the complex conjugate.) The error signal is
$$E = X - Y$$

The digital compensator produces
$$M^s = D^s E^s = D^s(X^s - Y^s)$$
The control signal is
$$C = M^s H$$
And finally the output is
$$Y = GC$$
$$Y = GM^s H = GHD^s(X^s - Y^s)$$

At this point we see a new development because we cannot solve for Y/X as usual. The occurrence of Y^s instead of Y prevents this solution. This essential feature of mixed discrete- and continuous-time systems distinguishes control theory from our previous discussions of pure digital systems.

But all is not lost. We start over and solve for Y^s/X^s, the overall system transfer function at the sample times. Thus

$$Y^s = G^s C^s = (GM^s H)^s$$

Now using our rule on sampling in Eq. (7.8) gives

$$Y^s = (HG)^s M^s = (HG)^s D^s (X^s - Y^s)$$

which can be solved for the transfer function

$$\frac{Y^s}{X^s} = \frac{D^s(GH)^s}{1 + D^s(GH)^s} \tag{7.10}$$

The term $(HG)^s$ also contains a sampled function because, as we have already observed, H is the product of the discrete operator $[1 - \exp(-i\omega T)] = (1 - Z)$ and the continuous integration operator $1/i\omega$. Thus our sampling rule of Eq. (7.8) gives

$$(HG)^s = \left[\frac{(1 - e^{-i\omega T}) G(\omega)}{i\omega} \right]^s$$

$$= (1 - e^{-i\omega T}) \left[\frac{G(\omega)}{i\omega} \right]^s$$

The closed-loop discrete transfer function then becomes

$$\frac{Y^s}{X^s} = \frac{(1 - e^{-i\omega T}) D^s [G(\omega)/i\omega]^s}{1 + (1 - e^{-i\omega T}) D^s [G(\omega)/i\omega]^s} \tag{7.11}$$

By evaluating this result in the Z domain, we can answer questions on stability, step function response, and other performance matters. Note that the term $[G(\omega)/i\omega]$ is just the sampled step function response of the plant since $G(\omega)$ is the Fourier transform of its impulse response and $(1/i\omega)$ corresponds to integrating this impulse to a step function. The step function response thus fits quite naturally into control theory, because of Eq. (7.11), and also because it is a logical command signal to use as a test of a control system.

Generally speaking, Eq. (7.11) can be used to evaluate control system response in two ways, using algebra or using computer-aided analysis via the DFT. In the first case, we must be given $g(t)$ analytically, integrate it, and find its Z transform. Since $D(Z)$ is given, the rest of Eq. (7.11) is in the

Digital Control Systems   155

Z domain ready for pole-zero analysis or the forming of a difference equation to generate the behavior of the control system at the sample times. On the other hand, perhaps the step function response of the analog system is measured by experiment, digitizing the result. Then the DFT can be used to estimate $[G(\omega)/i\omega]^s$ and to compute exactly the periodic spectra of the other terms in Eq. (7.11). The IDFT then can estimate the behavior of the system for any input. This same procedure could also be followed if $G(\omega)$ were specified analytically.

In either case, however, the results only give the behavior at the sample times. For many systems employing sufficiently high sample rates these results are good enough to describe system behavior. We have seen that there is no continuous-time system transfer function. However, we can in fact compute the output between sample times for any input by solving the same equations used above for $Y(\omega)$ in terms of $X(\omega)^s$, giving

$$Y(\omega) = \frac{D^s H G X^s}{1 + D^s (HG)^s} \qquad (7.12)$$

Once $Y(\omega)$ is known, the inverse Fourier transform can be estimated using the IDFT, giving a spatially aliased version of $Y(t)$ whose aliasing can be made as small as desired by choosing an appropriately long transform.

An example of using our results is now in order. We pick a special case of the control system represented by Fig. 7.13 and Eq. (7.11). In order to concentrate on the principles involved in combining sampled systems with analog systems, we let the actual digital transfer function be the simple proportional control $D(Z) = K$. And we will assume the plant has an exponential impulse response with $G(\omega) = 1/(\alpha + i\omega)$ (recall Problem 6.27). The main task at hand, then, is to evaluate $[G(\omega)/i\omega]^s$ for use in Eq. (7.11). That is to say, we need to integrate the exponential response and sample it with the Z transform as follows:

$$\text{IFT}[G(\omega)/i\omega] = \int_0^t e^{-\alpha t}\,dt = \frac{1}{\alpha}(1 - e^{-\alpha T}) \qquad t > 0 \qquad (7.13)$$

Because the plant is causal, the integration represented by $1/i\omega$ starts at $t = 0$, and the above result is zero for $t < 0$. Thus its Z transform is

$$[G(\omega)/i\omega]^s \rightarrow \frac{1}{\alpha}\left(\frac{1}{1-Z} - \frac{1}{1-e^{-\alpha T}Z}\right) = \frac{(1 - e^{-\alpha T})Z}{\alpha(1-Z)(1-e^{-\alpha T}Z)} \qquad (7.14)$$

by recalling the Z transform of the unit step function and of the exponential. Using Eq. (7.11) gives the discrete transfer function

$$\frac{Y^s}{X^s} = \frac{K(1 - e^{-\alpha T})Z}{\alpha + [K - (\alpha + K)e^{-\alpha T}]Z} \qquad (7.15)$$

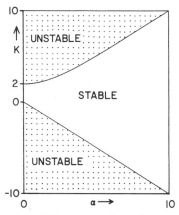

Figure 7.16 Plot showing the stability region of the discrete system transfer function of Eq. (7.15) in terms of the proportional gain K and the damping factor α.

This result has a zero at the origin and a pole at

$$Z_p = \frac{\alpha}{(\alpha + K)e^{-\alpha T} - K}$$

The region of stability is depicted in Fig. 7.16 by contouring the boundary of the region where $|Z_p| > 1$.

The response at the sample times due to any input can be determined by Eq. (7.15) using either the difference equation implied by the discrete transfer function or by using the DFT. In the latter case, zero-padded arrays are formed from the numerator and denominator coefficients of the discrete transfer function; they are transformed into the DFT frequency domain; the results are divided; and finally the IDFT is used to find the discrete impulse response. Time-domain convolution will then give the response due to any input.

The results of using Eq. (7.15) to determine the system step function response are shown in Fig. 7.17 for three quite different cases. In the first case, with $\alpha = 0.01$ (in our numerical work we use $\alpha T = \alpha$ where we have set the unit of time equal to the sampling period) and $K = 1.5$, a rapid rise time causes overshoot with fairly rapid settling to zero dc error. In the second case, with the same damping factor, decreasing the gain by a factor of three causes an overdamped response. The last case is particularly interesting with $\alpha = 0.001$ and $K = 1$. Now the control error is brought to zero in just one sample time and remains at zero with no subsequent oscillations. The case is called the *deadbeat response*, occurring in the limit $\alpha \to 0$ and $K = 1$. It is easy to see that the discrete system transfer function

Digital Control Systems

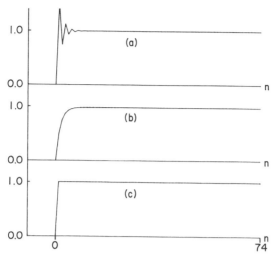

Figure 7.17 The step function response of a proportional digital controller for different damping factors α and gains K. (a) For $\alpha = 0.01$ and $K = 1.5$ overshoot occurs. (b) For $\alpha = 0.01$ and $K = 0.5$ overdamped response occurs. (c) For $\alpha = 0.001$ and $K = 1$ a deadbeat response occurs, the output reaching the command value in one sample time. The plots are solutions of the difference equation implied by Eq. (7.15) for 75 sample points.

given by Eq. (7.15) becomes in this limit $H \to Z$, the one-sample pure delay. This is a special case of deadbeat response where in general the error is driven to zero in a finite number of samples. This occurs, of course, when the discrete transfer function becomes an FIR operator. From Eq. (7.15) this happens for values of K and α related by

$$K - (\alpha + K)e^{-\alpha T} = 0$$

In general the deadbeat settling time depends on the sampling time since the error is zero in a finite number of samples. Thus the settling time can be made extremely small by using high sample rates. But this implies extremely high control signals giving the deadbeat controller another disadvantage in practice, in addition to the high sampling rate.

So far we have only explored the behavior of our control system at the sample times. Naturally we are curious about the state of the analog output in between these times; could it misbehave badly then? The answer to this question is given by application of Eq. (7.12), providing an instructive example of mixed discrete-time continuous-time systems. In this equation both the continuous-time and the sampled versions of HG are needed. We know that the product HG is

$$HG = \frac{1 - e^{-i\omega}}{i\omega(\alpha + i\omega)} = \frac{1 - Z}{i\omega(\alpha + i\omega)}$$

and multiplying Eq. (7.14) by $(1 - Z)$ gives

$$(HG)^s = \frac{(1 - e^{-\alpha T})Z}{\alpha(1 - e^{-\alpha T}Z)}$$

Using $D = K$ gives from Eq. (7.12):

$$Y(\omega) = \frac{\alpha K[1 - (1 + e^{-\alpha T})Z + e^{-\alpha T}Z^2]}{\alpha + [K - (\alpha + K)e^{-\alpha T}]Z} X(Z) \frac{1}{i\omega(\alpha + i\omega)} \quad (7.16)$$

The spectrum of Y is thus composed of the periodic spectrum of an IIR operator written in the Z domain, the periodic spectrum of $X(Z)$, and the nonperiodic spectrum $1/(i\omega)(\alpha + i\omega)$. So $y(t)$ is a continuous time-domain signal. And, in fact, its computation is really quite simple: We work entirely in the time domain. Using a step function for X, implementation of the difference equation implied by the IIR operator gives a discrete time series for all the sample times of interest. Multiplication by $1/(i\omega)(\alpha + i\omega)$ directs us to convolve the resulting discrete time series with the exponential function $(1/\alpha)[1 - \exp(-\alpha T)]$. Recall Eq. (7.13). This exponential function is in continuous time so we can compute $y(t)$ at as many points as we please. The results of using this scheme are shown in Fig. 7.18 in which we have evaluated $y(t)$ at eight intervening points between the sample times. The values at these sample times, computed from the discrete Eq. (7.15), are depicted by the vertical lines, showing how continuous-time response interpolates between the sample times.

Nothing bad happens between sample times. Rather $y(t)$ there is simply an interpolation from the monotonic exponential function, so intervening

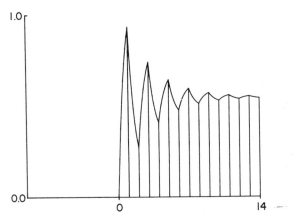

Figure 7.18 The step function response of a digital controller showing behavior between sample times for 15 samples. The plot is a solution of Eq. (7.16) with $\alpha = 1.5$ and $K = 1.8$, plotting eight of the continuous-time values between sample points. Note in passing that this combination of α and K leads to a large dc error.

Digital Control Systems 159

oscillations are impossible. Furthermore, we note that the poles of the discrete IIR operator of Eq. (7.16) are identical to the poles of the discrete operator of Eq. (7.15) so their stability regions are the same. Thus $y(t)$ has no instabilities that are not manifest at the sample times. Clearly, however, this is not necessarily always true because the behavior between sample times depends on the plant transfer function. If it has oscillatory behavior of periods less than the sample times, $y(t)$ may have also. In any case, our analysis procedure has been quite general. For any given control system, $y(t)$ can be computed for any input to verify satisfactory behavior.

We have used the proportional digital controller $D(Z) = K$ in the above discussion to provide a simple example, concentrating on the way continuous-time and discrete-time systems can be analyzed when combined. But any digital algorithm can be contemplated. For example, proportion, derivative, and integral control might be used with various gains, called a *PID controller*. A simple digital form of a PID compensator is

$$D(Z) = K_\mathrm{P} + \frac{K_\mathrm{I}}{1-Z} + K_\mathrm{D}(1-Z)$$

Of course other digital expressions, such as the bilinear transform, may be used to express differentiation and integration. More importantly, the digital controller can use complicated algorithms, nonlinear ones, time-varying ones, or reprogrammable ones all quite easily with the use of modern digital electronics and computers.

The analysis of more complicated LSI controllers follows the same general course outlined above. Digital and/or analog subsystems may be included in complicated feedback or feedforward loops. The location of the samplers within closed loops greatly effects the system behavior, as demonstrated in Problem 7.23. Additionally, some systems—even simple ones—may not have a continuous or even discrete system transfer function, as shown by Problem 7.30. This is, in effect, because the sampler, with its on–off states, is not an LTI device viewed in continuous-time theory. Nonetheless, we can always compute the output of the system, given the input, which is sufficient for most practical applications of digital controllers.

The greater challenge is the design of optimal digital controllers. Given the plant behavior $G(\omega)$ and a specific configuration such as Fig. 7.13, what digital algorithm $D(Z)$ will best meet a given set of control specifications according to some criterion? The hard part is specifying the criterion. For example, if the system is optimized for a step function input, it will not necessarily be optimal for other inputs. Clearly, further discussion of control theory would lead us well beyond the purpose of this chapter. However as a final remark, we point out that if the required control behavior were specified at the sample times in terms of the systems impulse response, the solution is readily available. In the example of the control configuration discussed above, Eq. (7.10) tells us that

$$\frac{Y^s}{X^s} = \frac{D^s(GH)^s}{1 + D^s(GH)^s} = R(Z)$$

where $R(Z)$ is the required control action. The equation is simply solved for D giving

$$D(Z) = \frac{R}{(1-R)(HG)^s}$$

If we pick $R(Z) = Z$, for example, we get the digital controller $D(Z)$ for a deadbeat response. Clearly, perfect fidelity $R(Z) = 1$ is impossible, leaving the question of the next best controller open.

This chapter completed our discussion of Fourier transforms by adding the fourth possible form—the continuous-time, discrete-frequency Fourier series. Most important, by using the sampling function, we have shown the relationship between continuous and discrete signals—a relationship which is indispensable for the correct use of the DFT in signal processing. Then, manipulations of data and their spectra led to the processes of sinc interpolation, decimation, and multiplexing, useful in many fields of communication and signal processing. Because digital controllers almost always interface to analog systems, digital control theory has provided us with another arena for discussing the relationship between continuous-time functions and discrete-time ones. Even though control theory is an extensive subject in its own right, we have seen how to design and analyze digital control systems using the concepts developed in previous chapters. Next we shall see how to design and analyze digital systems that attempt to meet a given spectral response. And again we shall see demands made on our understanding of the relationships between continuous-time and discrete-time systems.

Problems

7.1 Calculate the Fourier series coefficients for the square wave, the triangle wave, and the full wave rectified sine wave shown in Fig. 7.1.

7.2 In minimizing the total squared error given in Eq. (7.6), we minimized with respect to a_m^*. Repeat the calculation minimizing the total squared error by varying a_m.

7.3 Write out the Fourier series of Eqs. (7.4) for the special case that $f(t)$ is real. Show that they can be written as

$$f(t) = a_0 + \sum_{n=1}^{\infty} [a_n \cos(n\omega_0 t) + b_n \sin(n\omega_0 t)]$$

where

$$a_0 = \frac{1}{T}\int_{-T/2}^{T/2} f(t)\,dt$$

$$a_n = \frac{2}{T}\int_{-T/2}^{T/2} f(t)\cos(n\omega_0 t)\,dt$$

$$b_n = \frac{2}{T}\int_{-T/2}^{T/2} f(t)\sin(n\omega_0 t)\,dt$$

7.4 The sampling function of Eqs. (7.7), shown in Fig. 7.4, violates the requirements for possessing a Fourier transform. But, does the sampling function and its transform shown in the figure satisfy the Wiener–Khintchine theorem?

7.5 Interpret and discuss the statement: In the discrete-time theory, all functions can be considered band limited, which is also consistent with our experience. Conversely, in continuous-time theory, a one-sided function must have a broadband spectrum that is not consistent with our experience.

7.6 In this chapter, we have explored the fundamental relationships between various types of Fourier transforms. To take advantage of the FFT in frequency domain processing, the relationship of the DFT spectrum to the spectra of various signals must be thoroughly understood. Complete the table below, describing the spectra of the indicated signals, the way that the DFT could be used to compute these spectra, and the significance of the results.

DFT Relationships to Various Signals

$f(t)$	$F(\omega)$	DFT method	DFT results
1. Discrete, finite length	Continuous, periodic	Zero pad $f(t)$	Exact
2. Discrete, infinite length, periodic			
3. Discrete, infinite length, nonperiodic			
4. Continuous, finite length			
5. Continuous, infinite length, periodic	Discrete, nonperiodic; is Fourier series	Fit exactly one period of $f(t)$ into DFT	Aliased if $f(t)$ broadbanded
6. Continuous, infinite length, nonperiodic			

7.7 Show that the Fourier series coefficients of the repetitive signal consisting of portions of one-half of a sine wave, shown in Fig. 7.8(a), are $(-1)^n 8n/\pi(4n^2 - 1)$.

7.8 Write a computer program to calculate the Fourier series coefficients of Problem 7.1 using the DFT and observe that your results approach the correct coefficients in the limit of an infinite time domain sampling rate, that is, as the number of points used in the DFT become large.

7.9 Write a computer program to compute the DFT spectrum of a Gaussian time domain signal. Compare the DFT spectrum with the exact Gaussian spectrum given in Fig. 6.3 for different sampling rates and window lengths.

7.10 By using the DFT with time domain zero padding, display the spectrum of the discrete MA smoothing operators of Fig. 3.5.

7.11 The above three problems demonstrate three common uses of the DFT and the appropriate interpretation of the DFT spectrum. As a demonstration of the fourth case discussed in the text, concoct a periodic band-limited function and write a computer program to compute its DFT spectrum. Observe that for the special case where the period of the function just spans the DFT samples, the exact spectrum is produced by the DFT.

7.12 A security analyst claims that the stock market quotations from a certain corporation exhibit three cyclic patterns: a monthly cycle, a four-month cycle, and a quarterly cycle. Given that he used the weekly closing quotes for a one-year period for his analysis, which of these cyclic patterns do you believe in?

7.13 Describe sinc interpolation in both the time and frequency domains for 3:1 and 4:1 upsampling.

7.14 Repeat the zoom processing example shown in Fig. 7.11 using frequency band number two, showing that the resulting spectra for even-numbered bands are the frequency-reversed version of the original band spectrum. State the significance of this result in terms of the DFT circle.

7.15 Discuss how the spectrum recovered from FDM or in zoom processing, shown in Fig. 7.11, is really the result of aliasing.

7.16 A certain digital frequency analyzer device has its memory limited to 1000 data samples, and it digitizes analog signals every 1/400 sec. Make a comparison of frequency resolution before and after 4:1 zoom processing. Assume in each case that we fill the memory with the maximum number of data points subsequent to the zoom band-pass filtering. Calculate the width of the sinc function produced by the data window in each case to make this comparison.

Problems

7.17 Show that the spectrum resulting from zero interlacing in Fig. 7.12. must have its amplitude halved in order to satisfy Parseval's theorem.

7.18 Using the phase shift theorem, verify that the spectrum shown in Fig. 7.12(d) is correct for the $\frac{1}{2}$-shifted sampling function.

7.19 As an alternate to using the $\frac{1}{2}$-shifted sampling function of Fig. 7.12(d), develop the TDM spectrum of Fig. 7.12 by simply using subtraction. That is, find the B-channel result of Fig. 7.12(e) by subtracting the unshifted zero-interlaced result from the original channel B data.

7.20 Formulate the discrete version of the modulation theorem and draw sketches in both the time domain and the frequency domain, showing the FDM encoding process that leads to the spectrum shown in Fig. 7.11(a).

7.21 Use ample zero padding and display the magnitude spectrum of the 15-point zero-phase Spencer smoothing operator below. Be sure to display the results on both linear and decibel scales.

$$\frac{1}{320}(-3, -6, -5, 3, 21, 46, 67, 74, 67, 46, 21, 3, -5, -6, -3)$$

7.22 For samplers preceding and following a continuous system $H(\omega)$, as in Fig. 7.14(b) and 7.14(c), make two sets of sketches of spectra, one depicting sampling before convolution $h*x^s$ and the other sampling after convolution $(h*x)^s$. Show how $(h*x)^s \neq h^s * x^s$ because of cross-products produced by the wide band spectrum of h.

7.23 In the figure below two analog systems are cascaded with and without a sampler between them. Find the digital transfer function in each case, showing that $G(Z)H(Z) \neq GH(Z)$.

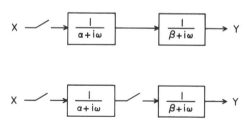

7.24 Use Eq. (7.10) along with the other equations describing the control system of Fig. 7.13 to derive Eq. (7.12).

7.25 Consider the digital control system of Fig. 7.13 with $D(Z) = K$, the D/A zero-order hold, and the plant $H(\omega) = 1/i\omega$. Find the digital transfer function and show that it is stable for $0 < KT < 2$. Show that at $KT = 1$, the transfer function is a pure delay of one time unit, a deadbeat response.

7.26 For the proportional control used in Problem 7.25 substitute derivative control $D(Z) = K_D(1 - Z)$ and show that the resulting control system is stable only for $K_D KT < 1$.

7.27 In the above problem use the bilinear transform for the derivative controller and show that the system is now stable only for $-1 < KT < 0$. What is the meaning of this result?

7.28 Show that the deadbeat response is impossible for the system described by Eq. (7.15) with nonzero α.

7.29 Write a computer program that implements the difference equation implied by Eq. (7.15). Study the behavior of the system for various values of K and αT by plotting the response due to a unit step function input.

7.30 Calculate the output $Y(Z)$ for the two closed-loop systems shown below, showing that they do not have a discrete transfer function $H(Z) = Y(Z)/X(Z)$ because $X(Z)$ cannot be separated out.

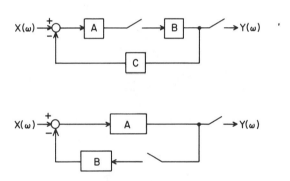

8  Digital Filter Design

All LSI systems can be thought of as filters. Generally though, the term is reserved for those situations that are most easily thought of as a selective screening or filtering of certain frequency components of a signal. All radio receivers, for example, contain a bandpass frequency filter that allows only the signal of the desired station to enter the set for further processing and amplification. The signals from other stations are rejected. A signal from some scientific experiment may contain an objectionable amount of 60-Hz noise from power lines. It can be filtered out with a notch filter. A radio astronomer may be trying to detect weak spectral lines from a distant galaxy. Passing this signal through a narrow bandpass filter will enhance the signal-to-noise ratio and therefore greatly improve the detection limit. These three examples are most easily thought of as a multiplication in the frequency domain of the signal's spectrum by the desired filter's frequency response. However, we know that the same result can be achieved by convolving the signal in the time domain with the Fourier transform of the desired frequency response curve. Both of these statements are about mathematical operations. The actual implementation of the filter could be accomplished by an analog device, such as an electrical circuit, or by carrying out mathematical operations on a digitized signal. Frequently, as in all three of these examples, the desired frequency response has discontinuities as in an ideal lowpass or bandpass filter. Neither analog devices nor digital operations can exactly produce the discontinuities of these ideal filters. Therefore, specifications of a digital filter often take the form of a tolerance scheme, as shown in Fig. 8.1(b), using a lowpass filter as an example. Digital filter design addresses the problem of computing the time

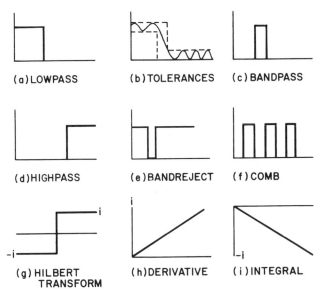

Figure 8.1 Examples of ideal filters. Frequently, filters are specified in terms of tolerances based on the ideal filter, as shown in (b) for the lowpass filter.

domain terms, called filter coefficients or the filter operator, for use in convolving with the input data to achieve a specific frequency response. Frequently, no constraints are imposed on the phase response except those arising from stability and, in some cases, causality. In other applications, phase distortion cannot be tolerated and the pure delay produced by linear phase behavior would then be of paramount interest. In all practical cases, digital filter design becomes a problem in approximation. In the case of an FIR filter, we are searching for the best polynomial in Z while for an ARMA filter, the quest is for the best rational fraction in Z. Because of the rapid changes possible near the poles of this rational fraction, the IIR filter generally achieves excellent amplitude response (particularly in the transition bands). But, it occurs at the expense of a nonlinear phase response dictated by stability requirements. On the other hand, FIR filters can trivially achieve exact linear phase, but their amplitude response is not as easily controlled by the polynomial in Z.

The purpose of this chapter is threefold: (1) to develop insight into the basic nature of the filter design problem, (2) to give two practical methods of designing FIR filters—windowing and the Parks–McClellan algorithm, and (3) to introduce the problem of IIR filter design. This last subject was really begun in the early chapters where we considered digitizing continuous systems. To finally complete the discussion on IIR filter design, the next two chapters on inverse theory and spectral factorization will be needed.

To meet these three objectives, we shall take the lowpass filter as the archetypical design objective. This is not really a limitation because other filters—bandpass, bandreject, and highpass—can be easily generated from them by simple operations. For example,

$$H(Z)_{highpass} = 1 - H(Z)_{lowpass}$$

gives the transfer function of a highpass filter. Also, the phase-shift theorem provides the mechanism for sliding the lowpass spectrum all around the periodic spectrum of digital operators to form bandpass and highpass filters. Bandreject filters can be formed from the bandpass filter by subtraction, as noted above. Finally, filters with multiple passbands or reject bands can be formed by linear combinations of the more primitive ones. So the discussion will use the lowpass filter as our prime example.

Digital Filter Design—The Problem

Usually, in the arena of digital filter design, a desired frequency response is specified and the job is to determine the time domain operator that will achieve this desired response. Most always, this desired frequency response is unobtainable. The reason for this nasty situation is of a fundamental nature and therefore is quite worthy of elaboration before we continue with actual design methods. A knowledge of the Fourier transform from the last chapters will be indispensable for our discussion. Consider that we would like to perform ideal lowpass filtering. The frequency response of such a filter is shown in Fig. 8.1, along with other ideal filters.

There are many applications, such as electronic circuits, communications, and signal enhancing where such a filter response is desired. Analog components cannot produce this ideal lowpass filter; neither can digital operations. Let us see what happens when we try to implement this ideal lowpass filter with a digital moving average filter. From the convolution theorem, we know that multiplying the spectrum of the input signal by this frequency response is identical to convolving in the time domain with the sinc function. This convolution process represents a nonrecursive moving average filter with an infinite number of coefficients. The desired operation can, of course, be attempted in either domain, but a fundamental problem arises in either case.

If convolution in the time domain is chosen, the sinc function will have to be truncated in order to actually carry out the computation. That is to say, in the time domain the desired filter operator becomes the sinc function, multiplied by the boxcar, which yields the truncated sinc function, as shown in Fig. 8.2. In the frequency domain, the operation equivalent to

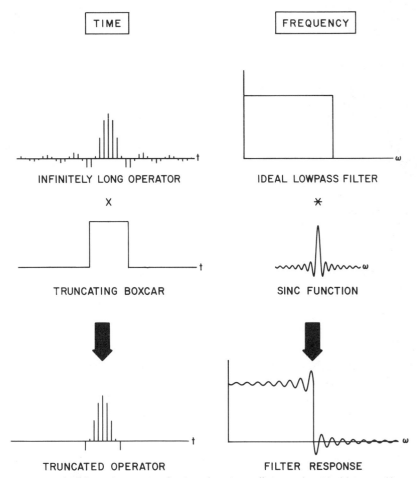

Figure 8.2 The effect of truncating the time domain coefficients of an ideal lowpass filter.

this boxcar truncation is convolution—convolution with the boxcar's Fourier transform. Its transform is a sinc function. This effect on the ideal lowpass filter of truncating the time domain operator is sketched in Fig. 8.2. The nonzero width of the truncating boxcar's sinc function smears out the sharp transition band of the ideal lowpass filter; it produces ripples in its passband; and it causes nonperfect attenuation in the stopband. In the limit of an infinitely long time domain operator, this sinc function becomes the Dirac delta function. Only then does the ideal lowpass filter's spectrum remain unscathed.

Thus, by viewing the operation in the frequency domain, we see that the ideal lowpass filtering is not achieved. It is true that some control over the

Digital Filter Design—The Problem

undesirable wriggles in the frequency response is afforded by using a smoother truncation of the time domain coefficients. However, the transition band would then be broader. This approach to FIR filter design, of selecting a tapering function (called a window) that gives an acceptable compromise between the wriggles and the transition bandwidths, is called *windowing*. This idea of windowing clearly points out the MA digital filter problem: any window that truncates the time domain operator to a manageable length will alter the frequency response from the desired shape.

It may seem that we could achieve any filter response curve that we wished by using multiplication in the frequency domain. However, this does not work either. For example, if we choose to perform this ideal lowpass operation in the frequency domain, it is true there is no distortion of the rectangular filter shape. However, the DFT of the time signal will be imprecise. Consider a sinusoidal component of the signal in the filter stopband located between DFT frequencies, as shown in Fig. 8.3. Some DFT components of the signal fall into the rectangular passband of the filter; complete rejection is not achieved. From this viewpoint, in the frequency domain, the filtering problem is recast into the fundamental problem of spectral analysis. But, as we saw in Chapter 7, usually it is

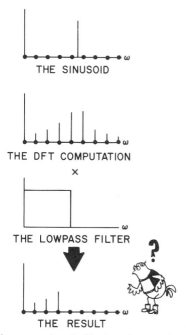

Figure 8.3 The result of frequency domain filtering of a pure sinusoid whose frequency lies between DFT computation points. The ideal lowpass function is not achieved.

impossible to determine the exact spectrum of any real data set. Fundamentally then, the digital filter design problem exists because the finite bandwidth filter requires an infinitely long operator in the time domain. It makes no difference if the actual implementation is attempted in the time or in the frequency domain.

In some special cases, we could argue that we know the exact spectrum of our data. For example, if the data were discrete-time data of a finite length, then the DFT of this data, amply zero padded, yields points that lie on the exact spectrum. Does point-by-point multiplication of this spectrum with an ideal lowpass filter response then give the ideal result? The answer is no. The reason is that the exact spectrum is really continuous and the multiplication of the ideal lowpass function at discrete points does not fix the filtered spectrum on the continuum. It is not confined between the DFT computation points. Clearly, the more dense the DFT computation points, the closer this filtering operation approaches the ideal. But nonetheless, the ideal result would only be obtained with a DFT of infinite length.

You may ask about filters that are less demanding than the ideal lowpass filter. What if we specified some $H(\omega)$ that has finite-width transition zones and less than infinite attenuation in its stopbands? Can we then find a time domain operator that has exactly $H(\omega)$ as its spectrum? The argument in the preceding paragraph still prevails. The desired frequency response is really continuous. But, multiplication in the frequency domain only fixes it at a finite number of points. Most always the unconstrained portion, between DFT points, will not coincide with the desired response.

You may ask about a recursive scheme. Why not use an ARMA filter with its infinite impulse response to achieve the required time domain operator? This is a good idea and ARMA filters do, in fact, achieve good amplitude response in the frequency domain with rather limited control over phase. However, recursive filters do not circumvent the fundamental problem. They have transient effects that propagate infinitely far along the data set. That is, the use of an ARMA filter still requires an impossible infinitely long data set to achieve exact results. When using such a filter, we have to be sure that the data set is sufficiently long compared to the transient decay time to ensure tolerable results.

A summary of the discussion is this: in order to filter according to some prescribed frequency response, we would need to know the spectrum of our data exactly. For continuous data, this is never exactly possible, complicating the discussion. On the other hand, the spectrum of finite-length discrete data can be computed exactly. The problem of digital filter design then lies in the fact that most desired filter responses are not time limited, producing aliasing when sampled in the frequency domain. Under this aliased condition, no finite number of coefficients, in either the time or the frequency domain, can completely control the continuous frequency response of the filter. The situation is analogous to the digitization of a signal having an infinite bandwidth. Aliasing is then inevitable, and subsequent faithful recovery of the original continuous signal is impossible. The problem of digital filter design is thus significant, worthy of understanding and worthy of pursuing.

Designing FIR Filters Using Windows

The windowing approach to designing FIR filters follows directly from the simple notion of truncating the impulse response function to a manageable length. As we have consistently seen in our discussions, it is a central point in digital signal processing that a digital LSI system response is a continuous function of frequency:

$$H(\omega) = \sum_{n=-\infty}^{\infty} h_n e^{-i\omega n}$$

while its time domain operator is, of course, discrete:

$$h_n = \frac{1}{2\pi} \int_{-\pi}^{\pi} H(\omega) e^{i\omega n} \, d\omega$$

For most filters of interest, the h_n are infinitely long and therefore unmanageable. Filters with sharp transition bands, such as the ideal lowpass filter, are examples. The window method consists of selecting a finite length window in time w_n whose truncations of h_n produces an acceptable frequency response. That is, we are searching for a w_n such that

$$\hat{h}_n = h_n \cdot w_n$$

is an operator of the desired, or reasonable, length such that

$$\hat{H}(\omega) = H(\omega) * W(\omega)$$

is the best frequency response. Of course, "best" is meaningless without some criterion. If we selected a least-squares condition, for example, then the rectangular window would solve the window selection problem. This is so because of the property (that we showed in the section "The Least-Squares Convergence of the Fourier Series" in Chapter 7) of the Fourier series that makes

$$\hat{H}(\omega) = \sum_{-N}^{N-1} h_n e^{-i\omega n}$$

a least-squares fit to

$$H(\omega) = \sum_{-\infty}^{\infty} h_n e^{-i\omega n}$$

However, it turns out that for most filter applications, a least-squares fit is not the most desirable solution. To see why, we proceed with a presentation of the most commonly used windows and examine their effects on the ideal lowpass filter. Normally, for odd N, zero phase is desired in such a filter and is trivially achieved by using a symmetrical operator. Then, its FT is real, it has zero phase shift at all frequencies. (The operator will also be acausal by one-half the filter length. Alternatively, the filter could be used in a causal fashion by delaying the output one-half the filter length. Then, the filter has exactly linear phase.)

As alternatives to the sharp truncation of the rectangular window, multiplication of the time domain coefficients by smoother windows produces a gradual tapering of the coefficients with an associated improvement in stopband rejection at the expense of transition zone broadening. Some commonly used windows for odd N are listed here (also see Fig. 8.4):

Bartlett: $W_n = \begin{cases} 2n/(N-1) & 0 \le n \le (N-1)/2 \\ 2 - 2n/(N-1) & (N-1)/2 \le n \le N-1 \end{cases}$

Hanning: $W_n = \frac{1}{2}\{1 - \cos[2\pi n/(N-1)]\}$ $0 \le n \le N-1$

Hamming: $W_n = 0.54 - 0.46 \cos[2\pi n/(N-1)]$ $0 \le n \le N-1$

Blackman: $W_n = 0.42 - 0.5 \cos(2\pi n/(N-1)) + 0.08 \cos(4\pi n/(N-1))$

Kaiser: $W_n = \dfrac{I_0[\omega_a \sqrt{\bar{N}^2 - (n - \bar{N})^2}]}{I_0(\omega_a \bar{N})}$

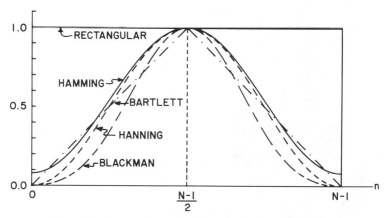

Figure 8.4 Commonly used windows for designating FIR filters.

Designing FIR Filters Using Windows

where $\bar{N} = (N-1)/2$. The Kaiser window is the ratio of two modified zeroth order Bessel functions, I_0. The Hanning and Hamming windows not only have similar names, but are of similar form. All of these tapered windows are named after workers who first introduced their use. Figure 8.5 shows the frequency responses of each of these windows along with the frequency response generated from their truncation of the coefficients of an ideal lowpass filter.

The observation made in the introduction of this section can be verified in these figures. The rectangular window has the sinc function for its frequency response. The sinc has the narrowest central peak and the largest side lobes of all the windows' spectra. Thus, the filter response produced by the rectangular window has the sharpest transition band at the expense of the poorest stopband rejection. At the other extreme, the Blackman window, with its broad central peak and small side lobes, produces a lowpass filter with a wide transition band but with superior rejection in the stopband. The Kaiser window is mathematically optimized in the sense of producing a minimum spectral energy beyond some specified frequency. But, because the spectral response of the window is convolved with the desired filter function, it is difficult to interpret what significance this optimization of the Kaiser window has for the final filter. In fact, all windowing procedures suffer from this same lack of optimization criterion. Windowing also has the disadvantage of little design flexibility. Exact passband and stopband frequencies cannot be specified *a priori*, because the convolution with the window spectral response smears out the discontinuity in an unpredictable way. Furthermore, we cannot proceed in a direct manner to calculate the required filter length for given filter specifications. The longer the operator, of course, the closer the results approach the ideal, as shown in Fig. 8.6. Cut and try is the only option using windowing. Nonetheless, because of its simplicity relative to more advanced optimal filter design techniques, windowing is an attractive approach for the casual user of digital filters. One frequently finds these limitations of windows to be quite tolerable for a particular application.

The general procedure would then be as follows: given an arbitrary desired filter spectral response, compute its N point IDFT to produce a symmetric operator N points long. This specification of the filter's response by digitizing it in the frequency domain is called *frequency sampling*. The length N of the DFT used can be quite large, much larger than the number of time domain coefficients desired. Next, this long time domain operator is truncated to the desired length by using one of the windows discussed. The resulting coefficients are the desired filter operator. Check the actual frequency response by computing the filter's DFT using ample zero padding. If the results are unacceptable, a different window and more points can be tried. When acceptable filter coefficients are found, they are convolved with the data to get the filtered results.

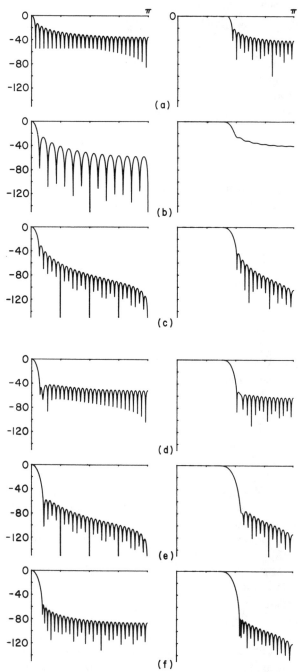

Figure 8.5 On the left is the frequency response of six common windows: (a) rectangular; (b) Bartlett (triangular); (c) Hanning; (d) Hamming; (e) Blackman; and (f) Kaiser. On the right is the result of truncating the infinitely long ideal lowpass filter coefficients to 57 terms using the windows on the left. The vertical scale is in decibels.

Frequency Sampling and the Parks–McClellan Algorithm

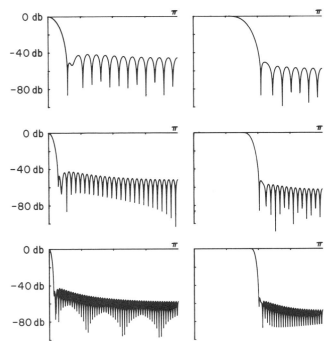

Figure 8.6 The dependence on operator length of the frequency response of a Hamming window on the left, and its corresponding lowpass filter on the right. $N = 51$, 101, and 201.

This procedure is described by the equation

$$r_t = d_t * (w_t \cdot f_t)$$

where d_t represents the data sequence, w_t is the window function, f_t are the filter coefficients, and r_t is the resulting filtered data. In the frequency domain, the above equation reads

$$R(\omega) = D(\omega) \cdot [W(\omega) * F(\omega)]$$

to give us an alternate prescription for windowing. Note that these double convolution/multiplication operations are not associative.

Frequency Sampling and the Parks–McClellan Algorithm

Because of the lack of a quantitative criterion in windowing, we are motivated to consider other design methods for FIR digital filters. As in the windowing method, we start by specifying the desired response in terms of N frequency points and perform the IDFT to find the corresponding N time domain filter coefficients. Then, rather than modifying the coefficients by

8/ Digital Filter Design

windowing, we would readjust the frequency points by some trial and error method, such as some iterative procedure, until a satisfactory frequency response is obtained according to a given criterion. Let us explain this idea with an example.

Suppose that we wish to design a lowpass filter with a cutoff at $\frac{1}{2}$ the Nyquist frequency. We start with a given N for use in our DFT algorithm. In this example, we will suppose $N = 32$. Therefore we specify the first eight real frequency terms to be one and the remaining real components to be zero. We also set all the imaginary terms equal to zero, to give a zero-phase filter. This picking of points from the desired filter frequency response curve is the same frequency sampling discussed in the previous section. It, of course, is not limited to our lowpass example. Such sampling could be done on any specified frequency response curved at any number of points.

In our example, the IDFT produces 32 time domain coefficients that we want to use to form a symmetrical operator under linear convolution. But the DFT relates to circular convolution, so these 32 coefficients must be linearized. Additionally, in this process, we wish to form an operator with an odd number of points. Only a symmetrical operator with an odd number of points can conveniently produce an exactly zero-phase FIR filter (symmetrical operators with an even number of coefficients lead to cumbersome half-unit delays). This required conversion of the 32 coefficients to an odd number will alter the response specified by the frequency sampling. Certainly, the scheme that least disturbs the original IDFT arrangement of coefficients will result in the least disturbed frequency response. We thus construct a 33-point operator from the 32 coefficients by placing the number one coefficient in the center, symmetrically flanked with the 15 paired coefficients on each side, and then adding the 17th coefficient to each end, using it twice. This linearized operator is shown at the top of Fig. 8.7. The frequency response of this new operator can now be displayed by zero padding and performing the DFT. The computer program of Fig. 8.8 shows how this may be accomplished using a 32-point IDFT, 8:1 zero padding, and then a 256-point DFT. The results are shown in Fig. 8.7. The curve shows the characteristic Gibbs phenomenon common in most FIR filters. As anticipated, the curve misses the original sampled points because we have been forced to change the operator to an odd number of coefficients. At any rate, so far we have a rather disappointing result with the first ripple in the stop band at about 0.1 or -20 db. For many practical applications, such a filter would be unsatisfactory.

Since large ripples near the transition are associated with the Gibbs phenomenon at this discontinuity, it follows that a reduction in ripple amplitude should accompany a broadening of the transition band. Figure 8.9 dramatically confirms this idea where we have set $R(8) = \frac{1}{2}$ in order

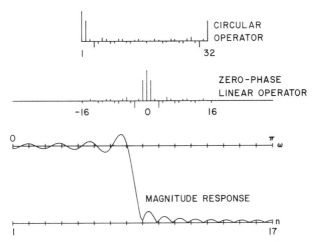

Figure 8.7 Top: Circular operator computed by using a 32-point IDFT to frequency sample an ideal lowpass filter. Middle: The symmetric zero-phase operator formed by linearizing the circular operator. Bottom: Zero padding the linear operator and computing the DFT produces the frequency response of the zero-phase operator.

1 REAL T(32), R(17), I(17)
2 REAL T0(256), R0(129), I0(129)
3 DO 4 J = 1, 8
4 R(J) = 1
5 CALL IFT(R, I, T, 32)
6 T0 = 0   T0(1) = 1
7 DO 9 J = 1, 17
8 T0(J) = T(J)
9 T0(256 − J + 2) = T(J)
10 CALL FFT(T0, R0, I0, 256)

Figure 8.8 Program to determine the frequency response of a 33-point operator formed by frequency sampling. Lines 1 to 4 generate the real and imaginary terms of an ideal, zero-phase, lowpass filter. In lines 6 to 9 a 256-point symmetrical operator is formed by zero padding 8:1 with the 33-point operator centered among the zeros so that I0 = 0, identically.

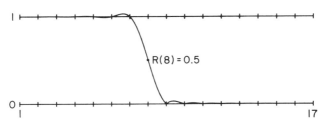

Figure 8.9 Frequency response of lowpass filter of Fig. 8.7 with the transition zone broadened by setting $R(8) = \frac{1}{2}$.

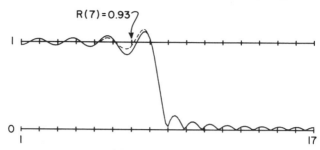

Figure 8.10 Frequency response of the lowpass filter of Fig. 8.7 with the passband ripples near the transition equalized somewhat by setting $R(7) = 0.93$. The maximum ripple error is now reduced from the original, shown by the dashed curve for comparison.

to broaden the sampled transition band. But what can be done to reduce ripple size without sacrificing the narrow transition band? Figure 8.10 suggests an answer where we have modified the sampled frequency response immediately before the transition by setting $R(7) = 0.93$. The change of this single point has reduced the size of the positive going ripple just before the transition at the expense of increasing the preceding negative going ripple. These two ripples now have about the sample amplitude, which is smaller than the largest ripple in the original design. We have succeeded in reducing the largest error.

Thus by fiddling with the points in the frequency domain, we have reduced the largest error—a step most would agree is an improvement. It also is clear that by adjusting these 17 points we will never be able to achieve our desired ideal lowpass filter, whose response is represented by a continuum.

Motivated by these examples, intuitively it would seem that if the error in meeting a given filter design specification were spread out uniformly in frequency, an overall better filter would result. One way of thinking of doing this would be to change the filter coefficients to reduce the largest errors at the expense of allowing the smaller errors to increase. That is, let us minimize the maximum error—a strategy thus called *minimax*. This philosophy seems very rational and appealing because we will have minimized the worst-case error. This measure of closeness of fit, which minimizes the maximum deviation from some desired result, is called the *Chebyshev criterion* and is used in other approximation problems as well as digital filter design. It, of course, is quite different than the least-squares criterion. If the Chebyshev criterion is satisfied for a given problem, we would normally expect that the sum of the squared error is not a minimum. Likewise, if the least-squares criterion is satisfied, we would expect that the largest error could have been made smaller.

Frequency Sampling and the Parks–McClellan Algorithm

As we have seen in the case of FIR filter design, these errors normally occur as ripples. If we imagine an iterative process whereby we continually reduce the largest ripples that pop up by adjusting the filter coefficients, all the ripples would finally end up the same size. Hence, for FIR filters, the Chebyshev criterion leads to an *equal ripple approximation*.

Now let us pursue this Chebyshev concept for zero-phase FIR filters. The frequency response of an FIR filter is

$$\sum_{-N}^{N} h_n e^{-i\omega n}$$

Since this $2N + 1$ long impulse response must be symmetric for zero-phase $h_n = h_{-n}$, the frequency response of the linearized filter is

$$h_0 + 2\sum_{1}^{N} h_n \cos(\omega n)$$

As suggested previously, we use an iterative procedure. It starts with the desired frequency response $H(\omega)$ and the filter length $2N + 1$ specified. Normally $H(\omega)$ will be piecewise linear, like an ideal bandpass or lowpass filter, but it could be any given function. For definiteness, we will be thinking of an ideal lowpass filter. We form an estimate of the impulse response h by some method (compute the IDFT of $H(\omega)$ for example) and then compute the actual frequency response from our first filter coefficient estimate:

$$\hat{H}(\omega) = h_0 + 2\sum_{1}^{N} h_n \cos(\omega n)$$

If the IDFT is used, this $\hat{H}(\omega)$ will agree with $H(\omega)$ exactly at some frequencies but will have ripples elsewhere. According to our minimax approach, we are searching for the maximum error between $H(\omega)$ and $\hat{H}(\omega)$:

$$H(\omega_j) - \left[h_0 + 2\sum_{1}^{N} h_n \cos(\omega_j n) \right] = E_j$$

where ω_j is the frequency at which maximum error E_j occurs. Now, the cosine terms in this equation can be thought of as an Nth-order trigonometric polynomial and therefore they have at most $N - 1$ extrema between $0 < \omega < \pi$. Parks and McClellan have shown that the error E has at least $N + 2$ extrema in the closed interval. This allows us to write the previous equation at each of $N + 2$ extrema ω_j and to demand the equiripple condition

$$H(\omega_j) - [h_0 + 2\sum h_n \cos(\omega_j n)] = -(-)^j E \qquad J = 0 \text{ to } N + 1 \quad (8.1)$$

where E represents an unknown equiripple amplitude. In practice, a weighting function can be applied to E by writing $w(\omega)E$ to specify different tolerances in various bands of the filter. Equations (8.1) form the basis of the Parks and McClellan algorithm for the Chebyshev approximation procedure for zero-phase FIR filter design.

Both the $N + 2$ ω_j and the $N + 1$ h_n are unknown, as is E. The equations are linear in the h_n, but nonlinear in the ω_j. After the first estimate of the h_n is selected, the error function is searched over a finely divided set of points to determine the location of the extrema ω_j. These extrema will not all have equal values (equiripple) so they are substituted in the $N + 2$ equations [Eqs. (8.1)], and they are solved as a linear set of equations for the $N + 2$ unknowns, giving a new set of h_n and E. Using the revised h_n, the error function is again formed and searched for a new set of extrema frequencies ω_j. The process is continued until E remains constant. Then the h_n are the impulse response coefficients for an equiripple zero-phase FIR filter.

To solve Eqs. (8.1) in a systematic fashion, we write them in matrix form:

$$\begin{pmatrix} 1 & 1 & 2\cos(\omega_0) & 2\cos(2\omega_0) & \cdots \\ -1 & 1 & 2\cos(\omega_1) & 2\cos(2\omega_1) & \cdots \\ & & & & \\ & & & & \\ & & & \cdots 2\cos(N\omega_{N+1}) \end{pmatrix} \begin{pmatrix} E \\ h_0 \\ h_1 \\ \vdots \\ h_N \end{pmatrix} = \begin{pmatrix} H(\omega_0) \\ H(\omega_1) \\ \cdot \\ \cdot \\ H(\omega_{N+1}) \end{pmatrix}. \quad (8.2)$$

or

$$\mathbf{M} \cdot \mathbf{T} = \mathbf{H}$$

where **M** is an $(N + 2) \times (N + 2)$ matrix containing the alternating error coefficients in its first column, all ones in the second column corresponding to the coefficients of h_0, and the appropriate coefficients of the other unknowns, h_n, in its remainder. The column vector of unknowns **T** contains the error E as its first element and the filter coefficients as the rest of its elements. The desired frequency response is specified numerically at $N + 2$ points in the column vector **H**. (If it is desired to apply the different error weights $W(\omega)$, they would appear in the first column of the matrix.) After the extremal points ω_j are found from a beginning trial filter, these linear equations are solved for the new filter coefficients by inverting M, and the procedure is repeated until E is constant.

The short computer program given in Fig. 8.11 is an instructive example

```
11  REAL W(18), M(18, 18), T(18), H(18), R0(129)
12  W(1) = 1  W(9) = 57  W(10) = 65  K = 2
13  DO 20  J = 1, 127
14  X = (R0(J + 2) - R0(J + 1)) * (R0(J + 1) - R0(J))
15  IF X .GT. 0 THEN 20
16  IF W(K) = 0 THEN 18
17  K = K + 2
18  W(K) = J + 1
19  K = K + 1
20  CONTINUE
21  DO 25  J = 1, 18
22  M(J, 1) = -1 ** (J + 1)
23  M(J, 2) = 1
24  DO 25  K = 3, 18
25  M(J, K) = 2 * COS((K - 2) * PI * (W(J) - 1)/128)
26  T = INV(M) MPY H
27  GOTO 6
```

Figure 8.11 A simple program that implements the minimax strategy for our 33-point lowpass filter example. The 129-point frequency response R0 is computed from the frequency sampling trial function in the program shown in Fig. 8.8. The array H contains the desired frequency responses specified at 18 points, two are at the limits of the transition zone, the other 16 are at the local extrema. The search on the 129-point grid for the extrema W(K) is performed in lines 13 through 20. Three extrema frequencies are specified initially in line 12: $\omega = 0$ and the transition zone edges at frequency points 57 and 65. The matrix M is filled in lines 21 through 25, and finally the unknown filter coefficients and E are computed in line 26. The process is repeated by returning to line 6 of Fig. 8.8, to compute a new zero-padded operator and then its 129-point frequency response via the DFT.

showing this strategy applied to our previous 33-point FIR filter. As expected, the results of this program, shown in Fig. 8.12, display the equiripple character, with a ripple amplitude smaller than the largest deviation shown in the trial filter of Fig. 8.7. This reduction of the maximum error has occurred without sacrificing the specified narrow transition zone, which, of course, is the goal of the minimax idea.

The short computer program, presented here, may serve as a basis for a working solution for equiripple FIR filter design in many applications. However, this program is not robust. Those that wish a more comprehensive algorithm may want to use the program published by Parks and McClellan in *Programs for Digital Signal Processing* (1979). An example of a 55-point multiband filter obtained by their algorithm is shown in Fig. 8.13. Different error weights applied to each stopband have produced different levels of rejection in those bands, while equiripples occur within a band.

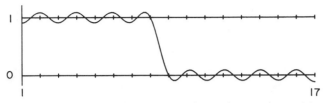

Figure 8.12 The 33-point equiripple filter generated from only one iteration of the program in Fig. 8.11. The figure shows the equiripple character of its frequency response. Note that the ripple size is considerably less than the largest ripple in the trial function of Fig. 8.6, at no sacrifice of transition zone bandwidth.

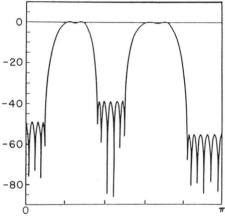

Figure 8.13 Magnitude response (in db) of a 55-point multiband equiripple FIR filter designed by the Parks–McClellan algorithm. Different rejection tolerances have been used in each stopband. (From *IEEE Programs for Digital Signal Processing*, 1979.)

More details on both the Parks and McClellan procedure and other FIR filter design methods may be found in specialized texts, such as those of Oppenheim and Schafer (1975) and Rabiner and Gold (1975). Our purpose has been to introduce the fundamental problem of FIR filter design and to present a simple version of its solution.

Recursive Filters

In many discussions we have stressed how AR operators correspond to an infinite number of MA coefficients—hence, recursive schemes have a lot of computational power. At first sight then, we might expect to be able to

Recursive Filters 183

design AR digital filters that exactly meet an arbitrarily specified frequency response. But this in general is also not possible, even though the AR operators are indeed quite powerful. Although at the beginning of this chapter we have discussed at some length the underlying reason for the existence of the filter design problem, it is worthwhile to present two more arguments that show AR operators cannot solve the problem exactly.

For the first argument, consider the inverse filter to the given AR design problem. Then, we are back to the problem of finding the finite term MA operator that meets the inverse frequency response; in general, we have seen that is impossible. In other words, we are saying that almost all desired frequency responses $F(\omega)$ have time domain coefficients that are not time limited, and the same is true of $F^{-1}(\omega)$.

Another way to see that AR filters cannot be the ultimate solution to the filter design problem is to observe that each pole has a specific frequency response,

$$|H(\omega)|^2 = \frac{1}{1 + \rho^2 - 2\rho \cos(\omega - \omega_0)}$$

whose flexibility is provided by only two factors ρ and ω_0. Since any AR filter of a finite number of poles is the cascaded effect of each individual pole, the filter design does not have the flexibility to meet an arbitrarily specified spectrum. Nonetheless, these poles can provide frequently desired sharp features in a filter's spectral response, making them quite attractive in many digital filter applications.

A common approach to the design of IIR filters is the digitization of known analog systems—either via their spectral response function or through the differential equations that govern their behavior. Viewed as the digitization of continuous systems, the IIR filter design problem becomes much wider; it becomes the problem of emulating continuous systems with digital techniques. As such, we have already addressed many approaches in our development of digital signal processing.

First, in Chapter 1, we saw how to transform differential equations into difference equations using forward, backward, and central difference operators. This method is called *mapping differentials*. In Chapter 3, we introduced another method of mapping differentials into the Z plane—the bilinear transform. The stability of the methods was a major concern that we discussed.

Another method (Problem 2.12) digitized the continuous-theory impulse response function. This method, called *impulse invariance*, preserves the impulse response upon digitization, resulting in a spectral response that is an aliased version of the analog system's response function. Likewise, any

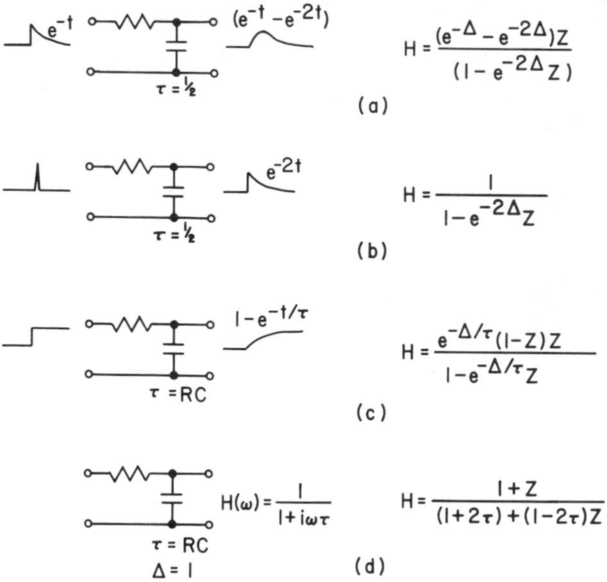

Figure 8.14 Summary of some methods of digitizing continuous systems using the simple lowpass RC filter as an example. The first three form the discrete transfer function by taking the ratio of digitized forms of known continuous inputs and outputs: (a) a nonstandard example of this, (b) impulse invariance, and (c) hold invariance. In (d), the bilinear transform is used. Each method makes a different approximation and therefore produces a different result.

(a) $H = \dfrac{(e^{-\Delta} - e^{-2\Delta})Z}{(1 - e^{-2\Delta}Z)}$, $\tau = \tfrac{1}{2}$

(b) $H = \dfrac{1}{1 - e^{-2\Delta}Z}$, $\tau = \tfrac{1}{2}$

(c) $H = \dfrac{e^{-\Delta/\tau}(1-Z)Z}{1 - e^{-\Delta/\tau}Z}$, $\tau = RC$

(d) $H(\omega) = \dfrac{1}{1 + i\omega\tau}$, $H = \dfrac{1+Z}{(1+2\tau) + (1-2\tau)Z}$, $\tau = RC$, $\Delta = 1$

known response of the analog system could be digitized along with its excitation function. The ratio of the digitized output function to the digitized excitation function then yields a digital transfer function. (Problem 2.11 is an uncommon example using an exponentially decaying pulse for input.) Impulse invariance is just one example of this approach; another is *step invariance* (sometimes called *hold invariance*), where the input is the unit step function (see Problem 2.15). Finally, there is the *direct pole–zero* placement that we explored in Chapter 3. While this method may be reasonable for a simple 2-pole notch filter, for example, it is not practical for higher order filters because of the unpredictable interaction between many poles and zeros.

Some of these methods and their results are summarized in Fig. 8.14, using the simple RC lowpass filter as an example. Because each method takes a different approach to approximating the analog system, each one gives different results.

In addition to this list of approaches to IIR filter design, we wish to discuss two more methods. One must wait for our discussion on inverses. The other, using the bilinear transform, we take up next.

Digitizing Rational Functions of ω

The frequency response of all LSI continuous systems are rational functions of ω. This will be clear to readers familiar with Laplace transform theory, for any system describable by a linear constant-coefficient differential equation will lead to a rational function of the Laplace transform variable. Thus the approach is quite general, but for definiteness, we will present four specific examples of power spectra of continuous systems. By specifying only the power spectrum, the phase spectrum is then left free to provide the stability required in the resulting ARMA operator. They are all derived from analog lowpass systems. They are as follows:

Butterworth filter: $\quad |H(\omega)|^2 = \dfrac{1}{1 + (\omega/\omega_0)^{2N}}$ (8.3a)

Chebyshev type I filter: $\quad |H(\omega)|^2 = \dfrac{1}{1 + e^2 T_N^2(\omega/\omega_0)}$ (8.3b)

Chebyshev type II filter: $\quad |H(\omega)|^2 = \dfrac{1}{1 + e^2 [T_N(\omega_0)/T_N(\omega_0/\omega)]^2}$ (8.3c)

Jacobi elliptic filter: $\quad |H(\omega)|^2 = \dfrac{1}{1 + e^2 R_N^2(\omega, L)}$ (8.3d)

In the above, ω_0 is the cutoff frequency and e relates to ripples in the response function. The $T_N(x)$ are Chebyshev polynomials defined by

$$T_N(x) = \cos(n \cos^{-1} x) \quad |x| < 1 \quad (8.4a)$$

$$T_N(x) = \cosh(n \cosh^{-1} x) \quad |x| > 1 \quad (8.4b)$$

with the recursion

$$2x T_n(x) = T_{n+1}(x) + T_{n-1}(x)$$

The first few polynomials are given by

$$T_0(x) = 1$$
$$T_1(x) = x$$
$$T_2(x) = 2x^2 - 1$$
$$T_3(x) = 4x^3 - 3x$$
$$T_4(x) = 8x^4 - 8x^2 + 1$$

The $R_N(\omega, L)$ are the Jacobi elliptic functions with the factor L controlling its ripple properties.

Each of these power spectra have denominators that are polynomials in ω of order $2N$. When factored into the magnitude spectra, they are of order N, the number of poles of the filter.

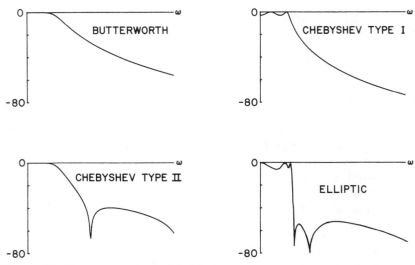

Figure 8.15 The power spectra (in db) of the four most common types of analog frequency selective filters. Each has special properties, making them attractive templates for IIR digital filter design.

Each of these filters has their own interesting characteristics. The Butterworth's power spectrum has its first $2N-1$ derivatives zero at $\omega = 0$, a property called maximally flat. The magnitude spectrum is monotonically decreasing with ω rolling off at $6N$ db/octave at high frequencies. The power spectrum is 3 db down at ω_0, independent of the filter's order.

The Chebyshev filters have equiripples in either the passband (type I) or the stopband (type II), minimizing the maximum error there. The elliptic filter has equiripple in both its passband and its stopband and achieves the sharpest transition possible for a given order filter. Examples of these four functions are plotted in Fig. 8.15.

The main point here is that these power spectra are rational functions of ω that can be, in principle, factored into their zeros. Then, the digital AR operator is formed by replacing ω with the bilinear transform and only using one-half of the roots of the polynomial—the half with minimum phase. The resulting digital transfer function will have the power spectrum of the continuous function, within the bilinear transform approximation, and a phase spectrum dictated by stability.

To see how this works, let us first assume that the power spectrum has been factored, producing a denominator polynomial of order N:

$$H(\omega) = \frac{K}{(\omega - \alpha_1)(\omega - \alpha_2) \cdots (\omega - \alpha_N)}$$

Digitizing Rational Functions of ω

where K is an unimportant constant. Using the bilinear transform to replace ω with $2(1-Z)/i(1+Z)$ gives

$$H(\omega) = \frac{K i^N (1+Z)^N}{[2(1-Z) - i(1+Z)\alpha_1] \cdots [2(1-Z) - i(1+Z)\alpha_N]}$$

Collecting powers of Z within each factor gives

$$H(\omega) = \frac{K i^N (1+Z)^N}{[(2 - i\alpha_1) - (2 + i\alpha_1)Z] \cdots [(2 - i\alpha_N) - (2 + i\alpha_N)Z]} \quad (8.5)$$

So the digital transfer function has N zeros at $Z = -1$ and N poles at

$$Z_p = \frac{2 - i\alpha}{2 + i\alpha}$$

By writing the zeros of the α polynomial in terms of their real and imaginary parts x and y, the requirement that the Z_p lie outside of the unit circle becomes

$$\left| \frac{2 - i(x + iy)}{2 + i(x + iy)} \right| > 1$$

or

$$\frac{(2+y)^2 + x^2}{(2-y)^2 + x^2} > 1$$

This is true only if $y > 0$. That is, the zeros of the ω polynomial must lie in the upper half ω plane.

So the prescription for forming digital filters from rational functions in ω, such as Eqs. (8.3), is to factor the ω polynomial into its zeros and use only the ones that lie in the upper-half ω plane to form the digital transfer function by Eq. (8.5).

In general, the required factoring can be messy business, depending on the form of the analog power spectrum. But the Butterworth function is quite tractable and will thus serve as both an instructive and a useful example. We are then looking for the zeros in the ω plane of the denominator of Eq. (8.3a):

$$1 + (\omega/\omega_0)^{2N} = 0$$

These zeros occur at

$$(\omega/\omega_0)^{2N} = -1$$

which has multiple roots of unity

$$(\omega/\omega_0)^{2N} = -1 = e^{i\pi + i2\pi\kappa}$$

or

$$\omega/\omega_0 = e^{i\pi(2\kappa+1)/2N} \qquad \kappa = 0, 2N - 1 \quad (8.6)$$

These roots lie on a circle of radius ω_0 in the ω plane, separated by the angle π/n from one another. For the minimum phase condition, we select only one-half of the roots, the ones lying in the upper-half ω plane. These are the ones for $0 < \kappa < N - 1$.

We will find it especially convenient to group these roots, pairing ones horizontally across the circle of radius ω_0. That is, we wish to pair

$$\kappa = 0 \quad \text{with} \quad \kappa = N - 1$$
$$\kappa = 1 \quad \text{with} \quad \kappa = N - 2$$
$$\vdots \quad \quad \vdots \quad \quad \vdots$$
$$\kappa = \kappa \quad \text{with} \quad \kappa = N - K - 1$$

and

$$\kappa = N/2 - 1 \quad \text{with} \quad \kappa = N/2$$

These pairs have a simple relationship among them that we see by writing

$$\alpha_{N-\kappa-1} = e^{i\pi[2(N-\kappa-1)+1]/2N} = -e^{-i\pi(2\kappa+1)/2N}$$

$$\alpha_{N-\kappa-1} = -\alpha_\kappa^*$$

Then, if we define the factors in the couplets occurring in Eq. (8.5) as

$$a_\kappa = 2 - i\alpha_\kappa \tag{8.7a}$$
$$b_\kappa = 2 + i\alpha_\kappa \tag{8.7b}$$

we get

$$a_{N-\kappa-1} = 2 - i\alpha_{N-\kappa-1} = 2 + i\alpha_\kappa^* = a_\kappa^* \tag{8.8a}$$
$$b_{N-\kappa-1} = 2 + i\alpha_{N-\kappa-1} = 2 - i\alpha_\kappa^* = b_\kappa^* \tag{8.8b}$$

Finally, we can write the transfer function of Eq. (8.5) as

$$H(Z) = \frac{K(i\omega_0)^N (1+Z)^N}{(a_1 - b_1 Z)(a_1^* - b_1^* Z) \cdots (a_m - b_m Z)(a_m^* - b_m^* Z)} \tag{8.9}$$

The last pair occurs at $m = N/2 - 1$. If the order N is odd, one root lying on the imaginary ω axis is unpaired. See Problem 8.11. In either case, the transfer function of Eq. (8.9) is seen to be real.

To summarize the method, we give the simplest example, that of the 2-pole Butterworth filter. There is just one zero in the first quadrant of the ω plane at

$$\alpha = \omega_0 e^{i\pi/4}$$

Digitizing Rational Functions of ω

which gets paired with the other in the second quadrant. Using Eq. (8.9), the desired transfer function is

$$H(Z) = \frac{-\omega_0^2(1+Z)^2}{(a-bZ)(a^*-b^*Z)}$$

$$H(Z) = \frac{-\omega_0^2(1+Z)^2}{|a|^2 - 2\operatorname{Re}(a^*b)Z + |b|^2 Z} \tag{8.10}$$

Using the definitions of Eqs. (8.7) for a and b gives by straightforward algebra the required three coefficients in the denominator of Eq. (8.10):

$$|a|^2 = 4 + 4\operatorname{Im}\alpha + |\alpha|^2 = 4 + 4\omega_0 \sin(\pi/4) + \omega_0^2 \tag{8.11a}$$

$$2\operatorname{Re}(a^*b) = 8 - 2|\alpha|^2 = 8 - 2\omega_0^2 \tag{8.11b}$$

$$|b|^2 = 4 - 4\operatorname{Im}\alpha + |\alpha|^2 = 4 - 4\omega_0 \sin(\pi/4) + \omega_0^2 \tag{8.11c}$$

The design is thus complete for any specified cutoff frequency ω_0. The results are plotted in Fig. 8.16, showing good low-frequency comparison to the analog Butterworth function.

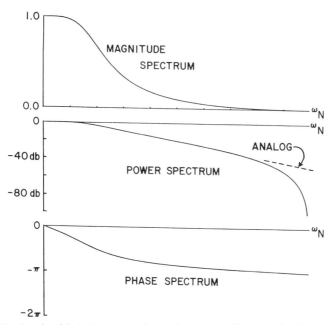

Figure 8.16 Results of designing a second-order Butterworth filter using the bilinear transform. This 2-pole, 2-zero, digital filter is given by Eqs. (8.10) and (8.11) for $\omega_0 = 0.2\pi$. On the linear scale (top), the results essentially match the analog function. On the db scale (middle), the magnitude spectrum matches the analog function out to about $\frac{2}{3}$ of Nyquist, where the zero at $Z = -1$ in the digital filter causes large attenuation. Note the fairly linear phase in the passband region.

In specifying ω_0, one needs to take into account the mapping of the bilinear transform. Remember that it maps analog frequencies from $-\infty$ to $+\infty$ onto the unit circle from $-\pi$ to π. This produces a distortion on the frequency axis that is zero at zero frequency and infinite at π. To compensate for this *frequency warping*, the design frequency must be *prewarped* by using

$$\omega_0 = 2\tan(\omega/2) \qquad (8.12)$$

where ω is the desired cutoff frequency and ω_0 is the one to use in the design equations. Figure 8.16 shows our 2-pole Butterworth design using the AR coefficients of Eqs. (8.11) and the frequency warping of Eq. (8.12) for a desired cutoff frequency of 0.2π.

DFT Frequency Filtering

In the beginning of this chapter, we saw how point-by-point multiplication of the DFT spectrum by some specified frequency response function does not produce the filtering action desired. Yet the speed of FFT algorithms makes DFT processing very attractive in many applications. The speed advantage depends on operator length. For short operators, direct convolution in the time domain will be faster. But for longer operators, DFT processing can be faster, having a speed that is independent of the operator length. Time domain operators have the advantage of continuous processing, whereas DFT filters require batch processing. Even so, with the available speed of computers—from signal processing chips to mainframes—batch processing frequently can be accomplished within one sample interval, depending on the application.

We know that the linear time domain convolution can be exactly emulated by DFT processing by zero padding to avoid the effects of circular convolution inherent in the DFT. The proposed DFT processing is then to (1) zero pad the data 2:1, (2) multiply the DFT spectrum by the desired digital filter frequency response curve, and (3) transform the filtered data back to the time domain with the IDFT.

But what is the actual frequency response of this type of processing? We saw the answer in our discussion of frequency sampling used both in the windowing method and the Parks–McClellan algorithm. This frequency sampling corresponds to the linear time domain convolution of an operator whose coefficients are determined by the IDFT. Thus, the true frequency response of DFT processing is obtained by the following steps: (1) take the IDFT of the specified frequency response sampled on the same grid to be used in the actual processing; (2) zero pad this equivalent time domain operator at least 2:1, or preferably more (remember that this operator is

DFT Frequency Filtering

on the DFT circle so the zeros go in the middle of the array); and (3) take the DFT of this zero-padded operator to get the true frequency response of the DFT processing.

The relationship between this true frequency response and the desired frequency response depends on the nature of the desired frequency response curve. Normally, the desired frequency response is such that its IDFT is spatially aliased. This effect then produces large departures from the desired frequency response, usually occurring as ripples between the frequency sampled points. On the other hand, if the frequency sampling is done on a response curve that is relatively broad, its IDFT may converge rapidly, resulting in minimal spatial aliasing. A special case occurs when the frequency sampled points lie on the actual spectrum of a short FIR operator. Then, its IDFT produces this short operator with many zeros already in the resulting array. Further zero padding does not change the result. The interpolated values then lie on a smooth curve with no ripples between the frequency sampled points. This is exactly the case of recovering a continuous band-limited function from its unaliased sampled points, but now the sampling is in the frequency domain.

A demonstration of DFT frequency filtering is now in order. Specifically, we propose the following processing example: (1) zero pad the data 2:1; (2) take its 128-point DFT; (3) multiply the resulting DFT magnitude spectrum by a Butterworth response curve, leaving the phase unchanged; and (4) take the IDFT to get the filtered result back in the time domain. The true frequency response of this operation, determined as discussed previously, is shown in Fig. 8.17 for Butterworth functions of order 8, 10, and 12. For 8th order, the Butterworth function is broad enough that little spatial aliasing occurs with the 128-point DFT used. The resulting frequency response is essentially equal to the specified Butterworth function. But, for higher orders, increasingly severe spatial aliasing produces ripples in the stopband, limiting rejection far above the specified Butterworth function. We see, as usual, that sharper transition zones are achieved at the expense of stopband rejection. DFT processing is thus a simple and effective means of frequency filtering, but the true frequency response of the operation should always be checked in order to appreciate what is really being done to the data.

The design of digital filters is a vast and challenging subject. We have only been able to introduce the problem and give a few approaches to solutions. Some of these, such as frequency sampling, followed by windowing, for FIR designs are particularly easy to put into practice for the scientist or engineer interested in quick solutions. Others, such as the Parks–McClellan algorithm and design of recursive filters just discussed, are more tedious to apply in general. However, comprehensive computer programs are available in the open literature, such as in *IEEE Programs*

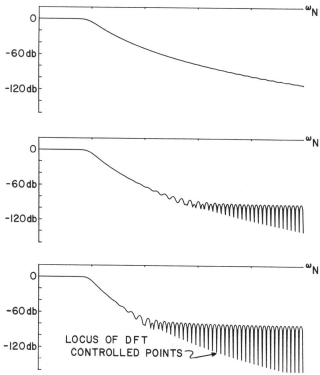

Figure 8.17 The true frequency response of using the DFT to lowpass filter by multiplying the DFT frequencies by Butterworth functions of order 8, 10, and 12 (top, middle, and bottom in figure, respectively). The DFT used was 128 points long. For sharp transition zones compared to the length of the DFT used, severe aliasing in the spatial domain occurs, producing results far less desirable than the Butterworth function.

for Digital Signal Processing (1979). Frequency filtering using DFT processing is particularly attractive because of its speed and ease of use. However, the true frequency response of this procedure can be greatly inferior to the multiplying function specified in the DFT frequency domain. After we discuss inverses and spectral factorization in the next two chapters, we will present yet another approach to IIR filter design—one that is not tedious and lets the computer do the work.

Problems

8.1 In using the frequency sampling approach to filter design, we start out with

$$R = 1, 0, 0, 1, 1, 0, 0, 0, 1$$

and

$$I = 0, 1, 1, 0, 0, 1, 1, 1, 0$$

as a first guess to the arrays forming the real and imaginary parts of a DFT algorithm. What basic category of filter are we trying to design? Carefully sketch both the magnitude and phase response of this filter from zero to Nyquist, labeling all values possible on the axis. Remember that $H(\omega)$ is a continuous function of frequency.

8.2 Imagine that we have a signal that is sampled every 3.125 msec and it is contaminated with 60 Hz line frequency noise. Try to design a sharp 60 Hz notch filter using the frequency sampling technique. We want the filter to be exactly zero phase with 33 coefficients. You can use the program of Fig. 8.8 as a guide.

After you have a program that displays the actual frequency response of the filter, try adjusting the sampled points in frequency to improve the filter by reducing the ripple without widening the notch.

When you get tired of that game, show how easily the ripple can be reduced at the expense of widening the notch by changing one of the sampled points. This exercise should give you a feel for the Gibbs phenomenon and an appreciation for the value of equiripple filters.

8.3 Experiment with windowing by using the 33-term lowpass filter of the section "Frequency Sampling and the Parks–McClellan Algorithm" as the starting point. That is, compute the IDFT of the ideal lowpass 16-point frequency function. Then, multiply by one of the windows of the section "Designing FIR Filters Using Windows," for example, the Hamming window, to obtain the windowed filter coefficients. Finally, display the filter's response via the DFT by zero padding.

8.4 Repeat Problem 8.3 for a different filter, for example, a multiband filter, such as in Fig. 8.13.

8.5 Write your own program to implement the minimax strategy by using Fig. 8.11 as a guide. Duplicate the result of Fig. 8.12.

8.6 Note that in our example of the minimax approach in Fig. 8.11 we have specified an extremal frequency at $\omega = 0$, which ignores the apparent extremal at $\omega = \pi$. Modify your program of Problem 8.5 to force an extremal frequency at $\omega = \pi$ and omit $\omega = 0$. Notice, carefully, in your results that now $\omega = 0$ is not, in fact, a maximal ripple. This flexibility arises in this example from the fact that there are 19 frequencies that may be convenient to use as extrema, but only 18 are needed for the iteration scheme. The published program of Parks and McClellan combines these cases into one algorithm.

8.7 The equiripple filter program of Fig. 8.11 is written specifically for a 33-point operator. Generalize the program to an operator of length N and improve the lowpass filter by using larger N.

8.8 The lowpass filter shown here is sometimes called an RC integrator:

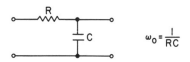

Using ac circuit theory, write down $H(\omega)$ and $|H(\omega)|^2$ for the filter and observe that it is a 1st-order Butterworth filter with a pole at $\omega = +i\omega_0$. Comment on the location of this zero in the ω plane. Next, use the bilinear transform to form a digital transfer function and show that its pole is necessarily outside the unit circle.

8.9 Write a computer program to implement the 2nd-order Butterworth filter discussed in the text. Apply a square wave to the input of your causal digital filter. There are three frequencies to keep in mind: the cutoff frequency ω_0, the square wave frequency ω, and the Nyquist frequency. By appropriate selection of these frequencies, show the following.

1. The output does indeed look like a lowpassed square wave.
2. For an appropriate selection of ω_0 compared to the input signal spectrum, the output does look like the integral of the input.
3. For frequencies near Nyquist, the filter behaves poorly.

8.10 Write a computer program that takes any FIR filter, such as those developed in the preceding problems, and slides the frequency response along the frequency axis. Thus, any of the lowpass filters discussed can be converted into bandpass and highpass filters. Demonstrate this by plotting their frequency spectra.

8.11 Show that an odd-order Butterworth function will produce a zero in the ω plane that lies on the imaginary axis at $i\omega_0$ and that this zero produces an additional factor in the transfer function equal to

$$H(Z) = \frac{i\omega_0(1+Z)}{(2+\omega_0)-(2-\omega_0)Z}$$

8.12 From our discussion of the Butterworth filter, in Eq. (8.10), the poles are located at $Z_p = a/b$. Show by writing $|Z_p|^2$ in terms of α that these poles necessarily lie outside of the unit circle when the α lie in the upper-half ω plane.

8.13 Derive the frequency prewarping expression in Eq. (8.12). Discuss the effect of frequency warping on the shape of desired frequency response functions.

8.14 A first-order Chebyshev type I analog filter ($T_1 = \omega$) with $\omega_0 = 1$ has a power spectrum given by

$$|H(\omega)|^2 = \frac{1}{1 + e^2 \omega^2}$$

Use the bilinear transform to design a 1st-order ARMA digital filter from this power spectrum. Show that your answer necessarily has its pole outside the unit circle.

8.15 A Butterworth filter is to be designed that has a constant passband within 1 db to $\omega = 0.2\pi$ and a stopband attenuation of at least 10 db for frequencies between 0.5π and π. Determine ω_0 and N that meet or exceed these specifications.

8.16 The phase-shift theorem can be used to generate various filters from a lowpass model filter, such as the four types listed in Eqs. (8.3). Assuming a high-order Butterworth digital filter with a cutoff frequency of 0.2π, sketch the resulting bandpass and highpass filters that could be generated by using the phase-shift theorem.

8.17 A certain causal analog LSI system has a frequency response given by

$$H(\omega) = \frac{1}{-i\omega^3 - 2\omega^2 + i2\omega + 1}$$

Show that the system is a 3rd-order Butterworth filter with $\omega_0 = 1$ rad/unit time. From the above expression, digitize the system directly by using the bilinear transform and make a plot of the frequency response of the discrete-time transfer function, showing that it is minimum phase.

8.18 To better appreciate the problem of digital filter design, write a computer program that plots the frequency response of a 4-pole, 4-zero ARMA filter for a given location of the poles and zeros. Remember that for a real filter the poles and zeros occur in complex conjugate pairs, so the inputs to the program should be the location of just two poles and two zeros. Use polar coordinates to specify their locations. You will need to develop equations that compute the five MA and five AR coefficients from the polar coordinates of the given poles and zeros. Then, use the DFT to compute the filter's spectrum. This direct pole–zero procedure results in a video game that is educational, challenging, and intriguing. Try to place the poles and zeros to give some prescribed filter response, such as a lowpass or a bandpass filter.

8.19 In our discussion of DFT frequency filtering, we assumed the data to be processed were pure digital data. Discuss the implications of using DFT processing on data formed by digitizing continuous signals.

8.20 In DFT frequency filtering of discrete data, a particular desired filter response can be approached to any degree of accuracy by using successively longer DFTs. Make a suite of true frequency response plots using 128-, 256-, and 512-point DFTs to perform 12th-order Butterworth lowpass filtering with a cutoff at one-fifth Nyquist.

8.21 Discuss the application of successively larger DFTs, as in Problem 8.20, applied to data digitized from continuous signals. What happens if the amount of data is limited?

8.22 Discuss zero-phase, linear-phase, and minimum-phase filters in the context of DFT processing.

8.23 Is DFT processing equivalent to MA, AR, or ARMA time-domain filtering? If so, explain how. If not, explain why not.

8.24 Discuss the application of DFT processing to allpass filters, such as the single-pole, single-zero allpass filter of Chapter 4 and the Hilbert transform.

8.25 A person tells you that certain data has been filtered using a 4-pole Butterworth filter in the frequency domain. What does this tell you about the frequency response of the filter used?

9

Inverse Filtering and Deconvolution

In the past chapter, we learned how to design convolution operators for use in the time domain that achieve a given desired frequency filtering of some data sequence. There are many data acquisition situations where a filter or convolution was unavoidably applied to the data. These require the inverse operation in order to recover the desired signal. This operation is, quite naturally, called inverse filtering or deconvolution. Inverse operators are extremely useful in all fields that use digital signal processing, from astrophysics and geophysics to biomedical studies and econometrics.

For example, a physicist may wish to remove some undesirable effects produced by a sensitive microwave receiver that detects signals from a distant galaxy. Likewise, reverberations produced by earth layers can contaminate the signal from a distant underground nuclear test; their removal will recover the original signal. In a different setting, acoustic waves patterned by the internal organs of a medical patient propagate according to LSI partial differential equations; if the effects of these propagating equations can be inverted, an image of the organs is recovered. As a final example, it turns out that inverses can be used to create prediction operators, operators that forecast future values of a time series from its past ones, suggesting intriguing applications to marketing and financing.

We have already encountered inverses in our discussion of Z transforms of couplets and recursive systems. Let us recall what we have discovered about inverses from thinking of the Z transform of one couplet. Since any operator of length N can be written as N factored couplets, studying just one couplet gave us insight to all finite-length operators. As we have seen in the section on inverse Z transforms in Chapter 2, if a causal couplet is minimum phase, it has a stable causal inverse. If it is a causal mixed phase couplet, the inverse can be made stable by allowing it to be anticipatory. If the couplet has a zero on the unit circle, the inverse will be unstable independent of causality considerations. All of these properties were seen from simply thinking of the expansion of the inverse into a geometric series. We thus anticipate the following problems with inverses. In general, inverse operators will be infinitely long and will have to be truncated somehow. If the operator has a zero near the unit circle, convergence of the inverse series will be slow; or worse yet, if the zero is on the unit circle, the inverse does not exist. Finally, in some applications, the luxury of using acausal operators may not be permitted. The causal requirement is frequently imposed by physics. Recall, for example, that both the impedance and admittance of a passive LSI one-port network must be causal and inverses of one another.

Furthermore, an additional problem arises in some applications as we have seen in our study of the Butterworth filter. Sometimes the operator whose inverse we are seeking is not specified completely. We might only know its autocorrelation function (or equivalently, in the frequency domain, its power or magnitude spectrum). In this case, the phase must somehow be determined. Frequently, the minimum phase condition, selected from physical considerations, leads to the problem of finding the one set of factors from all possible factors for the given autocorrelation function. Determination of the unique set of factors having a specific phase is called *spectral factorization*.

With this introduction behind us, we present our discussion on inverse operators by looking at four separate classes of problems with specific examples in each case. These problems will lead us to DFT inverses, Wiener least-squares filters, predication filters, and signal-to-noise-ratio enhancement filters.

Exact Inverses via the DFT

By now, we are used to thinking of operations in both the time and frequency domains. Thus, if we were presented with an operator for convolution in the time domain, and asked to find its inverse, we naturally would entertain the idea of computing its DFT, and then taking the IDFT of the reciprocal of the magnitude spectrum and the negative of the phase spectrum. That is, we are looking for f^{-1} such that

$$f_t * f_t^{-1} = 0, 0, 1, 0, 0$$

or in the frequency domain

$$F(\omega) \cdot F^{-1}(\omega) = \text{constant}$$

Exact Inverses via the DFT

Obviously, if

$$F(\omega) = |F|e^{i\phi(\omega)}$$

then

$$F^{-1}(\omega) = \frac{1}{|F|} e^{-i\phi(\omega)}$$

and

$$f_t^{-1} = \text{IDFT}[F^{-1}(\omega)]$$

But one must be careful. This line of reasoning is based on the DFT convolution theorem, which is only valid for circular convolution. In the DFT formalism, both the time domain and frequency domain functions are periodic. In some applications, this is exactly the case so that we can find the precise inverse via the DFT. Consider a surveillance radar whose antenna rotates about the vertical axis with a rather poor radiation pattern consisting of a wide central peak and large side lobes, as shown in Fig. 9.1. As far as azimuthal information goes, we imagine that both the targets and the antenna pattern are digitized at equal intervals. The received signal S is the circular convolution of the known antenna pattern A and the desired target sequence T:

$$S = A * T$$

Thus, by using the inverse operator A^{-1}, we can deconvolve the signal to get the desired target trace

$$T = A^{-1} * S$$

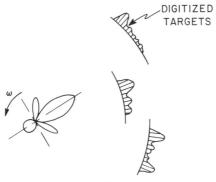

Figure 9.1 A rotating radar antenna. The digitized output of the system represents the circular convolution of the antenna radiation pattern with the target responses.

As a simple 4-long example, let

$$A = 1, 1, 0, 1$$

and

$$T = 1, 3, 1, 2$$

so that the received signal would be

$$S = A * T = 4, 6, 5, 6$$

We calculate A^{-1} via the DFT:

$$\text{DFT}(1, 1, 0, 1) = 3, 1, -1, 1$$

$$\text{DFT}(1, 1, 0, 1)^{-1} = \tfrac{1}{3}, 1, -1, 1$$

$$(1, 1, 0, 1)^{-1} = \text{IDFT}(\tfrac{1}{3}, 1, -1, 1) = \tfrac{1}{3}(1, 1, -2, 1)$$

and, as a check,

$$\tfrac{1}{3}(1, 1, -2, 1) * (1, 1, 0, 1) = (1, 0, 0, 0)$$

Now, we deconvolve using this inverse to recover T:

$$T = A^{-1} * S = (1, 1, 0, 1)^{-1} * (4, 6, 5, 6)$$
$$= \tfrac{1}{3}(1, 1, -2, 1) * (4, 6, 5, 6)$$
$$T = (1, 3, 1, 2)$$

the original target sequence. If any of the elements of the magnitude of the DFT vanish, then the inverse magnitude spectrum will blow up at those points resulting in the nonexistence of an inverse operator to A.

If our application involves one sequence that is periodic and one that is aperiodic, their convolution is still periodic. This was a fundamental result in our discussion on control theory; recall Eq. 7.8. Therefore in this case, inverses can be likewise obtained via the DFT. Before we continue, it would be beneficial to present some example cases: say that we excite some LSI system with a repetitive signal $f(t)$. The system impulse response function is $h(t)$ and the output is $S(t)$, as shown in Fig. 9.2. Such a situation might occur with the system as our object of study: we want to know $h(t)$, which completely describes the system. This requires the calculation of the inverse of a periodic sequence since

$$h(t) = f(t)^{-1} * S(t)$$

Of course, if an impulse were used for $f(t)$, the output would be the desired result with no need for an inverse computation. Another way of finding $h(t)$, which avoids an inverse calculation, is to excite the system with sine waves of different frequencies, one at a time, and record $S(\omega) = H(\omega)$

Exact Inverses via the DFT

Figure 9.2 An LSI system with a periodic input.

directly. If we still desire the impulse response, we can take the IDFT of $H(\omega)$, carefully considering leakage and the spectral width of $H(\omega)$, to estimate $h(t)$.

In some cases, however, neither impulses nor sine wave test signals are possible or practical. The appropriate excitation may not allow detailed control over $f(t)$. Air flow over an aircraft structure in a wind tunnel or over vocal cords in a person's voicebox causes periodic fluctuations that can be measured (admittedly, some easier than others). Velocity or acceleration of the structure may be taken as the input $f(t)$. The output may be the deflection of the aircraft structure, or in the case of speech, it may be the sound received at a distant microphone. The output can be related to the input by LSI systems having a nonperiodic system transfer function—a mechanical structure or an acoustic system of pharnyx, nasal cavity, and mouth.

Also, the known and unknown quantities depicted in Fig. 9.2 may be switched around. The input signal may be under study and the system may have a known response function. For example, $f(t)$ may be the optical (or radio) signal from a distant binary star system and $h(t)$ may represent the system response of the recording equipment used by an astronomer. In this case, we need the inverse of the aperiodic sequence $h(t)$ to deconvolve the received signal yielding the desired result:

$$f(t) = h(t)^{-1} * S(t)$$

This case, where one signal is periodic and the other is not, represents an important large class of problems that arise in many applications in science, engineering, and instrumentation. The examples given suggest only a few such applications. It is thus very convenient to realize that this class of problems can be cast into a form of circular convolution with exact inverses computed by using the DFT. Figure 9.3 shows schematically the convolution between a periodic sequence and one that is not periodic. There are two cases: the period of the periodic sequence is either greater or less than the length of the aperiodic sequence. In both cases, the convolution result is periodic with a period equal to the period of the periodic sequence. In both cases, circular convolution of two equal-length sequences holds by using appropriate padding, forcing the two sequences to be of equal length. As shown in Fig. 9.3, the nonperiodic sequence is zero padded and the periodic sequence is augmented with its replicas. Only whole replicas can

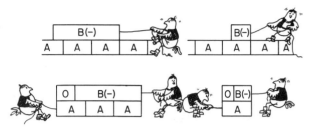

Figure 9.3 A schematic representation of the convolution between a periodic sequence and an aperiodic sequence. There are two cases: in the figures to the left, the periodic sequence is the shorter of the two. In the figures to the right, the periodic sequence is the longer of the two. In both cases, the convolution is periodic with a period equal to the length of the periodic sequence; it is reduced to circular convolution on the DFT circle with appropriate padding (bottom figures).

be used, resulting in DFTs with lengths equal to an even number, but not limited to a power of two. Other length DFT algorithms are readily available; see Elliott (1987). One, of course, can always use a direct DFT calculation (Problem 5.14). As an example of this procedure, suppose that we wish to deconvolve

$$C = A * B$$

where A is the 4-long repetitive sequence:

$$A = \ldots 1, 1, 3, 2 \ldots$$

and B is the nonrepetitive 6-long sequence:

$$B = (1, 2, 3, 4, 5, 6)$$

This is the first case shown in Fig. 9.3. The signal C, which we wish to deconvolve in order to recover either A or B, would be the repetitive sequence

$$C = \ldots 43, 35, 32, 37 \ldots$$

Padding appropriate for representing the circular convolution of A and B on the DFT circle replicates A once to make it 8-long and zero pads B to make it of equal length:

$$A \rightarrow \bar{A} = (1, 1, 3, 2, 1, 1, 3, 2)$$

$$B \rightarrow \bar{B} = (1, 2, 3, 4, 5, 6, 0, 0)$$

Now if C and B are given and we wish to recover A by deconvolution, we compute the DFT inverse of $\bar{B}$ and convolve it with C. This circular deconvolution will recover two replicas of A exactly.

On the other hand, if A and C are given and we wish to recover B, we would attempt a DFT inverse of $\bar{A}$. However, this inverse does not exist because the DFT of $\bar{A}$ will contain zero elements causing the inverse to blow up. The zero elements of the DFT of $\bar{A}$ are easily predicted by considering Fig. 7.12: replication in one domain produces zero interlacing in the other domain. The absence of an inverse in this case is not surprising; we would not really expect to be able to recover the 6 elements of B from only the 4 independent elements of the circular convolution in C.

In the second case shown in Fig. 9.3, where the periodic sequence is the longer of the two, no replication is required. Thus, barring accidental singularities, we could expect to find exact DFT inverses for use in circular deconvolution.

Linear Deconvolution—The Problem

As we saw in the preceding section, deconvolution problems with at least one periodic sequence are somewhat fortunate occurrences leading to exact inverses. Probably more frequently, we are faced with a situation where neither sequence is periodic. Then, the superposition process is linear convolution, which we shall see admits, almost always, only approximate inverses. Again, we will consider a definite example—again, we will take the radar reflection problem. But now consider an airborne radar set illuminating a series of ground-based reflectors along a straight traverse as shown in Fig. 9.4. Under the assumption of no aliasing, the digitized received signal at the aircraft S is the result of linear convolution of the discrete antenna patterns with the discrete target sequence:

$$S = A * T$$

Now we seek an inverse to A such that

$$A^{-1} * A = 0, 1, 0, \ldots \qquad (9.1)$$

Digitized ground-based reflecting targets

Digitized response of antenna

Figure 9.4 The linear convolution of two aperiodic sequences.

under linear convolution. Consequently, the target signal that would have been digitized with an infinitely sharp antenna response can be recovered by the linear deconvolution operation

$$T = A^{-1} * S$$

So we would think. However, there is a fundamental problem. Before we address this problem, observe that the computation of A^{-1}, to any desired precision in the sense of Eq. (9.1), is trivial via the DFT. All that is necessary is to append a sufficient number of zeros so that when A^{-1} is computed, there is sufficient room in the array to hold the large number of terms of A^{-1}. Then, the linear convolution of Eq. (9.1) has been emulated. For example, the inverse of the minimum phase sequence

$$A = (4, 4, 1)$$

computed via the DFT by appending 61 zeros gives an inverse that converges to 1 part in 10^{13} in 44 terms. This inverse satisfies $A^{-1} * A = 1$ under linear convolution to 1 part in 10^{13}. (In general, A may be mixed phase. So to leave space for anticipatory terms in A^{-1}, zeros should be prefixed to A as well as appended.)

However, the precision of the inverse may be necessarily severely limited for other reasons. Consider the pictorial representation of the linear convolution of the antenna pattern with the reflecting targets, as shown in Fig. 9.5. Clearly, the received signal S corresponding to some specified time t contains no information further than N away. How then can the inverse operator be longer than N? In our computed example, the 3-long operator has an inverse that is 44 long. The action of this inverse would have to be very tricky (i.e., mathematically sensitive) not to produce spurious information. One way to desensitize this effect is to content ourselves with an approximate inverse by truncating the computed DFT inverse so that it is no longer than the original operator length. This indeed is a fairly reason-

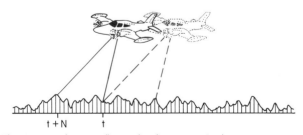

Figure 9.5 The response from a reflector that first occurs in the system output at flight time t remains in the output data until $t + N$. The system output at times outside of this interval cannot contain information about this reflector.

Linear Deconvolution—The Problem

able idea, but we know from our experience, from truncating Fourier series, for example, that if one wants to approximate an infinite series with a finite number of terms, one can do better than simply truncating the series. Thus, the question arises, which we treat shortly, as to how to obtain the best inverse of a given length.

It could be that you are still not satisfied that we must content ourselves with an approximate inverse. Let us explore the reason for this limitation from the time domain point of view. The equations for the full-transient linear convolution of a 3-long operator and a 5-long sequence are written below in Toeplitz matrix form:

$$\begin{pmatrix} x_1 & 0 & 0 \\ x_2 & x_1 & 0 \\ x_3 & x_2 & x_1 \\ x_4 & x_3 & x_2 \\ x_5 & x_4 & x_3 \\ 0 & x_5 & x_4 \\ 0 & 0 & x_5 \end{pmatrix} \cdot \begin{pmatrix} f_1 \\ f_2 \\ f_3 \end{pmatrix} = \begin{pmatrix} y_1 \\ y_2 \\ y_3 \\ y_4 \\ y_5 \\ y_6 \\ y_7 \end{pmatrix} \quad (9.2)$$

Suppose that we are given the 7-long result y and the known operator f. It is desired to invert this convolution thereby determining all the x's. Obviously, it is easy to accomplish this deconvolution exactly. For example, the very first equation of this set is $x_1 f_1 = y_1$, which gives x_1 directly. The next equation can then be solved for x_2, and so on through the whole set. In fact, the 7 equations only have 5 unknowns. We can invert the convolution because of the zeros at the beginning and end of the x-data stream. In applications where it is somehow possible to turn off the x-data stream, so that the end effect of the convolution process does indeed contain zeros as in Eq. (9.2), it is possible to obtain an exact deconvolution. It is significant that in this suggested procedure each current value of x is computed from two previous values of x and the current value of y. That is, the scheme uses a recursive computation so that the deconvolved value of x depends on all preceding values of y. In practice, there may be severe problems with this approach, such as numerical instability and noise contamination in a long recursion. End-effect values other than zeros will work just as well, of course; the only requirement is that they somehow be known.

However, in the majority of deconvolution applications, this end-effect information is lacking. Equation (9.2) then contains two additionally unknown x's before x_1 and two more unknown x's after x_5 in the array. Now, the set of seven equations has nine unknowns, preventing a solution. This is another way of seeing that we cannot find an exact inverse in applications, like our radar example, where the complete transient convolution is unavailable.

Notice, in passing, that in the case of circular convolution, the array in Eqs. (9.2) would have the end-effect zeros replaced with elements from the set x_1 to x_5, thereby introducing no new unknowns. The set of seven equations would then be overdetermined with only five unknown x's. Barring singularities and inconsistencies from experimental error, the equations would then be inverted exactly using matrix algebra.

Wiener Least-Squares Filters

Now we know that for most applications we must seek an approximate inverse of a predetermined length. Normally this length will be much shorter than would be dictated by convergence considerations of the Z transform for a given accuracy. It turns out that the procedure we are going to present is much more general than computing just inverses; so, we will simply start by considering a convolution equation in the time domain:

$$A * X = D \qquad (9.3)$$

Consider that D and A are given and that we are to determine X. Thus, we are considering a deconvolution problem, but not necessarily restricting ourselves to a simple inverse where $D = 1$. First, we use the matrix form of the convolution where A is an $m \times n$ maxtrix and x is n long:

$$m \begin{pmatrix} \overset{\leftarrow n \rightarrow}{A} \end{pmatrix} n \begin{pmatrix} x \end{pmatrix} = m \begin{pmatrix} D \end{pmatrix}$$

The numbers of rows and columns are indicated. In the case we are considering, $m > n$, there are more equations than unknowns, and we cannot, in general, expect to find an exact solution for the n x's. Two surprisingly simple steps give us a formal solution. First we premultiply by A transposed:

$$n \begin{pmatrix} \overset{\leftarrow m \rightarrow}{A^T} \end{pmatrix} m \begin{pmatrix} \overset{\leftarrow n \rightarrow}{A} \end{pmatrix} n \begin{pmatrix} X \end{pmatrix} = n \begin{pmatrix} \overset{\leftarrow m \rightarrow}{A^T} \end{pmatrix} \begin{pmatrix} D \end{pmatrix}$$

Then, we premultiply by the square $n \times n$ matrix $(\mathbf{A}^T\mathbf{A})^{-1}$, giving

$$\mathbf{X} = (\mathbf{A}^T\mathbf{A})^{-1} \cdot \mathbf{A}^T\mathbf{D} \qquad (9.4)$$

Wiener Least-Squares Filters

Thus, the solution for **x** is obtained using well-known matrix operations of transposes and inverses. Yet, what kind of solution is this? We show in the appendix to this chapter that it turns out that this simple solution, surprisingly enough, minimizes the sum square error from the original convolution equation. That is, if x is computed from Eq. (9.4), then

but

$$\sum_j A_{ij} x_j \neq d_i$$

$$\sum_j A_{ij} x_j - d_i = e_i$$

and our solution makes

$$\sum e_i^2$$

a minimum. Equation (9.4) is called the normal solution, because it makes vector x normal to the error vector e. From the point of view of statistics, this least-squares solution is quite pleasing. However, other criterion can be imagined, such as the Chebyshev approach, which may make as much or more sense for some filter applications. It's worth emphasizing that like so many problems in digital data processing the fundamental limitations can be traced to the finite length of the data window. So is the case with inverses. We could recast the problem of approximating the inverse as a problem in estimating the data beyond the acquisition window. Could the zeros of the matrix in Eq. (9.2) be replaced with a better estimate, for example? They represent the edge effect. We will pursue this approach in a matter related to inverse approximations when we discuss spectral estimation techniques. One of the greatest rationales for using the least-squares criterion is that it leads, as we have seen, to very simple equations. Let us pursue them.

The operator x, which satisfies Eq. (9.3) in the least-squares sense, is given by Eq. (9.4) and is called a Wiener filter after Norbert Wiener, the famous MIT mathematician. In order to calculate the filter, x, from Eq. (9.4), we need to understand the nature of the matrix $\mathbf{A}^T\mathbf{A}$ and the column vector $\mathbf{A}^T\mathbf{D}$. Once these two quantities are calculated, we simply invert the matrix by any common procedure and multiply it by the column vector. Let us look at the matrix first. If we write out the matrix in some detail it looks like this:

$$\begin{pmatrix} a_1 & a_2 & a_3 & \cdots \\ a_0 & a_1 & a_2 & \\ a_{-1} & a_0 & a_1 & \\ \vdots & & & \end{pmatrix} \begin{pmatrix} a_1 & a_0 & a_{-1} & \cdots \\ a_2 & a_1 & a_0 & \\ a_3 & a_2 & a_1 & \\ \vdots & & & \end{pmatrix} = \begin{pmatrix} \phi_0 & \phi_1 & \cdots & \phi_n \\ \phi_1 & & & \\ \vdots & & & \\ \phi_n & & & \end{pmatrix}$$

By inspecting the individual matrix products, we see the matrix ϕ_{ij} is formed by the autocorrelation of a_t with zero-lag values along the diagonal and autocorrelations of successively higher lags off the diagonal. The autocorrelation function is even; the autocorrelation matrix is symmetrical. It is of order n, equal to the specified length of the Wiener filter. Similarly, by writing out the product that forms the column vector,

$$n\left(\begin{array}{ccc} & \xleftarrow{\quad m \quad}\rightarrow & \\ & \vdots & \\ a_{-1} & a_{-2} & a_{-3} \\ a_0 & a_{-1} & a_{-2} \\ & a_1 & \\ & \vdots & \end{array}\right) m\left(\begin{array}{c} \vdots \\ d_{-1} \\ d_0 \\ d_1 \\ \vdots \end{array}\right) = n\left(\begin{array}{c} \vdots \\ c_{-1} \\ c_0 \\ c_1 \\ \vdots \end{array}\right)$$

we see that C is formed from the cross-correlation of the elements of a_t with the desired result D. The zero-lag cross-correlation is in the center with other lags above and below, the negative lags are on top, and the positive ones are on the bottom. This cross-correlation vector is also n long. With this insight into the composition of ϕ, the autocorrelation matrix, and c, the cross-correlation vector, we can write

$$\begin{pmatrix} \phi_0 & \cdots & \phi_n \\ \vdots & & \\ \phi_n & & \phi_0 \end{pmatrix} \begin{pmatrix} \\ x \\ \\ \end{pmatrix} = \begin{pmatrix} c_{-(n-1)/2} \\ c_0 \\ c_{(n-1)/2} \end{pmatrix} \tag{9.5}$$

This form of the normal equations can be readily solved by inverting the autocorrelation matrix. For applications requiring frequent solutions, particularly for large N, an algorithm by Levinson should be used. This recursive scheme, exploiting the symmetry of the autocorrelation matrix, is more efficient both in computing time and memory than matrix inversion. Here our intention is to concentrate on the fundamentals of least-squares filters and their applications; we discuss the Levinson recursion for the zero-delay inverse filter later in this chapter.

Application to Inverses

In order to gain familiarity with Wiener least-squares filtering, we now will explore some simple applications to inverse filtering where we can solve the normal equations by hand. We start by finding the 3-long inverse to the

Application to Inverses

minimum phase couplet $(2, 1)$. The autocorrelation of $(2, 1)$ is
$$(2, 1) \otimes (2, 1) = (2, 5, 2)$$
We pick a zero-delay spike as the desired output:
$$D = (2, 1) * (f_1, f_2, f_3) = (1, 0, 0, 0)$$
Then the cross-correlation between the input and the desired output becomes
$$(2, 1) \otimes (1, 0, 0, 0) = (1, 2, 0, 0, 0)$$
The normal equations then become
$$\begin{pmatrix} 5 & 2 & 0 \\ 2 & 5 & 2 \\ 0 & 2 & 5 \end{pmatrix} \begin{pmatrix} f_1 \\ f_2 \\ f_3 \end{pmatrix} = \begin{pmatrix} 2 \\ 0 \\ 0 \end{pmatrix}$$
Notice two things concerning delay. First, we can ask for a least-squares inverse with any delay. Second, we have selected a zero-delay inverse because the causal minimum delay couplet will have a causal inverse. The cross-correlation vector C used in the normal equations only requires the n-long center section of the full cross-correlation. Solving the three equations for three unknowns by any method (Cramer's rule, matrix inversion, the Levinson algorithm, etc.) gives
$$f = \tfrac{1}{85}(42, -20, 8)$$
This inverse filter applied to the original couplet gives
$$f * f^{-1} = (2, 1) * \tfrac{1}{85}(42, -20, 8) = \tfrac{1}{85}(84, 2, -4, 8)$$
while the exact result would be
$$f * f^{-1} = (1, 0, 0, 0)$$
giving a squared error of
$$\sum e^2 = \left(\frac{1}{85}\right)^2 + \left(\frac{2}{85}\right)^2 + \left(\frac{4}{85}\right)^2 + \left(\frac{8}{85}\right)^2 = 0.012$$

It is interesting to compare this error to that obtained by simply truncating the exact series inverse of $(2, 1)$:
$$f_s^{-1} = \frac{1}{f(Z)} = \frac{1}{2 + Z} = \frac{1/2}{1 + (Z/2)} = \frac{1}{2}\left(1 - \frac{Z}{2} + \frac{Z^2}{4} + \cdots\right)$$
$$f * f_s^{-1} = (1, 0, 0, 1/8)$$

with

$$\sum e^2 = \left(\frac{1}{8}\right)^2 = 0.016$$

As anticipated, the Wiener inverse is better, in the least-squares sense, than the truncated inverse. Parenthetically, note that the maximum error in the Wiener inverse is also less than that of the truncated inverse, but this maximum error, of course, is not minimized. Programs for similarly computing Wiener filters are straightforward; a 13-line example is shown in Fig. 9.6.

In Fig. 9.7, a comparison of inverses to the couplet $(1, 0.5)$, computed by three different methods, shows the significance of the least-squares approach: the 16-long inverses, computed by the DFT, least-squares method, and truncation of the geometric series, show reasonable agreement among themselves. They all give excellent results in the sense that these inverses, convolved with the couplet $(1, 0.5)$, give the unit sequence to quite high accuracy, typically to better than 1 part in 10^5. The major point is that 16 coefficients are an unreasonably large number for use as a deconvolution operator for a couplet. The only reasonable deconvolution operator in Fig. 9.7 is the 3-long least-squares version, which differs markedly from 3-long inverses produced by the other two methods. Remember that the inverse, formed by zero padding and then using the DFT method, inverts the padded operator exactly under circular convolution. For large padding, the result approaches the geometric series inverse; for small padding, the DFT inverse bears no relation to a linear deconvolution operator.

```
          M = 2    N = 3
          REAL A(M), C(N), D(M + N − 1), P(N, N), X(N)
    3     DATA A,D/1, 0.5, 1, 0, 0, 0/
          DO 6 J = 1, N MIN M
          DO 6 I = 1, M − J + 1
    6     P(1, J) = P(1, J) + A(J + I − 1) * A(I)
          DO 12 J = 1, N
          DO 10 I = J, N
          P(J, I) = P(1, I − J + 1)
   10     P(I, J) = P(J, I)
   11     DO 12 I = 1, M
   12     C(J) = C(J) + A(I) * D(I + J − 1)
   13     X = INV(P) MPY C
```

Figure 9.6 A simple program to compute Wiener filters to deconvolve the equation $A * X = D$ using inversion of the autocorrelation matrix. This example computes the 3-long inverse to A by setting $A = (1, 0.5)$ and $D = (1, 0, 0, 0)$ in line three. Lines 4 through 10 compute the autocorrelation matrix P. Lines 11 and 12 compute the cross-correlation vector; line 13 solves for X.

Filter Delay Properties   211

First 16 terms of $1/(1 + 0.5) = \sum(-0.5)^n$:
1	−0.5	0.25	−0.125
0.0625	−0.03125	0.015625	−0.0078125
0.00390625	−0.001953125	9.765625E−4	−4.8828125E−4
2.44140625E−4	−1.220703125E−4	6.103515625E-5	−3.051757813E−5

16-long DFT inverse:
1.00001525902	−0.500007629511	0.250003814755	−0.125001907378
0.0625009536889	−0.0312504768444	0.0156252384222	−0.00781261921111
0.00390630960556	−0.00195315480278	9.765774014E−4	−4.882887007E−4
2.441443503E−4	−1.220721752E−4	6.103608759E−5	−3.05180438E−5

16-long least-squares inverse:
0.999999999825	−0.499999999563	0.249999999083	−0.124999998145
0.0624999962783	−0.0312499925512	0.0156249850997	−0.00781247019813
0.00390619039558	−0.00195300579082	9.763240815E−4	−4.878044129E−4
2.431869507E−4	−1.201629639E−4	5.722045899E−5	−2.28881836E−5

3−long least-squares inverse:
0.9882   −0.4706   0.1882

Figure 9.7 Inverses to the couplet (1, 0.5).

Filter Delay Properties

To elaborate on the delay behavior of Wiener inverses, we consider another example composed of a set of three operators, all 3 long and all with the same autocorrelation. We select the desired output of the convolution of each of these to match their phase delay.

Operator		Inverse		Desired output	Phase delay
(441)	*	$(f_1 f_2 f_3)$	=	10000	Minimum
(252)	*	$(f_1 f_2 f_3)$	=	00100	Mixed
(144)	*	$(f_1 f_2 f_3)$	=	00001	Maximum

The corresponding cross-correlations between the inputs and the desired outputs with the 3-long center sections underlined are shown below. To the right is shown the mean-square error of each inverse calculated with the desired output and, subsequently, with the cross-correlation reflecting the desired delay.

Operator		Desired output		Cross-correlation	Squared error
(441)	⊗	(10000)	=	144̲0000	0.097
(252)	⊗	(00100)	=	0025̲200	0.076
(144)	⊗	(00001)	=	0000441̲	0.097

It turns out that the errors shown are a minimum for each inverse among all possible delays. Of course, an inverse can be calculated for any given operator for any desired delay; but we claim, without proof, that the squared error will be larger than shown above. As one example, the squared error for the 3-long inverse to the minimum phase sequence (4, 4, 1) for one unit of delay increases from 0.097 to 0.210.

It is apparent from looking at the center section of the cross-correlation vectors above that the minimum phase inverse (also the maximum) requires only a knowledge of the operator's autocorrelation to determine its inverse within a scale factor. This is so because the required cross-correlation in that special case is zero except for one element. Consequently, the normal equations in the form

$$\begin{pmatrix} \phi_0 & \cdots & \phi_{n-1} \\ \vdots & \ddots & \vdots \\ \phi_{n-1} & & \phi_0 \end{pmatrix} \cdot \begin{pmatrix} f_1 \\ \vdots \\ f_n \end{pmatrix} = \begin{pmatrix} 1 \\ 0 \\ \vdots \\ 0 \end{pmatrix} \quad (9.6)$$

define an n-long minimum phase inverse to the operator whose n lags of its autocorrelation is ϕ. Note that in this special case only the operator's autocorrelation is required to find its inverse. This is equivalent to only needing the operator's power spectrum. The phase information is contained in the fact that we are assuming the operator has minimum phase.

A very useful application of this approach has been in seismic reflection prospecting where the received signal from some seismic source is the result of convolution with a series of reflecting surfaces, as shown in Fig. 9.8. We would like to deconvolve the received seismic trace T to re-

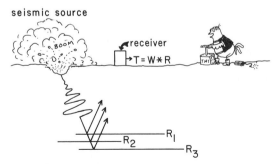

Figure 9.8 A one-dimensional model of the seismic reflection experiment. The reflection series R, representing the desired geological data, is convolved with the seismic source wavelet, $W(t)$. The received trace T can be approximately deconvolved to recover R if sufficient information is known about W.

cover the reflection series

$$R = W^{-1} * T$$

But, in practice, it is frequently impossible to measure the seismic signal, W, delivered to the subsurface and thereby compute its inverse. However, in many cases, it can be argued on physical grounds that W is a minimum phase sequence, and thus we need only its autocorrelation to compute its inverse by Eq. 9.6. If, in addition, we can assume that over sufficiently large depths the geological structure is such that the reflection surfaces form a statistically random series, then the autocorrelation of the wavelet is equal to the autocorrelation of the trace. To show this, we simply write out the autocorrelation of the seismic trace:

$$T \otimes T = (W * R) \otimes (W * R) = (W \otimes W) * (R \otimes R)$$

and

$$R \otimes R = \delta$$

because of the assumed random character of R. Thus,

$$T \otimes T = W \otimes W$$

In general, this approach, sometimes called spiking deconvolution because it sharpens up the seismic events to give better resolution of the reflecting surfaces, has been quite successful in the petroleum exploration industry. However, problems exist for sources that may not be minimum phase (recall the large errors in computing inverses using incorrect delays noted previously) and in certain areas where the geological structure is in fact not random but displays repetitive bedding.

Applications to Prediction Filters

One of the more interesting applications of the Wiener least-squares filters is prediction. We simply seek a filter such that its operation on a sequence produces the next element of the sequence. That is to say, we write the matrix form of the convolution process as

$$\begin{pmatrix} x_1 & 0 & 0 & \cdots \\ x_2 & x_1 & 0 & \cdots \\ x_3 & x_2 & x_1 & \cdots \\ \vdots & & & \\ x_m & & & \end{pmatrix} \begin{pmatrix} f_1 \\ f_2 \\ f_3 \\ \vdots \\ f_N \end{pmatrix} = \begin{pmatrix} x_2 \\ x_3 \\ x_4 \\ \vdots \\ x_{m+1} \end{pmatrix} \qquad (9.7)$$

for a filter that will produce the next element of the sequence x from the past N elements. Solving these equations by the methods of least squares, we premultiply both sides by the transpose of the $N \times M$ matrix and carry out the indicated matrix multiplication to give

$$\begin{pmatrix} \phi_0 & \phi_1 & \cdots & \phi_N \\ \phi_1 & & & \\ \phi_N & & & \end{pmatrix} \begin{pmatrix} f_1 \\ \vdots \\ f_{N+1} \end{pmatrix} = \begin{pmatrix} \phi_1 \\ \phi_2 \\ \phi_{N+1} \end{pmatrix} \quad (9.8)$$

where the ϕ_j are the familiar autocorrelations of lag j. These least-squares normal equations now contain the autocorrelation of the sequence on the right side as well as the left. By inverting the $N \times M$ matrix (by any standard means) these equations can be solved for the prediction operator f. Then, it would be hoped that the sequence x is sufficiently predictable that when f operates off its end it will predict the unknown x_{m+1} value to some useful accuracy. The extent to which this is possible depends on the inherent deterministic qualities of x compared to its random features. For example, the complex sequence

$$x_n = \exp(i\omega_0 n \, \Delta t)$$

is completely deterministic and thus it can be exactly predicted one time step ahead with the 2-long real linear operator

$$x_{n+2\Delta t} = 2\cos(\omega_0 \Delta t) x_{n+\Delta t} - x_n$$

On the other hand, if x contains large amounts of random effects, the prediction of the $m + 1$ term will, of course, be much less reliable. The really interesting cases lie somewhere in between. For example, a sequence of stock market closing quotations for just about any given cooperation will display what appears to be large random fluctuations. Yet, it is reasonable to think that known and unknown underlying business cycles and market pressures force certain predictabilities into the stock trading price that may not be completely apparent from the casual visual inspection of its past performance.

The simple computer program in Fig. 9.9 computes a 3-long prediction filter from the previous 37-week Friday closings for use in forecasting the closing price of Control Data Corporation stock one week in the future. The operator coefficients are

$$-0.9543, \; 0.02426, \; 0.06831$$

and the predicted closing price is $37.17 compared to the actual price on that date of $37.13. A prediction within $0.04 of the actual value from a history that varied from a $43 high to a $27\frac{5}{8}$ low is deceptively encourag-

Applications to Prediction Filters

```
      M = 37   N = 3
      REAL A(M), C(N), D(M + N − 1), P(N, N), X(N)
      DATA A/33.500, 33.500, 34.250, 37.000, 39.000, 43.000, 41.500,
            39.250, 38.750, 38.250, 38.125, 38.625, 37.500, 33.000,
            32.750, 30.500, 28.825, 29.375, 28.750, 32.375, 31.625,
            29.750, 28.750, 27.625, 28.000, 26.875, 28.625, 29.500,
            30.000, 29.125, 32.875, 28.500, 28.375, 30.125, 30.750,
            34.625, 37.250
   10 S = SUM(A)/M   A = A − S
      DO 12 I = 1, M − 1
   12 D(I) = A(I + 1)
      DO 15 J = 1, N
      DO 15 I = 1, M − J + 1
   15 P(1, J) = P(1, J) + A(J + I − 1) ∗ A(I)
      DO 21 J = 1, N
      DO 19 I = J, N
      P(J, I) = P(1, I − J + 1)
   19 P(I, J) = P(J, I)
      DO 21 I = 1, M
   21 C(J) = C(J) + A(I) ∗ D(I + J − 1)
      X = INV(P) MPY C
```

Figure 9.9 A program to compute a 3-long operator to predict stock market closings one week in advance. The program is identical to the Wiener filter program in Fig. 9.6 except that line 12 asks for a desired output that is advanced one week from the input. The average value, removed from the market quotes in line 10, would have to be applied to the predicted value.

ing. We selected a 3-long prediction operator with hindsight—any other length gives poorer results. It is natural to measure how good a job the operator does in predicting the known part of the sequence by using a normalized mean square error:

$$E = \frac{\sum_1^m (\text{predicted values})^2}{\sum_1^m (\text{actual past values})^2} = \frac{\sum_1^m (f \ast x)_j^2}{\sum_1^m (x_j)^2}$$

Of course, the longer the prediction operator, the better the least-squares solution fits the past data. Unfortunately, E is not necessarily related to how well future unknown values will be predicted. For our 3-long example $E = 0.2202$ and decreases for longer prediction operators.

Returning now to the normal equations for the prediction operator, we note that the same operator would result for the whole class of sequences with the same autocorrelation. We also know that these sequences correspond to different phase characteristics and that they would look very different from one another. Naturally, we are curious about a prediction operator that has identical values for a large class of different sequences (all those with the same autocorrelation).

We need a new way of looking at Eq. (9.7). A simple trick with the matrix form of the convolution equation leads to a useful way of using and understanding prediction operators. We augment Eq. (9.7) by using the fact that we can treat elements of matrices as matrices themselves. That is, we write

$$\begin{pmatrix} x_1 & 0 \\ \hline x_2 & x_1 & 0 & \cdots \\ x_3 & x_2 & x_1 & 0 & \cdots \\ & x_3 & x_2 & x_1 & 0 & \cdots \end{pmatrix} \begin{pmatrix} 1 \\ \hline -f_1 \\ \vdots \\ -f_N \end{pmatrix} = \begin{pmatrix} x_1 \\ \hline 0 \end{pmatrix}$$

as a 2×2 block matrix form of Eq. (9.7). You can see that this form is correct by writing out the first few equations. The first equation is just an identity, $x_1 = x_1$, while the general equation is

$$x_j - x_{j-1}f_1 - x_{j-2}f_2 - \cdots - x_{j-N}f_N = 0$$

This, of course, is the same equation for the one-step prediction from the N past values that we had in our original formulation. However, the operator

$$f_e = 1, -f_1, -f_2, \ldots, -f_N$$

has a new interpretation. Instead of predicting the next value, it yields the difference between the predicted next term and its actual value. Therefore, we call it a *prediction error operator* of unit span. By placing an appropriate number of zeros between the unit term and $-f$, we can generate a *gapped* prediction error operator of any span.

Generating the normal equations for this $N+1$ long prediction error filter by the usual techniques of premultiplying both sides by the transpose of the $M \times (N+1)$ matrix gives

$$\begin{pmatrix} \phi_0 & \phi_1 & \cdots & \phi_n \\ \vdots \\ \phi_n \end{pmatrix} \begin{pmatrix} 1 \\ -f_1 \\ -f_2 \\ -f_3 \\ \vdots \\ -f_N \end{pmatrix} = \begin{pmatrix} x_1^2 \\ | \\ 0 \\ | \\ | \end{pmatrix} \qquad (9.9)$$

If these equations are solved for the $N + 1$ terms of the prediction error filter where the first term is normalized to unity, the remaining terms in f_e will be identical to the solution of Eq. (9.8). Now, however, we recognize the $N + 1$ long prediction error filter is the solution, within a scale factor, to the minimum phase inverse of the given sequence by comparing Eq. (9.9) to Eq. (9.6). That is, f_e is attempting, in the least-squares sense, to cancel each x_t by operating on past values $(x_{t-1}, x_{t-2}, \ldots)$; or

$$x * f_e \simeq 1, 0, 0, \ldots$$

and will do its best job if x itself is a minimum phase sequence. So, our curiosity is satisfied; this least sequences approach to prediction assumes that the underlying sequence is minimum phase.

An application where we have reason to believe that the sequence is minimum phase is marine reflection seismology. Here, strong reverberations are produced in the received seismic signal from multiple reflections between the ocean bottom and the air–water interface. It is possible to show that if the seismic source is minimum phase, then so are the reverberations. The total received signal is a superposition of three components: the wavelet; reverberations, which contain a great deal of predictability because of the way in which they are generated; and reflection coefficients from geological strata, which are assumed mostly randomly spaced and therefore unpredictable. The output of a prediction error filter designed from the autocorrelation of the received seismic signal would then consist largely of the unpredictable components, the geological information in the form of reflection events.

Matched Filters and Output Energy Filters

In order to keep our discussions as simple as possible, we have for the most part assumed our signals are completely deterministic with no random behavior. The results obtained are useful then in situations where the signal-to-noise ratio is high. On the other hand, there are many interesting applications of digital signal processing where the challenge is to recover a signal deeply buried in noise. The reception of radio signals from distant interplanetary space probes; the detection of seismic signals from distant sources, such as deep reflectors, earthquakes, or even nuclear explosions; and the recording and decoding of information from noisy communication channels are just three examples of situations where the desired signal is covered by noise. In many of these cases, the signal-to-noise ratio may be -20 db or less.

There are several ways to deal with signals that are weak relative to contaminating noise. Our approach in this section is to imagine treating a signal with an LSI filter in the time domain. We will define a signal-to-noise ratio and then seek the filter coefficients that maximize this ratio. We consider only two cases. The first leads to an operator called a matched filter, and the second is called an output energy filter. The first requires a knowledge of the signal to be detected, while the second needs only the autocorrelation of the signal. Both require the autocorrelation of the noise present.

For the matched filter, we wish to design an LSI operator that tells us when it finds the hidden signal buried in noisy data. Imagine this operator convolving along the data looking for the hidden signal. When there is complete overlap between the filter and the hidden signal, we want the output of the filter to be maximum, telling us it has found the signal. Clearly, the length of the filter should be equal to the length of the signal. The convolution of a signal $(S_0, S_1 \ldots S_n)$ and a filter $(f_0, f_1 \ldots f_n)$ is $2n+1$ long. Complete overlap of the two then occurs at time $t = n$, and at that time, we choose to require maximum power output of the filter. Thus, for the matched filter, we are led to define a signal-to-noise ratio:

$$\alpha = \frac{\text{power of filter output at } t_n \text{ when only signal is present}}{\text{measure of power of filter output when only noise is present}}$$

For the numerator, we write the filter output at time n and square it to get the instantaneous power

$$\left(\sum_j S_{n-j} f_j\right)^2$$

The denominator of α requires more care in evaluating. When only noise is present at its input, the instantaneous output from the filter changes randomly. However, if the noise has statistical properties that are not changing in time (such a sequence of random numbers is called a stationary times series), we can use some kind of average value for noise output power. We choose to write the convolution representing this noise output in matrix form:

$$P_i = \sum Q_{ij} f_j$$

or

P = Qf

where **P** is a column vector containing the output over the time length of **f**, and **Q** is the input noise sequence written as a convolution matrix. The vector dot product

$$P^T P = P_1^2 + P_2^2 + \cdots + P_n^2$$

Matched Filters and Output Energy Filters

is a sum of squares of the noise output values or the total noise power over the time of the filter. Substituting for **P** gives

$$\mathbf{P}^T\mathbf{P} = (\mathbf{Q}\mathbf{f})^T\mathbf{Q}\mathbf{f} = \mathbf{f}^T\mathbf{Q}^T\mathbf{Q}\mathbf{f} = \mathbf{f}^T\mathbf{\Phi}\mathbf{f}$$

where we have recognized $\mathbf{Q}^T\mathbf{Q}$ as the autocorrelation matrix $\mathbf{\Phi}$ of the noise out to n lag values. If the noise represents a stationary phenomenon, this value will be the same for all time. Using these two parameters, the filter output signal-to-noise ratio becomes

$$\alpha = \frac{\left(\sum_j S_{n-j} f_j\right)^2}{\sum_{ij} \phi_{ij} f_i f_j}$$

It is worth pausing at this point to emphasize that this definition of α, although quite reasonable, is an arbitrary selection from many possibilities. We shall see that the value of this particular choice is the simplicity of the resulting filter. The filter that maximizes α is found by setting the derivatives of α with respect to each filter coefficient f_k equal to zero:

$$\frac{\partial \alpha}{\partial f_k} = \frac{2\left(\sum_j S_{n-j} f_j\right) S_{n-k}}{\sum_{ij} \phi_{ij} f_j f_i} - \frac{2\left(\sum_j S_{n-j} f_j\right)^2 \sum_i \phi_{ik} f_i}{\left(\sum_{ij} \phi_{ij} f_j f_i\right)^2} = 0$$

or

$$S_{n-k}\left(\sum_{ij} \phi_{ij} f_j f_i\right) = \left(\sum_j S_{n-j} f_j\right) \sum_i \phi_{ik} f_i$$

These k equations are satisfied by requiring that the filter coefficients obey

$$\sum_i \phi_{ik} f_i = S_{n-k}$$

This is our solution. Writing it out in matrix form gives

$$\begin{pmatrix} \phi_0 & \cdots & \phi_n \\ \vdots & & \vdots \\ \phi_n & & \phi_0 \end{pmatrix} \begin{pmatrix} f_0 \\ \vdots \\ f_n \end{pmatrix} = \begin{pmatrix} S_n \\ \vdots \\ S_0 \end{pmatrix} \quad (9.10)$$

which can be solved directly for the filter coefficients by inverting the noise autocorrelation matrix. Frequently, the noise is very wide band, or white, in which case the autocorrelation matrix approaches the identity. That is to say, purely uncorrelated noise has zero autocorrelation except at zero lag. Under these conditions, Eq. (9.10) yields a filter that is exactly equal to the time reversed version of the signal that we are seeking to detect. Since

```
REAL S1(N), A(N, N), S(512), F(N), S3(N + 512 − 1)
F = INV(A) MPY S1
DO 4 I = 1, N
4 FO = FO + F(I)**2
F = F/SQR(FO)
DO 8 I = 1,512
DO 8 J = 1, N
8 S3(I + J − 1) = S3(I + J − 1) + F(J) * S(I)
```

Figure 9.10 Program for matched filtering in colored noise. The inputs are S1, the time reversed signal; A, the noise autocorrelation matrix; and S, a 512-long data stream. The outputs are F, the matched filter coefficients, and S3, the filtered output. The filtered coefficients are normalized to unit total energy.

convolution with the signal's time reversed terms is the same as correlation without reversing the order, the desired filter operation can be performed by simply forming the cross-correlation between the incoming signal and the signal we are trying to detect. This filter is thus called a *correlator*. The output of such a correlator, when the signal is present, is the signal's autocorrelation function plus the action of the filter on the noise. If the noise is not white, sometimes called colored, it is said to be autocorrelated noise because its autocorrelation is nonvanishing at lags other than zero. In this case, the filter is still, by convention, called a correlator but its coefficients must be determined by Eq. (9.10). Regardless of the noise characteristics, the correlator design is based on a complete knowledge of the sought-after signal. That is to say, the correlator is matched to the signal and for that reason it is frequently called a *matched filter*.

The correlator, or matched filter, is particularly useful in echo ranging problems, such as seismic work and radar. In these situations, the shape of the transmitted signal is known, and it is desired to know only when the reflected signal is returned. Clearly, the greatest range resolution occurs for the transmitted signal with the sharpest autocorrelation function, that is, a wide bandwidth signal.

The matched filter is relatively easy to implement, as shown by the simple program in Fig. 9.10. As expected, a large signal in white noise will produce its autocorrelation at the output of the matched filter, as shown in Fig. 9.11. In the presence of colored noise, the output differs markedly from the signal's autocorrelation because of the noise contribution. This is shown in Fig. 9.12, where a signal has been detected under low signal-to-noise conditions in colored noise. The matched filter has produced rather dramatic results considering that it is impossible to visually detect the signal under the given conditions. A price paid for this very welcomed

Matched Filters and Output Energy Filters

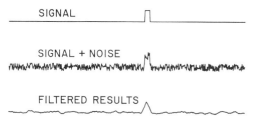

Figure 9.11 Results of the matched filter acting on a large signal in white (broadband) noise, generated by a random number computer code. The bottom trace shows that the output approximates the autocorrelation of the signal under the given conditions.

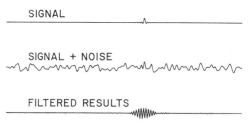

Figure 9.12 Results of the matched filter acting on a weak signal buried in colored noise. The colored noise was generated by passing white noise through Spencer's 15-point smoothing filter discussed in Problem 9.14. This spectra of the signal and of the noise have considerable overlap. The bottom trace shows the remarkable detection of a weak signal that is indistinguishable in the signal + noise trace.

signal-to-noise enhancement is that the matched filter produces an output somewhat remotely related to the original signal and usually about twice as long.

The other signal-to-noise enhancement filter that we wish to consider is the output energy filter. Again, we will define a particular signal-to-noise ratio and then maximize it to determine the filter's coefficients. This time we will attempt to maximize not the power output at a single time but the total energy over the length of time of the signal. So motivated, we write for the signal-to-noise ratio

$$\lambda = \frac{\text{sum of squares of output when signal only is present}}{\text{sum of squares of output when noise only is present}}$$

A filter that maximizes this ratio is called an *output energy filter* because it maximizes the signal energy output relative to the noise energy output. In our discussion of the matched filter, we saw that the sum of the squares of the output due to noise could be written as

$$\mathbf{f}^T \Phi \mathbf{f}$$

where Φ is the noise autocorrelation matrix, and $\mathbf{f}$ is the filter column vector. Likewise, then, the sum of the square of the output due to an input signal would be written as

$$\mathbf{f}^T \Psi \mathbf{f}$$

where Ψ is the autocorrelation matrix of the signal. The output energy filter's signal-to-noise ratio can then be written:

$$\lambda = \mathbf{f}^T \Psi \mathbf{f} / \mathbf{f}^T \Phi \mathbf{f}$$

Instead of the normal method of maximizing by setting derivatives equal to zero, we take advantage of the matrix character of this expression. From this definition, we write

$$\lambda \mathbf{f}^T \Phi \mathbf{f} - \mathbf{f}^T \Psi \mathbf{f} = 0$$

or

$$\mathbf{f}^T \{(\Psi - \lambda \Phi) \mathbf{f}\} = 0$$

This last expression is the vector dot product of the row vector $\mathbf{f}^T$, with the column vector in braces. The dot product can be zero for several reasons or combinations of reasons. For example, $\mathbf{f}^T = 0$ is obviously not the solution we desire. Likewise, $\mathbf{f}^T$ could be perpendicular to the other vector, not a particularly useful condition. On the other hand, if the vector in braces is set to zero,

$$(\Psi - \lambda \Phi) \mathbf{f} = 0 \tag{9.11}$$

we generate a particularly nice set of equations. This set of simultaneous linear equations represents the generalized eigenvalue problem that for a filter length n provides for the determination of n eigenvectors f and their corresponding n eigenvalues λ. By selecting the eigenvector f corresponding to the largest eigenvalue λ, we are assured of maximizing this signal-to-noise ratio among the class of signals of length n with the given autocorrelation of both noise and signal.

The solution of the generalized eigenvalue problem involves the diagonalization of two matrices, Φ and one associated with Ψ. Both Φ and Ψ are symmetric and real. We start by diagonalizing Φ to Φ_d via an orthogonal transformation $\mathbf{U}$ (if $\mathbf{U}$ is orthogonal, $\mathbf{U}^T = \mathbf{U}^{-1}$):

$$\Phi_d = \mathbf{U}^T \Phi \mathbf{U}$$

Since Φ_d is diagonal, it is possible to define

$$\Phi^{1/2} = \mathbf{U} \Phi_d^{1/2} \mathbf{U}^T$$

and

$$\Phi^{-1/2} = \mathbf{U} \Phi_d^{-1/2} \mathbf{U}^T$$

where $\Phi_d^{1/2}$ and $\Phi_d^{-1/2}$ have the square root, and the reciprocal square root respectively, of Φ_d along their diagonals. The square root notation is appropriate for these matrices because

$$\Phi^{1/2}\Phi^{-1/2} = \mathbb{1}$$

$$\Phi^{1/2}\Phi^{1/2} = \Phi$$

and

$$\Phi^{-1/2}\Phi^{-1/2} = \Phi^{-1}$$

follow from the above definitions of $\Phi^{1/2}$ and $\Phi^{-1/2}$.

The second diagonalization is performed on

$$\Psi' = \Phi^{-1/2}\Psi\Phi^{-1/2}$$

via the matrix **V** to give

$$\Psi_d = \mathbf{V}^T\Psi'\mathbf{V} = \lambda$$

the diagonal matrix containing the eigenvalues λ. We see that these two transformations do indeed satisfy the generalized eigenvalue problem by writing

$$\mathbf{V}^T\Psi'\mathbf{V} = \lambda$$

$$\Psi'\mathbf{V} = \mathbf{V}\lambda$$

$$\Phi^{-1/2}\Psi\Phi^{-1/2}\mathbf{V} = \mathbf{V}\lambda$$

$$\Psi\Phi^{-1/2}\mathbf{V} = \Phi^{1/2}\mathbf{V}\lambda$$

$$\Psi(\Phi^{-1/2}\mathbf{V}) = \Phi(\Phi^{-1/2}\mathbf{V})\lambda \quad (9.12)$$

Since each side of this equation is an $n \times n$ matrix and λ is diagonal, λ multiplies each column, f, of the matrix in brackets by one of its elements:

$$(\Phi^{-1/2}\mathbf{V})\lambda = \begin{pmatrix} | & | & \\ \lambda_1 f_1 & \lambda_2 f_2 & \cdots \\ | & | & \end{pmatrix}$$

Therefore, Eq. (9.12) can be written as n column vector equations:

$$\Psi\mathbf{f}_i = \lambda_i\Phi\mathbf{f}_i \quad i = 1 \text{ to } n$$

where λ_i is the scalar eigenvalue associated with its eigenvector f_i. If we have a matrix diagonalization computer routine available (see Press *et al.*, 1986), the above discussion leads to a quite simple procedure for finding the output energy filter. First, diagonalize Φ finding both Φ_d and **U**. Then form Ψ' and diagonalize it finding **V** and λ. The output energy filter is then the column vector of $\Phi^{-1/2}\mathbf{V}$ corresponding to the largest λ. A sample computer program for this procedure is shown in Fig. 9.13. Dramatic

```
      REAL AO(N, N), A1(N, N), R2(N, N), A(N, N), E(N), F(N)
      CALL EIGEN(AO, E, N)
      DO 4  I = 1, N
    4 R2(I, I) = 1/SQR(E(I))
      A2 = AO MPY R2 MPY TRN(AO)
      A = A2 MPY A1 MPY A2
      CALL EIGEN(A, E, N)
    8 DO 10  I = 1, N
      DO 10  J = 1, N
      F(I) = F(I) + A2(I, J) * A(J, N)
      DO 12  I = 1, N
   12 FO = FO + F(I)**2
      F = F/SQR(FO)
```

Figure 9.13 Subroutine to compute the output energy filter F from the noise autocorrelation matrix A0 and the signal autocorrelation matrix A1. A subroutine eigen is required that solves the regular eigenvalue problem leaving the eigenvalues, ordered from low to high, in E and their corresponding eigenvectors as columns in the matrix A.

results using an output energy filter generated by this program are shown in Fig. 9.14; a signal buried in noise beyond visual recognition is easily detected after filtering.

We have chosen to develop both of these signal-to-noise-ratio enhancement filters in the time domain because the definition of the signal-to-noise ratios and their subsequent optimization was most easily accomplished in that domain; it would be obscure in the frequency domain. Nonetheless, as suggested in Problem 9.14, we can gain a qualitative insight into the action of these filters by looking at them in the frequency domain. The spectral response of the filter can be compared to the noise power spectrum. Clearly, we would expect the most improvement in the signal-to-noise ratio in those cases where the spectral overlap is minimal. In the limiting case,

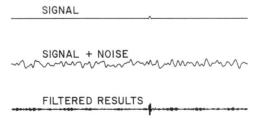

Figure 9.14 Results of the output energy filter acting on a weak signal that is completely buried in colored noise. The noise used is the same as that in the matched filter example. The traces are all 512 points long and the signal is 9 points long. In contrast to the matched filter, which is designed to peak the output at the center of the signal, the output energy filter tends to equalize the energy along the length of the signal.

where noise and signal spectra have no overlap, a bandpass filter would completely recover the signal. The matched filter and the output energy filter are the result of optimizing two different criteria for cases where the noise and the signal occupy the same frequency band.

The Levinson Recursion

In appendix B of Wiener's (1949) classic work on time series, Norman Levinson presented an algorithm for solving the normal equations, Eq. (9.6) or Eq. (9.9), that exploits their Toeplitz structure. Standard algorithms for solving an $n \times n$ system of linear equations require time proportional to n^3 and storage proportional to n^2. Levinson's method requires time proportional to n^2 and storage proportional to n.

The recursion starts with the 2×2 set of normal equations

$$\begin{pmatrix} \phi_0 & \phi_1 \\ \phi_1 & \phi_0 \end{pmatrix} \begin{pmatrix} 1 \\ f_1 \end{pmatrix} = \begin{pmatrix} v_2 \\ 0 \end{pmatrix} \quad (9.13)$$

where we have chosen to normalize the first element of f to 1; the two unknowns thus become f_1 and v_2. This is the form of Eq. (9.9). The solution to Eq. (9.13) is trivial:

$$f_1 = -\phi_1/\phi_0 \quad \text{and} \quad v_2 = \phi_0 - \phi_1^2/\phi_0$$

The idea of the recursion is to generate the solution to an $(n+1) \times (n+1)$ set of normal equations from an $n \times n$ set. The 2×2 set can be augmented by first writing

$$\begin{pmatrix} \phi_0 & \phi_1 & \phi_2 \\ \phi_1 & \phi_0 & \phi_1 \\ \phi_2 & \phi_1 & \phi_0 \end{pmatrix} \begin{pmatrix} 1 \\ f_1 \\ 0 \end{pmatrix} = \begin{pmatrix} v_2 \\ 0 \\ e_3 \end{pmatrix} \quad (9.14)$$

This contains the old known values f_1 and v_2 in an imbedded form of the original 2×2 set, and introduces no new unknowns since

$$e_3 = \phi_2 + \phi_1 f_1 \quad (9.15)$$

The main trick of the method is to make use of the fact that the Toeplitz structure allows this 3×3 augmented form to be also written as

$$\begin{pmatrix} \phi_0 & \phi_1 & \phi_2 \\ \phi_1 & \phi_0 & \phi_1 \\ \phi_2 & \phi_1 & \phi_0 \end{pmatrix} \begin{pmatrix} 0 \\ f_1 \\ 1 \end{pmatrix} = \begin{pmatrix} e_3 \\ 0 \\ v_2 \end{pmatrix} \quad (9.16)$$

Then, a 3×3 set of normal equations is formed by multiplying Eq. (9.16) by a yet to be determined constant c_3 and subtracting the result from

Eq. (9.14):

$$\begin{pmatrix} \phi_0 & \phi_1 & \phi_2 \\ \phi_1 & \phi_0 & \phi_1 \\ \phi_2 & \phi_1 & \phi_0 \end{pmatrix} \left\{ \begin{pmatrix} 1 \\ f_1 \\ 0 \end{pmatrix} - c_3 \begin{pmatrix} 0 \\ f_1 \\ 1 \end{pmatrix} \right\} = \begin{pmatrix} v_2 \\ 0 \\ e_3 \end{pmatrix} - c_3 \begin{pmatrix} e_3 \\ 0 \\ v_2 \end{pmatrix}$$

Next, collecting terms and writing this as a new 3×3 set of normal equations gives

$$\begin{pmatrix} \phi_0 & \phi_1 & \phi_2 \\ \phi_1 & \phi_0 & \phi_1 \\ \phi_2 & \phi_1 & \phi_0 \end{pmatrix} \begin{pmatrix} 1 \\ (1-c_3)f_1 \\ -c_3 \end{pmatrix} = \begin{pmatrix} v_2 - c_3 e_3 \\ 0 \\ e_3 - c_3 v_2 \end{pmatrix} = \begin{pmatrix} v_3 \\ 0 \\ 0 \end{pmatrix} \quad (9.17)$$

where we have had to introduce a second unknown, v_3. So, in going from the 2×2 set to the 3×3 set, two unknowns have to be introduced, c_3 and v_3. Equation (9.17) simply says that

$$c_3 = e_3/v_2 \quad (9.18)$$

and

$$v_3 = v_2 - c_3 e_3 \quad (9.19)$$

which solves the 3×3 problem.

It is easy to see how to generalize the recursion for forming any $n \times n$ set from an $(n-1) \times (n-1)$ set. Equations (9.15), (9.18), and (9.19) took us from the 2×2 to the 3×3 set. In general, Eq. (9.15) would be

$$e_n = \phi_{n-1} f_0 + \phi_{n-2} f_1 + \cdots + \phi_1 f_{n-2}$$

or

$$e_n = \sum_{i=0}^{n-2} \phi_{n-i-1} f_i^{n-1} \quad (9.20)$$

where f^{n-1} are the elements of the $n-1$ long vector. And in general, Eqs. (9.18) and (9.19) become

$$c_n = e_n/v_{n-1} \quad (9.21)$$

$$v_n = v_{n-1} - c_n e_n \quad (9.22)$$

At each stage the new n-long solution vector is formed from the $n-1$ long vector and the current value of c by

$$\begin{pmatrix} 1 \\ f_1 \\ f_2 \\ \vdots \\ f_{n-1} \end{pmatrix} = \begin{pmatrix} 1 \\ f_1 - cf_{n-2} \\ f_2 - cf_{n-3} \\ \vdots \\ f_{n-2} - cf_1 \\ -c \end{pmatrix}$$

The Levinson Recursion

```
      REAL P(N), F(N), G(N)
      F(1) = 1
      G(1) = 1
      F(2) = -P(2)/P(1)
      V = P(1) - P(2) * P(2)/P(1)
      DO 30 J = 3, N
      E = 0
      DO 10 I = 1, J - 1
   10 E = E + P(J - I + 1) * F(I)
      C = E/V
      V = V - C * E
      DO 20 I = 2, J - 1
   20 G(I) = F(I) - C * F(J - I + 1)
      G(J) = -C
      DO 30 I = 1, N
   30 F(I) = G(I)
```

Figure 9.15 A program that implements the Levinson recursion. The symbols follow the discussion in the text. The array P(N) is the given autocorrelation; F(N) is the solution. G(N) is a work-space array.

or

$$f_0 = 1$$
$$f_i = f_i - cf_{n-i-1} \quad i = 1 \text{ to } n - 2 \quad (9.23)$$
$$f_{n-1} = -c$$

Equations (9.20), (9.21), (9.22), and (9.23) constitute the recursion; the program of Fig. 9.15 implements it.

An important property of the Levinson recursion is that F is generated in a fashion that guarantees that it is minimum phase. It is possible to write Eq. (9.23) as

$$F'(Z) \rightarrow F(Z) - cZ^n F(1/Z)$$

and to show that the new $F'(Z)$ is minimum phase if F is minimum phase and if $|c| < 1$ (see Problem 11.4).

And we can see that $c^2 < 1$ by referring to the form of the normal equations in Eq. (9.9), showing that v_n is always positive (for example, $v_2 = \phi_0 - \phi_1^2/\phi_0$ is necessarily positive). Combining Eqs. (9.18) and (9.19) allows us to write

$$v_n = v_{n-1}(1 - c_n^2)$$

Since the v's are necessarily positive, $|c| < 1$.

The Levinson recursion is a powerful algorithm, useful whenever computer time or storage is a consideration. Moreover, the method can be extended to nonsymmetric Toeplitz structures as explained in the very useful book by Press et al. (1986). Their approach also allows for a general inhomogeneous vector V, so that general equations of the form

$$\sum_{j=1}^{n} R_{n+i-j} x_j = v_i \quad (i = 1, \ldots, n)$$

can be solved, with the Toeplitz symmetry of R being the only constraint.

Appendix: The Least-Squares Property of Wiener Filters

We have seen that the inverse problem reduces to solving a set of equations with more equations than unknowns. In the section on Wiener least-squares filters in this chapter, we have used the simple procedure of premultiplication by the nonsquare matrix to produce the normal equations containing a square matrix that can then be inverted by any common means to find the desired filter coefficients. This quick and dirty method gives no indication of the kind of solution produced by this matrix manipulation. We now show, as previously claimed, that the normal equations satisfy the convolution equation not exactly but, surprisingly enough, in a least-squares sense.

We start with the convolution

$$A * X = D$$

and seek an $(n+1)$-long filter $X = (x_{-(n-1)/2} \cdots x_0 \cdots x_{(n+1)/2})$ that satisfies this equation to some approximation for a given $(m+1)$ long sequence $A = (a_0, a_1 \cdots a_m)$ and a given desired $(m+n+1)$ long output $D = (d_{-(m+n)/2} \cdots d_0 \cdots d_{(m+n)/2})$. If D is selected to be the Kronecker delta sequence, then X is an approximate inverse to A, otherwise X is called a shaping filter and tries to produce the desired output D from A. We write the above convolution as $(m+n+1)$ equations for the $n+1$ unknown filter coefficients in matrix form:

$$\sum_{-(n-1)/2}^{(n-1)/2} a_{i-j} x_j = d_i$$

and realizing that the equations are not satisfied exactly, we write the sum square error

$$E = \sum_i \left(\sum_j a_{i-j} x_j - d_i \right)^2$$

The filter coefficients that minimize this sum squared error are found by setting the partial derivatives of this equation with respect to the co-

Appendix: The Least-Squares Property of Wiener Filters

efficients equal to zero:

$$\frac{\partial E}{\partial x_k} = 2 \sum_i \left(\sum_j a_{i-j} x_j - d_i \right) a_{i-k} = 0$$

or by rearranging terms:

$$\sum_j x_j \left(\sum_i a_{i-j} a_{i-k} \right) = \sum_i a_{i-k} d_i$$

Now, we observe that

$$\sum_i a_{i-j} a_{i-k} = \phi_{k-j}$$

is the autocorrelation of the input to the filter at the $(k-j)$th lag. Likewise, we identify

$$\sum_i^{n+m} a_{i-k} d_i = c_k$$

as the kth lag cross-correlation of A, the input, with D the desired output. Thus, the result may be written

$$\sum_j^n \phi_{k-j} x_j = c_k$$

which represents n equations for the n unknown filter coefficients. In matrix form, these equations become the familiar normal equations

$$\begin{pmatrix} \phi_0 & \phi_1 & \cdots & \phi_n \\ \phi_1 & & & \\ \cdot & & & \\ \cdot & & & \\ \cdot & & & \\ \phi_n & & & \phi_0 \end{pmatrix} \begin{pmatrix} x_{-(n-1)/2} \\ \vdots \\ x_0 \\ \vdots \\ x_{(n-1)/2} \end{pmatrix} = \begin{pmatrix} c_{-(n-1)/2} \\ \vdots \\ c_0 \\ \vdots \\ c_{(n-1)/2} \end{pmatrix}$$

where we have used the fact that $\phi_{-j} = \phi_j$ for real A.

Furthermore, we can compute the sum squared error resulting from these normal equations. We use matrix notation and write

$$\begin{aligned} \sum_i e^2 &= (\mathbf{D} - \mathbf{AX})^T (\mathbf{D} - \mathbf{AX}) \\ &= (\mathbf{D}^T - \mathbf{X}^T \mathbf{A}^T)(\mathbf{D} - \mathbf{AX}) \\ &= \mathbf{D}^T \mathbf{D} - \mathbf{X}^T \mathbf{A}^T \mathbf{D} - \mathbf{D}^T \mathbf{AX} + \mathbf{X}^T \mathbf{A}^T \mathbf{AX} \\ &= \mathbf{D}^T \mathbf{D} - \mathbf{X}^T \mathbf{A}^T \mathbf{D} - \mathbf{D}^T \mathbf{AX} + \mathbf{X}^T \mathbf{\Phi} \mathbf{X} \\ &= \mathbf{D}^T \mathbf{D} - \mathbf{X}^T \mathbf{A}^T \mathbf{D} - \mathbf{D}^T \mathbf{AX} + \mathbf{X}^T \mathbf{A}^T \mathbf{D} \\ &= \mathbf{D}^T \mathbf{D} - \mathbf{D}^T \mathbf{AX} \end{aligned}$$

since $A^TA = \Phi$ the autocorrelation matrix of the input and

$$\Phi X = A^T D$$

the normal equations in matrix notation. Now, we recognize

$$D^T D = c_0$$

as the autocorrelation of the desired output at zero lag and

$$D^T A X = (A^T D)^T X = \sum_j c_j x_j$$

as the inner product of the filter coefficients with the cross-correlation vector. By dividing the error by the positive number c_0, we obtain the *normalized mean squared error*:

$$E_n = 1 - \frac{\sum_{j}^{n+1} c_j x_j}{c_0}$$

Clearly, a zero-length filter would produce an $E_n = 1$. As the length of the filter increases, the least-squares solution will provide a better fit, and hence, the error must decrease monotonically with n. However, the error can never go negative because it is the sum of the squares of numbers divided by a positive c_0. Hence,

$$0 \le E_n \le 1$$

and thus provides a convenient criterion for filter performance. Sometimes another parameter

$$P = 1 - E$$

is used in describing the least-squares fit. It is called the *filter performance*.

Problems

9.1 It is proposed that the time domain operator $(3, -1, 2, 1)$ is the result of computing an inverse via a 4-long DFT calculation for the operator $(1, 1, 0, -1)$. Is it the DFT inverse? If not, what is the DFT inverse? What is the meaning of a 4-long exact inverse to $(1, 1, 0, -1)$?

9.2 Compute the 4-long DFT inverse to $(0, -1, 1, 1)$ and check your result by showing $(0, -1, 1, 1) * (0, -1, 1, 1)^{-1} = \delta$ exactly, under circular convolution.

Problems 231

9.3 Verify the statements made in the section "Exact Inverses via the DFT." That is, for

$$A = \ldots 1, 1, 3, 2 \ldots$$
$$B = (1, 2, 3, 4, 5, 6)$$

show that under circular convolution

$$A * B = C = \ldots 43, 35, 32, 37 \ldots$$

Calculate the appropriate DFT inverse of B and use it to recover A from C by circular deconvolution. Compute the DFT of the appropriately augmented A to verify that it has interlaced zero elements, preventing the calculation of its inverse operator.

9.4 In the section "Application to Inverses," we computed the 3-long least-squares inverse to $(2, 1)$. Compute by hand the 2-long least-squares inverse to $(2, 1)$ for zero delay and one unit of delay. Compute the normalized mean square error for each inverse and compare both to the 2-term truncation of the Z transform of $(2, 1)^{-1}$.

9.5 Compute the 3-long least-squares inverse to $(1, 2)$ for zero, 1 unit, and 2 units of delay. Compute the normalized mean square error in each case and observe that the maximum delay inverse has the least error.

9.6 Guided by the computer program of Fig. 9.6, explore least-squares inverses further. For example, write a program to compute 10-long inverses to the minimum phase sequence $(2.2, -4.2, 3.1, -1)$ with zeros at $Z = 1 \pm i$ and $Z = 1.1$ for all possible delays. Compare with the inverses to $(2.02, -4.02, 3, -1)$, which has zeros at $Z = 1 \pm i$ and $Z = 1.01$.

9.7 Simulate the discussion of Fig. 9.8 by first generating a reasonable facsimile of a seismic trace by convolving a minimum phase wavelet of your choice with a random reflection coefficient series R. Then, deconvolve this synthetic trace using only its autocorrelation to recover R.

9.8 Show that the LSI operator

$$x_{n+2\Delta t} = ax_{n+\Delta t} + bx_n$$

predicts $\sin(\omega_0 t + \phi)$ exactly for all t and ϕ when

$$a = 2 \cos(\omega_0 \Delta t) \quad \text{and} \quad b = -1$$

9.9 Using the program of Fig. 9.9 as a guide, write your own computer program to explore the dependency of the stock market prediction on prediction operator length. Compare the actual value of $37.13, known from hindsight, with your predicted values using various operator lengths from 1 to 37.

9.10 Make plots similar to Figs. 9.11 and 9.12 for the matched filter by writing your own computer program. Then, explore the properties of the matched filter by adding two cases not shown in these figures: a low signal-to-noise ratio in broadband noise and a high signal-to-noise ratio in colored noise.

9.11 Reproduce a plot similar to Fig. 9.14 by writing your own computer program based on the output energy filter program shown in Fig. 9.13. Run the case for white noise by setting the noise autocorrelation matrix equal to the identity matrix.

9.12 In the section "Matched Filters and Output Energy Filters," we computed the output energy filter by solving the generalized eigenvalue problem

$$(\Psi - \lambda\Phi)f = 0$$

by first diagonalizing Φ and then diagonalizing a matrix associated with Ψ. This procedure involved the reciprocal square roots of the eigenvalue of Φ. Alternately, we could premultiply the generalized eigenvalue equation by Φ^{-1} to give

$$(\Phi^{-1}\Psi - \lambda)f = 0$$

which immediately reduces the generalized eigenvalue problem to the ordinary eigenvalue problem requiring only the diagonalization of the matrix $\Phi^{-1}\Psi$. This solution requires only one diagonalization, but adds the calculation of a matrix inverse. You can very easily modify your program from Problem 9.11 to implement this simpler alternate procedure. What is the disadvantage of this second, apparently simpler, approach?

9.13 There is nothing that precludes the application of Wiener least-squares techniques to signal-to-noise ratio enhancement filters. You can do this by simply using the known signal that you are trying to detect for the desired output—that is, we would be asking for a filter that deconvolves the signal and noise into the exact signal. Write a computer program to do just this and compare the matched filter, the output energy filter, and the least-squares filter for the same combinations of type of noise, signal, signal-to-noise ratio, and operator length.

9.14 Inspect the frequency response of the 15-point Spencer smoothing filter (see Problem 7.21) used to generate the colored noise of the examples in Figs. 9.12 and 9.14. How do you think the spectra of the signals used in these two examples compare to the colored noise spectrum? Verify your answer by using signals with different spectra to study the performance of the matched filter in noise that is

Problems 233

generated by passing random numbers through the 15-point Spencer filter.

9.15 Use the bilinear transform to convert the second-order differential equation for sinusoids into a difference equation. Form the two-zero prediction error filter for sinusoids from the result. Show that the zeros of this filter are on the unit circle.

9.16 One application of inverse theory is recursive digital filter design. There are many approaches, but the simplest is to compute the least-squares inverse of the MA operator determined by frequency sampling on a dense grid. To demonstrate this idea, sample the Butterworth function $|H(\omega)|^2 = 1/(1 + \omega^{2N})$ on a dense grid and compute its IDFT to get an estimation of the filter's autocorrelation coefficients. Using these autocorrelations, solve the normal equations $\phi_{xx} B = 1$ to get the all-pole digital filter $H(Z) = 1/B(Z)$. Show that as the length of $B(Z)$ increases the frequency response of the AR filter approaches the specified function $|H(\omega)|^2$. How many poles are required to get a reasonable fit to an Nth order Butterworth function?

9.17 Apply the method of Problem 9.16 to other filter functions. In particular, develop digital filters from Chebyshev type I and type II functions. Comment on the AR order required to achieve a reasonable fit to the specified frequency response.

9.18 The approach to filter design taken in Problem 9.16 is limited by the all-pole model. Better filter performance can be expected if we include zeros in the filters transfer function $H(Z) = A(Z)/B(Z)$. In doing so, we must employ the least-squares inverse machinery in such a way to guarantee that $B(Z)$ is minimum phase. One scheme of doing so is as follows: (1) given $|H(\omega)|^2$, select a suitable $A(Z)$; (2) compute the frequency response $A(\omega)$ on a dense grid by taking the DFT of $A(Z)$ zero padded; (3) frequency sample $|H(\omega)|^2$ on the same dense grid and form $|B^{-1}(\omega)|^2 = |H(\omega)|^2/|A(\omega)|^2$; and (4) use the least-squares inverse procedure to find $B(Z)$ from the autocorrelation coefficients $|B^{-1}(\omega)|^2$.

Design Butterworth lowpass digital filters of various orders using this scheme and compare the results to the desired $H(\omega)$. Observe that a 2N-pole, 2N-zero digital filter duplicates the frequency response of an Nth-order Butterworth function quite well. Comment on this result.

9.19 A great advantage of the least-squares approach to IIR filter design used in the previous problems is that it can be used for any arbitrarily specified $|H(\omega)|^2$. To demonstrate this, use the procedure to develop bandpass filters from $|H(\omega)|^2 = 1/(1 + (\omega - \omega_c)/\omega_0)^{2N}$, where ω_c is the center frequency of the passband and ω_0 relates to the cutoff frequency. Compare your results to the specified $|H(\omega)|^2$.

Comment on the order of the digital filter required to get reasonable results.

9.20 Repeat Problem 9.19 but use a cosine taper in the transition bands. Make a suite of filters of different orders, comparing their frequency responses to the specified $|H(\omega)|^2$.

9.21 The frequency sampling required in the previous problems can require very long DFTs for high-order filters. Discuss limitations, if any, to this somewhat brute-force approach that can require very long DFTs. Discuss modifications to the procedure that might help control the required length of the DFT.

9.22 Show that, as expected, the Wiener least-squares formulization leads to $f = (1, 0, 0, \ldots)$ when one tries to filter a sequence into itself.

9.23 As pointed out in Problem 9.13, least-squares filtering can well be applied to signal-to-noise enhancement. Consider a received signal $y(t)$ to be composed of the desired signal $x(t)$ plus additive noise $n(t)$: $y(t) = x(t) + n(t)$. We desire an LSI operator $f(t)$ that gives an estimate $\bar{x}(t)$ for the true signal $x(t)$. In the frequency domain, $\bar{X}(\omega) = F(\omega)Y(\omega)$. We want to determine an optimal $F(\omega)$ by least-squares minimization of

$$\int_{-\infty}^{\infty} |\bar{x}(t) - x(t)|^2 \, dt = \int_{-\infty}^{\infty} |\bar{X}(\omega) - X(\omega)|^2 \, d\omega$$

Show that under certain assumptions on some cross-correlations, minimizing this integrated power in the frequency domain leads to

$$F(\omega) = \frac{|X(\omega)|^2}{|X(\omega)|^2 + |N(\omega)|^2}$$

Discuss the assumptions required of the cross-correlations in your derivation. But, regardless of these assumptions, make an argument that $F(\omega)$ as given above is a very reasonable filter, making it quite useful in practice.

9.24 The time domain formulization of the least-squares filter that filters noisy data y into a signal x leads to

$$\phi_{yy} f = \phi_{yx}$$

Show that these normal equations are roughly equivalent to

$$F(\omega) = \frac{\phi_{yx}(\omega)}{\phi_{yy}(\omega)}$$

and that under a suitable assumption about certain cross-correlations, this result is roughly equal to that of Problem 9.23.

10

Spectral Factorization

Frequently, in analyzing and processing signals, only their power spectrum is available, giving us no information on the signal's phase spectrum. We have seen such a case in the section "Wiener Least-Squares Filters" in Chapter 9, where under suitable assumptions the autocorrelation (or somewhat equivalently, the power spectrum) of a downgoing seismic source signal could be inferred from the received seismic trace. In the next chapter, we will discuss other cases where one must be content with estimating a signal's power spectrum at the expense of information on its phase spectrum.

Optics is another branch of science in which it is difficult to measure phase spectra; generally, it is only possible to measure the magnitude of an optical signal (as opposed to its instantaneous function of time) and to determine its Fourier magnitude spectrum. For example, coherent light shining on the baffel of Fig. 6.7 will produce the complete complex spectrum of the aperture, including magnitude and phase. Yet, normal detectors suitable for recording optical signals respond only to light intensity, revealing only the magnitude spectrum of the aperture.

To recover a time domain signal from its power spectrum, we also need to know its phase spectrum. But given a signal's power spectrum, there are a large number of possibilities for its phase spectrum. Of all these possibilities, a unique phase spectrum can be determined if sufficient additional information is known about the signal. A trivial example is a symmetrical time domain operator; it must have a zero-phase spectrum. Thus, this operator's power spectrum and a knowledge of its symmetrical behavior is sufficient information for recovery of the complete time domain signal. A more interesting and nontrivial case is the minimum phase signal. We will show how to recover this signal from its power spectrum alone. This is an important case because the minimum phase signal has several practical properties: remember that physical considerations may dictate that a signal must be causal and invertible—that is to say, it must be minimum phase; and, the one filter with the least group delay among all filters with the same magnitude spectrum has a unique phase spectrum—minimum phase.

The process of determining the minimum phase function belonging to a given power spectrum (or equivalently magnitude spectrum) is called *spectral factorization* because the power spectrum is the product of two factors, $F(\omega)$, and $F^*(\omega)$, either of which are to be determined. With the phase spectrum thus determined, $F(\omega) = |F(\omega)|\exp[-i\phi(\omega)]$ is completely known; an estimate of the time domain signal can then be found using the IDFT. Most important, it is just as easy to compute the IDFT of $(1/|F|)\exp(-i\phi)$, thus giving the signal's time domain stable causal inverse. Of course, if $|F|$ is zero anywhere (or nearly so) we would have difficulty computing the inverse. Our discussion in this chapter is an extension of the previous chapter's treatment of discrete inverses. In this chapter, we will treat three methods of spectral factorization and three related topics: causal functions, minimum-delay inverses, and applications of spectral factorization to the design of IIR filters. We start with the simplest approach first, the root method of spectral factorization.

The Root Method

Although the root method of spectral factorization is not practical for long data streams, it does provide a very useful method for short sequences, such as those used in digital operators. Equally important, the root method yields an invaluable conceptional insight into spectral factorization.

We work in the Z domain. The information on the magnitude spectrum is then contained in a sequence's autocorrelation. The spectral factorization problem is to find the minimum phase sequence given only its autocorrelation.

From Eq. (5.22), the autocorrelation of a sequence $(a_0, a_1 \cdots a_n)$ is

$$A^* \otimes A = A^*(1/Z)A(Z)$$

This $(n + 1)$ long sequence will have a real symmetrical autocorrelation of length $2n + 1$.

$$P(Z) = p_n Z^{-n} + \cdots + p_1 Z^{-1} + p_0 + p_1 Z + \cdots + p_n Z^n$$

The Root Method

If $P(Z)$ is multiplied by Z^n, we get a polynomial of degree $2n$ with real coefficients that can be factored into $2n$ terms:

$$Z^n P(Z) = P_n \prod_{k=1}^{2n} (Z - Z_k)$$

The roots of $Z^n P(Z)$ are either real or occur in complex conjugate pairs. Furthermore, it is well known from the theory of equations (and easy to show as well) that the symmetrical disposition of the coefficients forces the roots to occur in reciprocal pairs. Thus the complex roots of $Z^n P(Z)$ occur in groups of four. One root, Z_1, produces the associated factors

$$(Z - Z_1)(Z - Z_1^*)(Z - 1/Z_1)(Z - 1/Z_1^*)$$

in $Z^n P(Z)$. These four factors contribute to the autocorrelation $P(Z) = A^*(1/Z)A(Z)$ through the factors

$$(1/Z^2)(Z - Z_1)(Z - Z_1^*)(Z - 1/Z_1)(Z - 1/Z_1^*)$$

The division by Z serves to cast the Z polynomial back into the symmetrical form of an autocorrelation. Thus, we can see from the above expression that couplets can always be paired into ones that have complex-conjugate reciprocal zeros, $(Z - Z_1)$ and $(Z - 1/Z_1^*)$. To identify factors belonging to $A(Z)$ and $A^*(1/Z)$ in the autocorrelation, we can at first assign any factor to $A(Z)$, then its conjugate reciprocal pair must be grouped with $A^*(1/Z)$. This means that in collecting factors for assignment to $A(Z)$, we are always free to select the factors with zeros outside of the unit circle, leaving the other factors to $A^*(1/Z)$. Thus, an odd-length, real, symmetric sequence can always be factored into a product $A^*(1/Z)A(Z)$, where $A(Z)$ is minimum phase, unless, of course, there are zeros on the unit circle.

However, it is important to observe that, in fact, most odd-length, real, symmetric sequences are not autocorrelations of some other real sequence. This, perhaps startling, development occurs because of additional requirements among autocorrelation coefficients related to the Fourier transform autocorrelation theorem. As an example, even though the sequences presented in Figs. 3.4 and 3.5 can be factored into the form $A^*(1/Z)A(Z)$, none of them are autocorrelations because their Fourier transform (the power spectrum) has negative values, as shown in the figures, an impossible result for true autocorrelations. On the other hand, the sequence $(1, 3, 1)$ does have an all-positive Fourier transform, but the most general 3-long autocorrelation must be of the form $(a, 1 + a^2, a)$; thus $(1, 3, 1)$ is not an autocorrelation. These examples do not contradict our factoring theorem of symmetric polynomials. It simply means that for these cases when all the factors are collected having zeros outside of the unit circle, the result is not real. So, these sequences are not autocorrelations of real sequences.

As our last example, take the sequence $(-4, 0, 17, 0, -4)$, which we have constructed by computing the autocorrelation of a 3-long sequence. We thus know that it is a bona fide autocorrelation. It has roots at $2, \frac{1}{2}, -2$, and $-\frac{1}{2}$. From these four roots, we can form products of two factors to make up a sequence that has the given autocorrelation. The limitation is that $(Z - Z_1)$ and $(Z - 1/Z_1^*)$ cannot be used together. This leaves four possibilities:

$$A = (2 + Z)(2 - Z) = 4 - Z^2$$
$$B = (1 + 2Z)(2 - Z) = 2 + 3Z - 2Z^2$$
$$C = -(1 - 2Z)(2 + Z) = -2 + 3Z + 2Z^2$$
$$D = (1 - 2Z)(1 + 2Z) = 1 - 4Z^2$$

All of these sequences have the same autocorrelation, but only A is minimum phase. Sequence D is maximum phase; B and C are mixed phase.

The value of this discussion of the root method is that it points up the difficulty in identifying an autocorrelation sequence. If the sequence is really an autocorrelation, the minimum phase sequence associated with it can be determined from the zeros of the autocorrelation. And it does show how to find the minimum phase sequence $A(Z)$ that has the same autocorrelation of any given sequence $B(Z)$; just move all zeros inside the unit circle to their conjugate reciprocal position outside. This switching the roots around is, in fact, just

$$A(Z) \to \frac{Z - 1/Z_0^*}{1 - Z/Z_0} B(Z)$$

our allpass filter of Eq. (4.6), which leaves the magnitude spectrum unchanged. Furthermore, for sequences up to perhaps 100 terms, a computer program can be readily written to perform the factorization. Basically, the problem is one of finding the zeros of a polynomial, which in itself is a fairly difficult mathematical problem, but computer routines are available in most standard math libraries to do this task. Given such a subroutine, spectral factorization is performed by finding the zeros of the autocorrelation and then using only the ones outside of the unit circle to form the minimum phase sequence.

The Spectrum of a Real Causal Function

Next, we wish to extend the discussion of spectral factorization to the frequency domain where we can anticipate the usual insight and FFT

The Spectrum of a Real Causal Function

computational speed. First, however, we need to develop an important relationship between causality and the real and imaginary parts of a function's spectrum. This connection is surprisingly simple and interesting. The relationship occurs because of the basic symmetry properties of the Fourier transform. Any function can be written in terms of its even and odd parts (for now, our discussion will assume continuous functions): $f(t) = f_e(t) + f_0(t)$, where

$$f_e(t) = [f(t) + f(-t)]/2$$
$$f_0(t) = [f(t) - f(-t)]/2$$

From the symmetry properties of the Fourier transform, the even part of a real $f(t)$ has an even real spectrum and the odd part of $f(t)$ has an imaginary odd spectrum. That is, if

$$\text{FT}\{f(t)\} = F(\omega)$$

then

$$\text{FT}\{f(t)\} = \text{Re}\, F(\omega) + i\, \text{Im}\, F(\omega)$$

and

$$\text{Re}\, F(\omega) = F_e(\omega) \qquad (10.1\text{a})$$
$$i\, \text{Im}\, F(\omega) = F_0(\omega) \qquad (10.1\text{b})$$

where we have written

$$F_e = \text{FT}\{f_e\}$$
$$F_0 = \text{FT}\{f_0\}$$

Now, as shown in the example of Fig. 10.1, if $f(t)$ is real and causal,

$$f_e = \text{sgn}(t) f_0(t)$$

and

$$f_0 = \text{sgn}(t) f_e(t)$$

Taking the Fourier transform of these equations and using the convolution theorem gives

$$F_e(\omega) = (2/i\omega) * F_0(\omega) \qquad (10.2\text{a})$$
$$F_0(\omega) = (2/i\omega) * F_e(\omega) \qquad (10.2\text{b})$$

recalling that the Fourier transform of $\text{sgn}(t)$ is $2/i\omega$. Using Eqs. (10.1), we

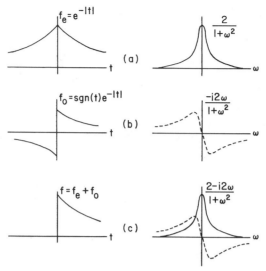

Figure 10.1 An example of the even and odd parts of a causal function. Its spectrum has real and imaginary parts that are Hilbert transform pairs.

can write these frequency domain convolutions as

$$\operatorname{Re} F(\omega) = \frac{1}{\pi} \int \frac{\operatorname{Im} F(\omega')}{\omega - \omega'} d\omega' \qquad (10.3a)$$

$$\operatorname{Im} F(\omega) = -\frac{1}{\pi} \int \frac{\operatorname{Re} F(\omega')}{\omega - \omega'} d\omega' \qquad (10.3b)$$

This is the desired result. Equations (10.3) show that if a real function is causal, the real and imaginary parts of its spectrum are Hilbert transforms of each other. Likewise, it is easy to trace these steps backward to show that if the real and imaginary parts of $F(\omega)$ are Hilbert transform pairs, $f(t)$ must necessarily be one-sided.

If we interchange the roles of time and frequency, the complex exponential discussed in Chapter 6 is an example of our general result: $\exp(\pm i\omega t)$ has Hilbert transform pairs as its real and imaginary parts in the time domain, producing a one-sided frequency spectrum.

In adopting the above discussion to discrete computations, we would naturally like to exploit the DFT, which immediately confronts us with two questions. The first is what shall we mean by causality when in fact DFT sequences cannot be zero for all $t < 0$ or all $\omega < 0$ since they are periodic? The answer is that we shall define causality in the DFT context to mean

The Spectrum of a Real Causal Function 241

that the sequence is zero in the first half of each period. The second question arises because of our band-limited/time-limited theorem of Chapter 6. Because of this theorem, we know that a sequence that is one-sided in one domain will have an infinite extent in the other domain—a causal time domain signal will necessarily contain frequencies to infinity, questioning how the Hilbert transform operations can be implemented via the DFT. The answer is that the ideal Hilbert transform, like the ideal lowpass filter and the ideal differentiator, just to name two other examples, is a valuable concept from continuous theory that can only be approximated in actual digital computations.

The design of digital Hilbert transformers uses the standard techniques that we have discussed in Chapter 8: windowing, frequency sampling, and equiripple approximations. The ideal discrete Hilbert transform operator can be obtained by interchanging the roles of time and frequency in the Fourier series expansion of the odd version of the square wave shown in Fig. 7.1. The time domain operator is infinitely long, of course, because of the discontinuity in the square wave; any practical digital computation will have to truncate the operator in some fashion, giving only approximate results. Figure 10.2 shows the ideal discrete Hilbert transform operator, its spectrum, and the actual magnitude spectrum resulting from truncation. The phase spectrum remains exact upon time domain truncation.

The alternative is to perform the Hilbert transform operation in the frequency domain accepting the errors introduced by the DFT spectrum computation. The actual operation—multiplication by sgn(ω)—gives a simple, and frequently adequate, approximation to the Hilbert transform.

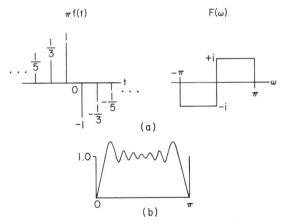

Figure 10.2 (a) The ideal discrete Hilbert transform operator and its spectrum. (b) The actual magnitude spectrum of a truncated finite-duration operator. Its phase spectrum is exact.

```
     COMPLEX A(N), B(N)
     CALL DFT(A, N)
     DO 10 I = 2, N/2
     B(I) = (0, -1) * A(I)
10   B(N + 2 - I) = (0, 1) * A(N + 2 - I)
     B(1) = B(N/2 + 1) = 0
     CALL IDFT(B, N)
```

Figure 10.3 A complex arithmetic computer program that computes the Hilbert transform of A via the DFT. The result is in array B.

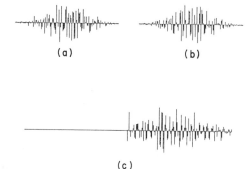

(a) (b)

(c)

Figure 10.4 An example of a causal DFT sequence. (a) Noise with a Gaussian envelope and (b) its Hilbert transform computed with the program of Fig. 10.3. The causal sequence (c) is formed by using (a) and (b) as its real and imaginary parts.

Figure 10.3 presents a short computer program that does exploit the DFT to do the Hilbert transform. The DFT type causality is demonstrated in Fig. 10.4 by using a discrete Hilbert transform pair for the real and imaginary parts of a DFT sequence. Its IDFT is zero (or nearly so) in the first half of the DFT period. The one-sided property is particularly dramatic for the example shown—a noise sequence with a Gaussian envelope.

This discussion on causality, the Hilbert transform, and their DFT behavior is not only interesting in itself, but it also has prepared us to investigate the frequency domain spectral factorization of Kolmogoroff.

Kolmogoroff Factorization

In the preceding section, we have found a relationship between the real and imaginary parts of a function's spectrum for a real causal time domain function. When describing complex spectra, we either use their real and

Kolmogoroff Factorization

and imaginary parts or their magnitude and phase; it is natural then to seek whatever similar relationships exist between magnitude and phase spectra. We start in the Z domain by writing an identity for the power spectrum

$$P(Z) = e^{\ln P(Z)} = e^{U(Z)} = e^{1/2 \ln P - i\phi} e^{1/2 \ln P + i\phi} \quad (10.4)$$

Thus we have immediately made the spectral factorization problem explicit by introducing the unknown phase spectrum ϕ; either exponential on the right gives the complete spectrum of a function whose power spectrum is P. Given this power spectrum, we would like to compute a phase spectrum from it in analogy with relating the real and imaginary parts of spectra of causal signals.

In Eq. (10.4), we have also introduced a sequence called U for $\ln P$. We divide this sequence into its past and future parts by writing

$$U(Z) = U^-(Z) + U^+(Z)$$

where

$$U^-(Z) = \sum_{-\infty}^{-1} u_n Z^n + \frac{u^0}{2} \quad (10.5a)$$

and

$$U^+(Z) = \sum_{1}^{\infty} u_n Z^n + \frac{u^0}{2} \quad (10.5b)$$

Next we equate each of these one-sided U's with the corresponding exponent or the right side of Eq. (10.4). That is, we choose to make an identification:

$$U^-(\omega) = \tfrac{1}{2} \ln |P(\omega)| - i\phi(\omega) \quad (10.6a)$$

and

$$U^+(\omega) = \tfrac{1}{2} \ln |P(\omega)| + i\phi(\omega) \quad (10.6b)$$

where we have evaluated Z on the unit circle. We are free to make such an identification because it merely defines $\phi(\omega)$, which was arbitrary at the start in Eq. (10.4). From our insight into Hilbert transforms in the preceding section, it is easy to see what this selection forces ϕ to be. Since U^+ is one-sided, its real and imaginary parts of its spectrum must be Hilbert transforms pairs, that is,

$$\phi = \mathcal{H}[\tfrac{1}{2} \ln |P(\omega)|] \quad (10.7a)$$

and

$$\tfrac{1}{2} \ln |P(\omega)| = -\mathcal{H}[\phi(\omega)] \quad (10.7b)$$

The spectral factorization that has been made by dividing U into U^- and U^+ has determined a unique phase spectrum that can be computed from a magnitude spectrum. This phase is the Hilbert transform of the logarithm of the magnitude spectrum.

So far, so good. But the factorization appears to have been rather arbitrary, motivated—if by anything—by simplicity. But let us see what has been wrought. First of all, the sequence determined by this phase function is

$$A(Z) = e^{U^+(Z)} = 1 + U^+ + \frac{(U^+)^2}{2!} + \frac{(U^+)^3}{3!} + \cdots \qquad (10.8)$$

and since U^+ is one-sided, containing only positive powers of Z, so is A; the time sequence determined by this factorization is causal.

The great significance of this determination of ϕ is that it is the one minimum phase function among all functions with the given magnitude spectrum. An easy way to see that $A(Z)$ must be minimum phase is to consider $A(Z) = \exp[U^+(Z)]$ inside the unit circle. If $A(Z)$ had zeros there, $U^+(Z_0)$ would have to diverge to minus infinity at those proposed zeros. But, this could not happen because $U^+(Z)$ was constituted from the causal convergent part (for $|Z|<1$) of $U(Z)$ by our selection in Eqs. (10.5). Thus, $A(Z)$ must be minimum phase. This is further verified by observing that U^+ evaluated on the unit circle is

$$U^+(\omega) = \frac{u_0}{2} + \sum_1^\infty u_n (e^{-i\omega})^n$$

and its imaginary part

$$\operatorname{Im} U^+(\omega) = u_1 \sin \omega + u_2 \sin(2\omega) + u_3 \sin(3\omega) + \cdots$$

is the phase spectrum of the sequence A. Clearly, this is a minimum phase function because it is periodic in ω; that is, it is not augmented by 2π as ω goes from 0 to 2π.

If the above discussion, including the conclusion on the minimum phase behavior of ϕ, appears obtuse, a practical example might be satisfying. We remember, however, that our discussion addressed a discrete sequence $U^+(Z)$ and its continuous Fourier spectrum; any practical digital Hilbert transform will only yield approximate results. The simplest approach is to use frequency sampling (as we used in Fig. 10.3) for the Hilbert transform, as shown in the simple program of Fig. 10.5. In this example, a 16-point DFT finds the minimum phase sequence corresponding to $(2, 3, -2)$ to be the correct result of $(4, 0, -1)$ within about one part in 10^4. Zero padding the original sequence and using a corresponding denser frequency sampling for the Hilbert transform produce an increasingly more accurate result. A 32-point DFT gives the correct answer to one part in 10^{10}.

Least-Squares Zero-Delay Factorization

```
        COMPLEX A(16) MAG(16)
        M = 16
        DATA A/2, 3, -2/
        CALL DFT(A, M)
        DO 10 I = 1, N
     10 MAG(I) = .5 * CLOG(A(I) * CONJG(A(I)))
        CALL IDFT(MAG, M)
        DO 20 I = 2, N/2
        A(I) = (0, -1) * MAG(I)
     20 A(N + 2 - I) = (0, 1) * MAG(N + 2 - I)
        A(1) = A(N/2 + 1) = (0, 0)
        CALL DFT(A, M)
        DO 30 I = 1, N
     30 A(I) = MAG(I) * CEXP((0, 1) * A(I))
        CALL IDFT(A, M)
```
Figure 10.5 A complex arithmetic computer program to find the minimum phase sequence corresponding to (2, 3, −2) using Kolmogoroff spectral factorization. The result is (4, 0, −1), accurate to one part in 10^4.

Of course, once the complete spectrum of the minimum phase sequence has been determined, it is child's play to compute the sequence's stable causal inverse; negate the phase spectrum and invert the magnitude spectrum in line 30 of Fig. 10.5.

The Kolmogoroff method is useful for thinking about the spectral factorization of continuous functions, for providing a framework for using approximations exploiting the DFT, and for giving a practical and simple method of computing an approximate inverse to a minimum phase sequence whose autocorrelation is known.

Least-Squares Zero-Delay Factorization

The concept of the Wiener least-squares shaping filter was developed early in Chapter 9 and its delay properties were discussed later in that chapter. There we saw that the convolution $A * X = D$ could be inverted to give

$$X = (A^T A)^{-1} \cdot A^T D$$

If the desired output D is selected to be the unit spike of zero delay, then the cross-correlation vector $A^T D$ contains the first element as the only nonzero one. The matrix $A^T A$ is generated from the autocorrelation of A. In the section "Filter Delay Properties" in Chapter 9, we demonstrated, but did not prove, that the zero-delay solution

$$X = (\boldsymbol{\phi})^{-1} \begin{pmatrix} 1 \\ 0 \\ 0 \\ \vdots \\ 0 \end{pmatrix}$$

is the best least-squares inverse to the minimum phase sequence whose autocorrelation is ϕ. The phase properties of x itself are not obvious, but we have shown in Chapter 9 that because of the Levinson recursion development, x itself must be a minimum phase sequence. Because X is a minimum phase approximate inverse to the minimum phase sequence A belonging to ϕ, we can apply zero-delay inverse filtering to spectral factorization. In this context, the spectral factorization problem is: given a $2N + 1$ long autocorrelation ϕ, find its minimum phase sequence A. The solution is: first find an M-long approximate minimum phase inverse x to the desired A. Then invert X to get an N-long A, using the zero-delay inverse again. The larger the intermediate length M is, the better the result (see Problem 10.14).

Both the least-squares zero-delay method and the Kolmogoroff method of spectral factorization give approximate results. The first method requires a matrix inversion in the time domain that can be computationally very efficient using the Levinson alogrithm; the second requires several DFT calculations that can be quite efficient using the FFT alogrithm. As usual, regardless of which domain is selected for computation, the fundamental problems of operator length and bandwidth cannot be eliminated; only the way in which they are dealt with is different.

But, in the factorization problem, a new, interesting, and practical question arises. One cannot always argue that the sought-after sequence (or its inverse) must be minimum phase. The question then is: are there methods that can factor a given sequence into appropriate components? In the next section, we will see how this can be done for the case of an ARMA system, with a noninvertible moving average and an invertible autoregressive component.

Iterative Least-Squares Factorization

We have stressed at the very beginning (beginning in Chapter 2) that the most general LSI operator that can be implemented in an actual computing scheme is a rational fraction, $B(Z)/A(Z)$. Unfortunately, complicated

Iterative Least-Squares Factorization

processes are often not of this form. Examples are the behavior of the national economy, speech, and seismic waves. The list is extensive, but all have one point in common. In order to emulate these processes in a recursively computable scheme, they must be approximated by the rational fraction model. The idea, so common in complex phenomenon, is to describe the process by a small number of parameters, in this case by the poles and zeros of a rational fraction. We shall see that this parsimonious parametric signal modeling has diverse applications, unifying topics in spectral estimation, predictive deconvolution, and IIR digital filter design. The one requirement imposed on the rational fraction model $B(Z)/A(Z)$ is that $A(Z)$ must be minimum phase.

Thus, we are led to consider the general ARMA expression $B(Z)/A(Z)$, where $A(Z)$ is minimum phase but $B(Z)$ has no restrictions. The problem to be solved is: given $S(Z) = B(Z)/A(Z)$, find both B and A. This is a type of spectral factorization problem reminding us of the previous section where we found the minimum phase inverse to an operator given only its autocorrelation. In this case, $A^{-1}(Z)$ is convolved with yet another sequence $B(Z)$. But, perhaps quite surprisingly, the minimum-phase condition on A permits a solution for both factors, $A(Z)$ and $B(Z)$.

Before proceeding with the solution, perhaps a more detailed example of the practical relevance of this problem is in order. Many fields such as radar, sonar, and seismology use echoes of minimum phase sources. Repeated echoes of the source signal between the same reflectors can produce annoying ghosts or reverberations that may severely interfere with the system's performance. These reverberations are, in principle, an infinitely long sequence of delays

$$1 + a_1 Z + a_2 Z^2 + \cdots$$

Since this reverberation train is minimum phase (Problem 10.9), it has a stable causal inverse $A(Z)$. Furthermore, in some cases, such as in marine seismic reverberations, the above series sums to a single-pole polynomial, $1/(1 + aZ)$ (see Problem 10.18). This reverberated signal is reflected from a series of deep interfaces with reflection coefficients B, yielding an upgoing signal of $B(Z)/(1 + aZ)$. On the return trip, reverberation again occurs in the water layer, producing a total upgoing signal

$$S(Z) = \frac{B(Z)}{(1 + aZ)^2}$$

at the surface receivers. The denominator is necessarily minimum phase because $|a| < 1$. But there is no physical reason to expect that the deep reflectors producing $B(Z)$ are a minimum phase sequence.

This marine seismic case is but one example of rational fraction modeling; another, which we shall pursue after this section, is the problem of IIR filter design.

The proposed technique hinges on the observation of the previous section that the solution to the least-squares zero-delay inverse problem is necessarily a minimum phase sequence. The problem is given:

$$S(Z) = B(Z)/A(Z)$$

find B and A. The procedure is iterative. We start with a guess for $B(Z)$ (which if nothing else is available, could be the unit zero-delay spike). We solve the least-squares problem $S*A = B$ for A, given an estimate of B. This solution for A is not minimum phase, but the minimum phase version of A can be formed by calculating the least-squares inverse of the least-squares inverse of A. That is to say, given an estimate of A, we find the zero-delay least-squares inverse A^{-1}. This is necessarily minimum phase. Then, we compute the zero-delay least-squares inverse of A^{-1}. This is the minimum phase component of A. Finally, a new estimate of B is found from $B = A*S$. The procedure is perhaps better illustrated graphically in Fig. 10.6. The Wiener least-squares computer routine of Fig. 9.6 (or the Levinson recursion, Fig. 9.15) is called three times for each iteration, using appropriate inputs. We hope that at the end of each iteration, the estimates of A and B approach their true values. If, after a reasonable number of iterations, the estimates for A and B stabilize, the process has been successful. If they do not stabilize, the assumed ARMA model does not fit the given sequence $S(Z)$ very well.

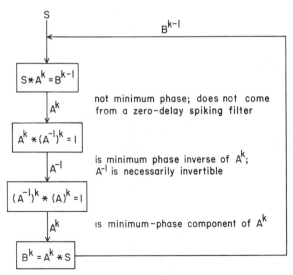

Figure 10.6 The iterative least-squares procedure factors S into its MA component B and its minimum phase AR component A. The initial trial for B can be the unit impulse.

Iterative Least-Squares Factorization   249

For a tutorial example of this iterative least-squares procedure, we take an ARMA sequence from digital filtering, a 6th-order lowpass Butterworth recursive transfer function,

$$H(Z) = \frac{0.0007378(1+Z)^6}{1 - 3.183Z + 4.6211Z^2 - 3.7785Z^3 + 1.18322Z^4 - 0.4800Z^5 + 0.0544Z^6}$$

This function does fit our data model. The numerator is noninvertible, with its multiple zeros on the unit circle, and the denominator is minimum phase, as is required for recursive filtering. To carry out the example, this $H(Z)$ is expanded via the DFT to form a long sequence representing the ARMA process. The least-squares factorization procedure acting on this long ARMA data stream will recover A and B. The procedure is outlined in Fig. 10.7. In this example, A and B can be recovered accurate to five digits, in 50 iterations using 800 terms of S.

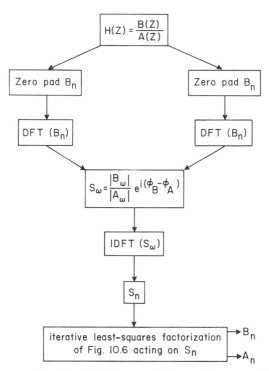

Figure 10.7 The generation of an ARMA data model for tutorial purposes. The long sequence S_n is generated by a DFT expansion of $H(Z)$. The separate MA and AR components of S_n are then recovered by the least-squares iterative factorization procedure.

Spectral factorization and ARMA data models have brought us to quite advanced concepts in digital signal processing. We will develop them more in the next chapter on modern methods of power spectral density estimation. Next, however, we will put these ideas to practical use in the problem of IIR digital filter design.

Applications to IIR Filter Design

In Chapter 8, we introduced IIR filter design based on analog filters. In that approach, the filter's frequency response is controlled by mathematical functions that bear a relationship to the desired frequency response but are not freely specified. Additionally, massive algebra is required dealing with the zeros of Butterworth functions, Chebyshev polynominals, or Jacobian elliptic functions. The results are not entirely predictable because some transformation (such as the bilinear transform) must be used to go from a function of frequency to a rational polynominal in Z.

Our approach here is stimulated by the ideas of the preceding sections on spectral factorization. We want to exploit computing algorithms, such as the FFT and least-squares inverses, letting the computer do the work. Furthermore, we would like to start with frequency sampling because this allows an arbitrary definition of the filter's desired response for meeting requirements of any given application. We seek an ARMA system function $H(Z)$ with the AR component providing the sharp features of the spectral response and the MA component providing the broader aspects.

By demanding that the computer do our work, the central feature of the scheme that emerges is brute force; we simply generate a long MA sequence that closely approximates the desired IIR filter. That is, given any desired $H(\omega)$, we frequency sample it on a dense grid, take its IDFT, and use the resulting long sequence as the input to the iterative least-squares procedure of Fig. 10.6. The number of poles and zeros are specified *a priori*; then, the iterative procedure determines their best location according to the least-squares criterion. The frequency sampling grid, and hence the length of the DFT used, should be sufficiently adequate that the resulting approximation to $H(Z)$ is limited only by the least-squares fit.

In the least-squares process, $B(Z)$ is found that best satisfies $S * A = B$. If we contemplate a causal form for B, then S should also be causal because A must be. Therefore, as a matter of convenience, we will develop a causal form for S. The simplest way to do this is to find the minimum phase sequence that corresponds to the frequency sampled determination of $|H(Z)|^2$. This can be done by any of the spectral factorization methods just

Applications to IIR Filter Design

discussed. With S minimum phase, $A(Z)$ operating on $S(Z)$ attempts to reproduce $B(Z)$ with zero delay; then, $A(Z)$ operating on the remainder of S attempts to produce zeros.

Now, an example is in order to demonstrate these ideas. We choose a 6th-order Butterworth lowpass function

$$|H(\omega)|^2 = \frac{1}{1 + (\omega/\omega_0)^{12}}$$

with $\omega_0 = 0.2\pi$. Next, we frequency sample this power spectrum with 128 points and compute the 256-point IDFT to get an estimation of the autocorrelation coefficients out to 128 lags. Then we use spectral factorization to find the first 128 points of the corresponding minimum phase sequence. This 128-point sequence fits the desired power spectrum very well, but we wish to develop a pole–zero model of it; so, we feed it to the iterative least-squares procedure. The frequency responses of the resulting rational fractions are shown in Fig. 10.8. First, we try a 6-pole, 6-zero model and find a very good Butterworth response after about 10 iterations. Further iterations move the poles and zeros about, but do not noticeably change the frequency response on the scale of Fig. 10.8. The iteration converges slowly after about 10 steps and stabilizes to 5 or 6 digits after hundreds of iterations. The unnormalized transfer function corresponding to Fig. 10.8(a) is

$$H(Z) = \frac{2 + 1.0019Z - 0.8402Z^2 - 0.4243Z^3 + 0.0165Z^4 + 0.0916Z^5 + 0.01729Z^6}{1 - 4.1790Z + 7.6580Z^2 - 7.8323Z^3 + 4.7005Z^4 - 1.5661Z^5 + 0.2258Z^6}$$

When a lower order pole–zero model is tried, the procedure again produces its best fit, but now there are an insufficient number of parameters to do the job, so the iteration converges rapidly; the location of the poles and zeros are now much more critical. The result of a 3-pole, 3-zero fit is shown in Fig. 10.8(b). The resulting transfer function

$$H(Z) = \frac{2 + 4.6433Z + 5.0091Z^2 + 3.3534Z^3}{1 - 2.3584Z + 2.0619Z^2 - 0.6546Z^3}$$

stabilizes to 12 digits after only 25 iterations. Interestingly, the lower order fit has maintained the sharpness of the 6th-order transition zone at the expense of ripples and a lower rejection at Nyquist, producing a very reasonable and useful low-order filter. In the attempt to make this fit to a higher order process, the procedure has exploited its flexibility by allowing $B(Z)$ to be nonminimum phase.

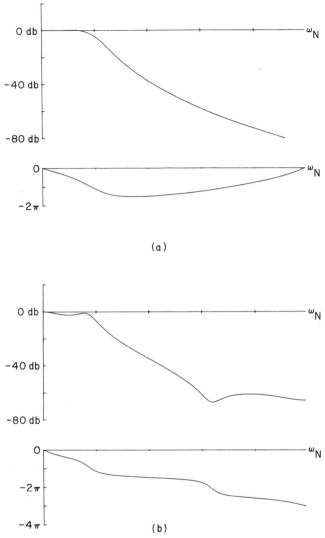

Figure 10.8 The result of designing an IIR filter by iterative least squares. (a) The frequency response of a 6-pole, 6-zero IIR digital filter formed by an iterative least-squares fit to a minimum phase 6th-order Butterworth function. (b) The frequency response of a 3-pole, 3-zero fit to the same 6th-order Butterworth filter. Note how the lower order digital filter develops ripples, similar to Chebyshev filters, in attempting to fit the higher order Butterworth function. Rejection at Nyquist is also reduced compared to the 6-pole, 6-zero filter, but the falloff in the transition zone is maintained.

No digital filter design method is superior to all others or is suitable for all applications. The same is true for this iterative least-squares method. It does have the advantage of easily accommodating any specified magnitude spectrum and of avoiding algebraic manipulations, letting the computer do all the work. But, it also has the disadvantage of distributing errors demanded by the least-squares fit, rather than according to the preference of the designer. Trial and error experimenting with various numbers of poles and zeros is one control available over error distribution; another is the initial selection of $|H(\omega)|^2$.

We started out discussing spectral factorization almost as a mathematical curiosity, using the root method to gain understanding of the problem. Then, by looking at the conditions imposed on the spectrum of a real causal function, we found that the requirement is simply that the spectrum's real and imaginary parts are Hilbert transform pairs. Next, we logically asked how this condition would be represented in terms of the magnitude and phase spectra, and found the log magnitude spectrum and the phase spectrum to be Hilbert transform pairs for a real causal signal. Surprisingly, this condition also necessarily produces a minimum phase time sequence, giving us the Kolmogoroff spectral factorization scheme. Alternately, least-squares zero-delay inverses gave us another method of spectral factorization, for just taking the inverse of the inverse (via zero-delay least squares) brings us back to the minimum phase version of the original sequence. This concept then led to the iterative least-squares procedure, which models a sequence as a rational fraction. To be recursively computable, the rational fraction must have minimum phase poles. Because many complex sequences that arise in the natural world are not rational fractions, approximations must be made in order to represent the processes of interest. Representing these high-order processes by low-order pole–zero models is an attempt to describe complicated processes by a parsimonious number of parameters; one goal of science is to reduce the complex to the simple.

Problems

10.1 Find all the sequences that have the same autocorrelations as $(2, 5, 2)$ and identify the phase of each one.

10.2 If a sequence is n long, how many other sequences have the same autocorrelation?

10.3 The energy distribution in a minimum phase sequence has a special behavior. Compare the total cumulative energy in two sequences that are identical except that they have one zero switched in the conjugate reciprocal fashion, producing a zero outside the unit circle for $A(Z)$ and inside for $B(Z)$. That is, write

$$A(Z) = (Z_1 - Z)P(Z)$$
$$B(Z) = (1/Z_1^* - Z)P(Z)$$

where $|Z_1| > 1$, and $P(Z)$ is any causal sequence. By writing out $\sum_{t=0}^{t} (A_t^2 - B_t^2)$ for each time, show that the accumulated energy of A is greater than B. Extend this argument to show that the minimum phase sequence has the greatest accumulated energy at each time of all sequences with the same magnitude spectrum.

10.4 The previous problem shows that the minimum phase sequence has the sharpest rise time of all causal sequences with the same magnitude spectrum. If we wanted to sharpen the attack of a minimum phase sequence even further, how should we modify the magnitude spectrum?

10.5 Explain why a function and its Hilbert transform have the same autocorrelation.

10.6 If the roles of time and frequency are interchanged, what becomes the significance of a minimum phase function?

10.7 Prove that the convolution of two minimum phase sequences is again a minimum phase sequence.

10.8 Show that the sum of two minimum phase sequences does not necessarily result in a minimum phase sequence.

10.9 Reverberations occur in many fields amenable to digital signal processing, such as radar, speech, sonar, and seismology. Show that a minimum phase sequence plus its echo is also minimum phase.

10.10 Show how exponential damping can always be used to stabilize the inverse of any causal system. That is, for any causal B_n, find γ such that

$$A_n = \gamma^n B_n$$

are the terms of a minimum phase sequence.

10.11 Is the function of Fig. 10.1(c) a minimum phase function?

10.12 The causal continuous function $y(x) = xe^{-x}$ (for $x \geq 0$) has the Fourier transform $(1 + i\omega)^{-2}$. Is $y(x)$ a minimum phase function?

10.13 For simplicity, the Kolmogoroff factorization computer code shown in Fig. 10.5 uses complex arithmetic. Write a similar computer program that uses only real arithmetic and uses a DFT subroutine that is limited to taking a real sequence into the complex frequency domain, outputting only $N/2 + 1$ points for the real and imaginary parts.

Problems 255

10.14 Adopt the computer code of Fig. 9.6 to implement least-squares spectral factorization. Verify the discussion of the section "Least-Squares Zero-Delay Factorization" by finding the minimum phase sequences that have the same autocorrelation as $(2, 3, -2)$ and $(2, 5, 2)$. Compare the results to the correct values and to results produced by the procedure of Problem 10.13.

10.15 The frequency response of the simple analog RC integrator is $1/[1 + i(\omega/\omega_0)]$. Compute the Hilbert transform of the real part of this spectrum, showing that the real part and the imaginary parts form Hilbert transform pairs. The system is thus causal. Does the system have a stable causal inverse? (Mathematical alert: the Hilbert transform required here is best done by contour integration in the complex plane.)

10.16 What do Eqs. (10.7) say about (1) the phase of a minimum phase allpass filter and (2) the magnitude spectrum of a minimum phase filter with symmetrical time domain coefficients?

10.17 Generally, we have considered the problem of determining the minimum phase function given a system's magnitude spectrum. But, as Eqs. 10.7 show, we can clearly reverse the procedure just as easily. Give a practical example of a situation in which a system's minimum phase function is known and we wish to determine the system's magnitude spectrum.

10.18 Short period reflections occur in many situations, such as in thin film optics, sonar, and marine seismology. To have a definite example in mind, consider marine seismic reverberations occurring between the air–water interface and the ocean–bottom interface, as shown schematically below. Because of the strong acoustic impedance contrast between air and water, the top reflection coefficient is nearly -1; let the bottom reflection coefficient be a. Show that the downgoing pulse is $X(Z)/(1 + aZ)$.

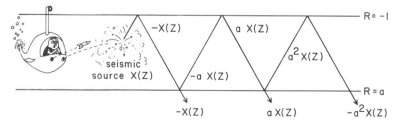

10.19 Write a computer program that implements the iterative least-squares procedure of Fig. 10.6.

10.20 Use the computer program of Problem 10.19 to factor the following sequence into its moving average and its invertible autoregressive parts:

(.125, .250, .40625, .0625, .1015625, .0156250, .0253906, .003906, .0063477, .0009766, .0015869, .0002441, .0003967, .0000610, .0000992, .0000153, .0000248, .0000038, .0000062,

.0000010, .0000015, .0000002, .0000004) $\simeq (1, 2, 3) * \dfrac{1}{(4, 0, -1)}$

10.21 In the iterative least-squares procedure of Fig. 10.6, an order must be selected for the intermediate inverse calculations; that is, the length of A^{-1} must be chosen. What determines how this order should be selected?

10.22 Write a program that implements the iterative least-squares algorithm using Kolmogoroff spectral factorization. How does this compare to the procedure of Fig. 10.6?

10.23 What is an example of a data model that does not fit the assumptions of the section "Iterative Least-Squares Factorization"? Create such a data sequence and attempt to use the computer program of Problem 10.19 to factor the signal. What happens?

10.24 Using least-squares inverse spectral factorization, write a computer program to generate the long, minimum phase, time domain operator corresponding to a specified power spectrum.

10.25 Repeat Problem 10.24 using Kolmogoroff spectral factorization.

10.26 Reproduce the results of Fig. 10.8 using the programs developed in Problems 10.19 and 10.24.

10.27 Using the ARMA filter design program of Problem 10.26, select a 6-pole model and generate a suite of results using Butterworth orders from 3 to 9. Comment on your results.

10.28 Experiment with the iterative least-squares ARMA filter design of Problem 10.26 by modifying the specified power spectrum. For example, try to sharpen the transition zone of the 6th-order Butterworth filter of Fig. 10.8 by placing a zero in the specified power spectrum at 0.3π.

10.29 Use the iterative least-squares ARMA filter design method to develop a passband filter based on the Butterworth-type function

$$|H(\omega)|^2 = \dfrac{1}{1 + [(\omega - \omega_c)/\omega_0]^{2N}}$$

where ω_c is the center of the passband and ω_0 relates to the cutoff frequencies. Comment on the digital filter order required to get a reasonable approximation to $|H(\omega)|^2$.

10.30 In the iterative least-squares IIR digital filter design method described in the text, the specified power spectrum was used to estimate a long, minimum phase operator. It was stated that the minimum phase condition on this operator was only a matter of convenience, not necessity. Discuss this point. Consider sampling $|H(\omega)|^2$ so as to produce a zero-phase time domain sequence. How would this then affect the subsequent least-squares modeling?

10.31 The design of minimum phase FIR frequency selective filters is considered an advanced topic in digital signal processing. Use least-squares spectral factorization to develop minimum phase FIR filters from FIR filters discussed in previous chapters. Do not hesitate to move their zeros slightly off the unit circle if you think it desirable. Be sure to check both the magnitude and phase spectra of your results.

11

Power Spectral Estimation

Power spectral estimation, with its wide variety of applications, can be easily understood in terms of the phase-shift theorem from Fourier transforms. To see why this is so and to give insight to practical applications, we will approach the topic by considering signal averaging of a deterministic signal.

Imagine that we have a controlled laboratory experiment. The experiment works well, is repeatable, is reliable, but for some reason or another, it produces a poor signal-to-noise (s/n) ratio. Perhaps technically the s/n ratio could be easily increased, but improving that aspect of the apparatus is precluded by economics. We are, however, fortunate to have a modern digital data acquisition and storage system at our disposal. In this situation, we can easily improve the s/n ratio using repeated sums of the experiment's results.

Provided that the experiment has a synchronizing signal available, we can perform *coherent signal averaging*. This is done by simply summing repeated results from the experiment. Each result is a noisy function of time that contains a deterministic signal, synchronized to the experiment's trigger signal, but imbedded in noise. Because the sums are synchronized to the signal while the noise is random, the signal strength builds faster than the noise. Using superposition, we can reckon the results of this summation on the signal and on the noise independently: after N coherent sums, the signal will be N times stronger than the single original result. However, the noise will only grow as $\sqrt{N}$ because at each point in time we are summing N signed random numbers with zero means. Thus, the signal-to-noise ratio becomes

$$\text{s/n} = N/\sqrt{N} = \sqrt{N}$$

11 / Power Spectral Estimation

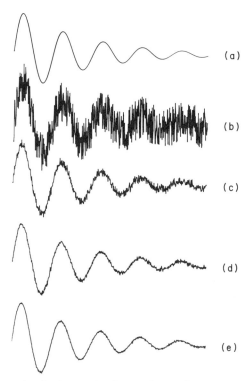

Figure 11.1 An example of coherent signal averaging used to extract a deterministic signal from random white noise: (a) the signal, (b) the signal contaminated by noise. Toward the tail end of the signal, the s/n ratio is quite low, less than one. (c), (d), and (e) Successively increased numbers of averages of data samples containing the same signal but different noise. Coherent averaging builds the signal as N but the noise as only $\sqrt{N}$, giving s/n improvement as $\sqrt{N}$.

As shown in Fig. 11.1, where each summation has been divided by the current value of N to prevent the result from continually getting larger, this signal averaging can be quite effective. Even though this signal averaging is effective, it is not nearly as effective as having a stronger signal to begin with—it takes four sums to reach the same s/n as obtained by only doubling the original signal strength.

The key to this coherent signal averaging is the synchronization of the desired signal, using some external timing trigger, or relationship. But what if this timing coherency is lost or is inherently unavailable? Clearly, without the timing reference, a signal like that of Fig. 11.1(a) will sum to zero under repeated summations containing random shifts of the time origin. But Fourier transforms save the day. There is one obvious function that is independent of these random shifts in the origin of time; it is the magnitude spectrum. Our old friend, the phase-shift theorem, tells us that under a translation in time, the signal's phase spectrum is rotated, but its magnitude spectrum is invariant.

Thus, if no timing reference is available for coherent signal averaging, all is not lost; some information still can be extracted in the presence of noise. For each run of the experiment (or acquisition of a data window), we compute the magnitude spectrum. Then, we average these magnitude spectra from all the runs. Since the magnitude spectrum is independent of time shifts, an estimation of the magnitude spectrum will improve with the number of averages. This is sometimes called *incoherent signal averaging*. The price paid is the loss of phase information; without it the original signal can never be recovered. Nonetheless, the magnitude spectrum provides very useful information in many applications.

As just one example, envision a U.S. submarine engaged in underwater reconnaissance. Its sonar picks up very weak acoustic signals that build and fade from a distant ship. It is known that U.S. submarines use 60-Hz power while certain foreign ships use 50-Hz primary power. There is no method for synchronizing averages from data windows made available when the distant ship comes within receiving range. Therefore, the incoherent method of averaging is used, and it yields an estimation of the magnitude spectrum of the signal emitted from the ship in question—information that is sufficient to determine whether the ship's primary power is 60 Hz or 50 Hz. An example of this sort is shown in Fig. 11.2, where a dramatic improvement has been made in the magnitude spectrum using this method of averaging.

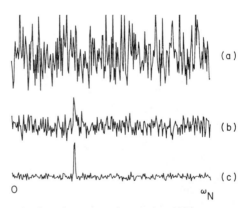

Figure 11.2 An example of incoherent signal averaging. (a) The power spectrum of a sinusoid imbedded in white noise with a s/n of −20 db. If no synchronization with this sinusoid is available, averages of the power spectrum still give useful information about the signal. In (b) and (c) averages of 20- and 200-fold of the DFT power spectra are shown. The price paid for incoherent averaging is the lost information on the signal's phase spectrum.

In the above application, two considerations emerge: one is adequate frequency resolution for discriminating between two closely spaced frequencies, and the other is statistical resolution for separating the ac primary power spectrum from that of the background noise. We need to look next at these considerations in more detail.

Signal-Like and Noise-Like Processes— The PSD Problem

The two examples of the discussion in the preceding section involved signals of very differing characteristics—deterministic signals and noise signals. As discussed in Chapters 1 and 9, deterministic signals contain a great deal of predictability. A sine wave, for example, can be predicted from just two terms as indicated in Eq. (1.1). Sometimes such signals are called *signal-like*. Random signals, on the other hand, are unpredictable. This does not mean that they lack information. Random signals, sometimes called noise-like signals, indeed contain a great deal of useful information, information that is tied up in their statistical properties. Examples of statistics that contain information are the mean, the variance, and higher order moments. One way of thinking about random signals is to imagine that there exists some source or process that is generating the numbers. Although this process produces random numbers, the way that it does it is time invariant in such a way that all statistical properties of the random signal are constant in time. Such a process is called a *stationary* process; its output is called a *stationary time series*.

This concept of stationarity allows us to take a straightforward approach to estimating the statistics of a time series given by J. W. Gibbs, the same mathematician that we met in connection with truncating Fourier series. In Gibbsian statistics, averages of random variables are computed over ensembles. To appreciate what ensembles are, imagine that we have a random number generating machine. We can turn it on and off, obtaining a sample run of data each time. This record is called a *realization* of the random process. The collection of all possible realizations is the ensemble. (In comparison with a random process in which each realization is different, a completely deterministic process will produce identical realizations.) Figure 11.3 plots realizations $x(t)$ from a certain random process. A computation of an average is done over the ensemble of all possible realizations. For example, the mean value of x at time t_1 would be

$$m_x = \overline{x(t_1)} = \lim_{N \to \infty} \frac{1}{N} \sum_{k=1}^{N} x_k(t_1) \qquad (11.1)$$

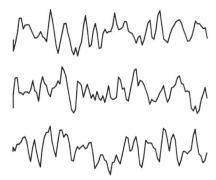

Figure 11.3 Three samples, or realizations, of an innovative process. Each plot is 100 data points produced by passing white noise through the 9-point lowpass filter of Fig. 3.5. According to the ergodic theorem, the statistical properties of each realization will converge to the same values as their length grows, reflecting the stationary nature of their source.

Because this computation requires an infinite number of realizations, we are led to postulating that a plausible estimate can be made from a finite number of realizations

$$\hat{m}_x(t_1) = \frac{1}{N} \sum_{k=1}^{N} x_k(t_1) \qquad (11.2)$$

Hopefully, in the limit of large N, this estimate approaches the true value of $m_x(t_1)$ in a well-behaved fashion. The branch of statistics that addresses this procedure is called *estimation theory*. Our excursions into the formal aspects of that branch of mathematics will be limited as much as possible.

Let us now apply the above discussion to our topic of interest, the power spectrum. We have available one realization of a random process yielding the Z transform of n random numbers:

$$X(Z) = x_0 + x_1 Z + x_2 Z^2 + \cdots x_{n-1} Z^{n-1}$$

We compute a *power spectrum density (PSD) estimate* by first forming the *autocovariance*

$$\hat{P}(Z) = (1/n) X^*(1/Z) X(Z) \qquad (11.3)$$

This autocovariance function differs from the autocorrelation function by the division of n. This divisor of n is necessary to keep the result from increasing linearly with the somewhat arbitrary record size n.

There are two fundamental problems with using Eq. (11.3) as an estimate for the autocovariance. First, as the finite record is shifted off of zero lag in the autocovariance computation, a portion of the record overlaps a region of nonavailable data that is necessarily treated falsely as though they were zero.

Signal-Like and Noise-Like Processes—The PSD Problem 263

The second problem concerns the behavior of the power spectrum computed from the autocovariance estimate of Eq. (11.3). It may seem plausible that a reasonable estimate could be made by computing the Fourier transform of the autocovariance estimate, expecting the estimate to improve with larger and larger n. Somewhat surprisingly, this is not the case. To see this, consider a real stationary time series with zero mean and a known constant variance σ^2. [The variance is the squared deviation from the mean $(1/n)\Sigma(m_x - x_i)^2$.] Then, the Fourier transform of Eq. (11.3) gives for our power spectrum density estimate

$$\hat{P}(\omega) = \sum_{k=-(n-1)}^{n-1} p_k e^{-i\omega k}$$

or

$$\hat{P}(\omega) = p_0 + 2 \sum_{k=1}^{n-1} \hat{p}_k \cos(k\omega) \qquad (11.4)$$

where each individual sinusoidal component contains the term

$$\hat{p}_k = \frac{1}{n} \sum_{t=0}^{n-k-1} x_t^* x_{t+k} \qquad (11.5)$$

Now let us look at the statistical properties of the p's. First look at the zero-lag autocovariance term

$$\hat{p}_0 = \frac{1}{n} \sum_{t=0}^{n-1} |x_t|^2 = \hat{\sigma}^2$$

which is an estimate of the variance σ^2. Since this sum is over n signed random numbers of variance σ^2, we expect the series to sum to σ^2 plus or minus the random fluctuations that will decrease with sample size as $1/\sqrt{n}$. That is, we expect

$$\hat{p}_0 = \sigma^2 \pm \sigma^2/\sqrt{n}$$

Likewise, for nonzero lags, because the random numbers are assumed to be completely independent of one another, we expect that $\hat{p}_1$ would be zero plus or minus its random fluctuation:

$$\hat{p}_1 = 0 \pm \frac{n-1}{n} \frac{\sigma^2}{\sqrt{n}}$$

and thus in general,

$$\hat{p}_k = 0 \pm \frac{n-k}{n} \frac{\sigma^2}{\sqrt{n}}$$

where we have decreased the fluctuation factor in proportion to the number of terms $(n - k)/n$. Now at any given frequency, Eq. (11.4) tells us that the power spectrum estimate $P(\omega)$ is a sum of n signed random numbers. This sum itself is a random variable, having a statistical fluctuation equal to the square root of the number of terms $\sqrt{n-1}$. Thus, the expected fluctuations of the power spectrum estimate using Eq. (11.4) is

$$\frac{\Delta P}{P} = \sqrt{n-1}\left(\frac{n-k}{n}\right)\frac{\sigma^2}{\sqrt{n}}$$

For large n, this fluctuation in the power spectrum does not approach zero, but rather it approaches a constant value σ^2. Fortunately, our estimate does not diverge with large n, but it does have the rather startling consequence that increasing the record length of our time series sample does not improve the estimate of the power spectrum. It is true that as the record length increases, the frequency resolution will increase, but the statistical resolution will still be of the order of the power spectrum itself. The estimate we have used in Eq. (11.4) is called the *periodogram*. In statistical language, we have found that the periodogram is not a consistent estimator of the power spectrum, but rather, independent of record length, it has a fluctuation from frequency to frequency and from record to record that is as large as the computed spectrum itself. This property of the periodogram is shown in Fig. 11.4, where successively longer records of a random digital

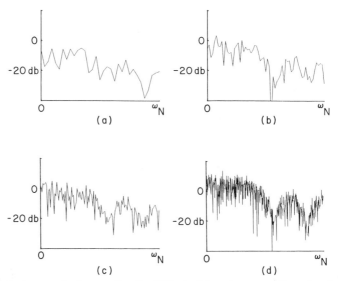

Figure 11.4 The result of computing the DFT of white noise passed through a 9-point MA filter. For (a) through (d), the DFT was 64, 128, 256, and 1024 points, respectively. The power output is plotted from zero frequency to Nyquist. As the length of the DFT is increased, frequency resolution increases, but statistical resolution does not.

signal are used to estimate the power spectrum, showing that there is improved frequency resolution without any improvement in statistical resolution. Basically, the reason for this result is that although the fluctuations in the autocovariance go as $1/\sqrt{n}$, the summation of the Fourier transform gives another factor of $\sqrt{n}$, preventing convergence with increasing n. When we increase the record size in the periodogram, frequency resolution is increased but statistical resolution remains constant.

To increase the statistical resolution, a different approach is needed. One approach, that of Gibbsian statistics discussed previously, is ensemble averaging. One way to do this, given a data sample of length N, is to break the sample into n equal length segments of length N/n, compute the periodogram of each, and then average these periodograms. In so doing, we are making the assumption that the statistical average is equal to the time average. A random process that has this behavior is called *ergodic*. The ergodic assumption seems eminently reasonable, but is really quite difficult to justify in most practical cases.

Clearly, in shortening our sample from N to N/n, we have gained our statistical resolution at the expense of frequency resolution. Thus, it becomes clear that in estimating power spectral densities we may be forced with a trade-off—one that can be quantified, even if only roughly. If the original record is of length ΔT and we break it up into n samples of length $\Delta T/n$, the uncertainty principle gives the frequency resolution

$$\Delta f(\Delta T/n) \geq \tfrac{1}{2} \qquad (11.6)$$

On the other hand, averaging the n samples will give a statistical resolution of

$$\Delta P/P = 1/\sqrt{n} \qquad (11.7)$$

for the n random numbers averaged at one frequency of the power spectrum $P(\omega)$. Combining Eqs. (11.6) and (11.7) gives a frequency–time–statistical resolution principle for estimating power spectra using averages over periodograms:

$$\Delta f\, \Delta T (\Delta P/P)^2 \geq \tfrac{1}{2} \qquad (11.8)$$

An example of this ensemble averaging technique is shown in Fig. 11.5, clearly showing the frequency resolution trade-off in favor of statistical resolution. The incoherent signal averaging of Fig. 11.2 is another example of ensemble averaging.

As in all digital signal processing schemes, a wide variety of computing technology can be used, from pencil and paper to supercomputers. An example using intermediate-scale technology is the modern compact DFT-based signal analyzers that are commercially available. Typically, these instruments can perform a variety of signal processing functions, including

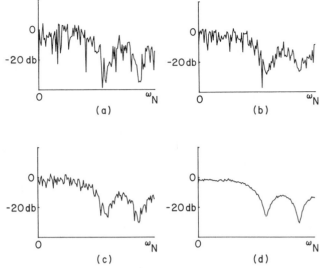

Figure 11.5 An example of ensemble averaging. (a) The DFT power spectrum of one 256-point-long sample of the output of the MA operator used in Fig. 11.4. In (b), (c), and (d), additional samples of the process have been used to compute power spectrum averages of 2-, 4-, and 96-fold, respectively. This averaging does not improve the frequency resolution, but the statistical resolution is increased, revealing the true spectral response of the MA operator.

Figure 11.6 A spectrum analyzer using DFT spectral averages. White noise is passed through the lowpass circuit under test and the HP-3562A signal analyzer averages the DFT power spectra of the output. This particular instrument collects 2046 samples in 8 msec and averages the number of DFT specta selected by the operator. (Photograph courtesy of Hewlett-Packard Company, reprinted with permission.)

DFT spectral analysis. Figure 11.6 shows an instrument of this type measuring the frequency response of a lowpass filter. The instrument applies white noise to the circuit under test and averages DFT spectra of the output until the desired statistical resolution is obtained according to Eq. (11.8).

This Gibbsian approach of ensemble averaging has bridled us with the limitation of the uncertainty principle of Eq. (11.8). A very interesting and important question is: can we do better than Eq. (11.8) by some other approach? This question has been the central topic of modern methods of spectral estimation. The answer to the question is yes, provided that additional, or prior, knowledge is available beyond that of the record itself. Unfortunately, in many practical applications the kinds of prior knowledge required by these methods of spectral estimation are usually not available, coloring spectral estimation with somewhat of the character of art, rather than science.

The power spectrum density analysis (PSDA) problem addresses how best to estimate power spectra. In PSDA, the available data are thought to represent a segment of an infinitely long sequence of random noise. The Fourier transform of the stationary noise does not exist in the strict sense because it does not have finite energy. However, as we observed in our discussion of the Wiener–Khintchine theorem in Chapter 6, such non-convergent signals can have a generalized Fourier transform yielding a spectrum of finite energy per frequency called the *power spectrum density*.

Even for signal-like processes, where the record appears rather deterministic, the concept of using additional information applies. Whenever the record does not contain the entire signal, some kind of assumption must be made concerning the behavior of the unmeasured portion of the signal outside of the data window. In the following sections, we will explore various types of these assumptions, which will enable us to estimate a corresponding power spectrum.

The MA Model: The Approach of Blackman and Tukey

The periodogram estimate of Eq. (11.4) computes the power spectrum by taking the Fourier transform of the autocorrelation (really the autocovariance, including the $1/n$ factor). This seems perfectly reasonable, reminding us of a special case of the convolution theorem, sometimes called the autocorrelation theorem, as we discussed in connection with both the DFT and the Fourier integral transform. However, the autocorrelation theorem needed to justify Eq. (11.4) is in the discrete-time, continuous-frequency formulation of Fourier transforms. Thus,

$$\hat{P}(\omega) = \text{FT}\{\hat{p}_k\} = \sum_k \frac{1}{n}(X^* \otimes X)_k e^{-i\omega k}$$

We expect that the appropriate form of the convolution is valid (leaving the proof for Problem 11.1), giving

$$\hat{P}(\omega) = 1/n |X(\omega)|^2 \qquad (11.9a)$$

as an equivalent form for the periodogram, where

$$X(\omega) = \sum x_n e^{-i\omega n} \qquad (11.9b)$$

That is, we can compute the periodogram by taking the discrete-time, continuous-frequency Fourier transform of the data and then form the power spectrum by using the squared modulus.

Although the periodogram of Eq. (11.9a) is a continuous function of frequency, we know that we can use the DFT to compute as many points as we wish in the frequency domain, all we need to do is zero pad in the time domain. This time domain zero padding is the same as assuming that x_n is indeed zero outside the record just as the autocovariance estimate of Eq. (11.5) does. Naturally, in terms of PSDA, we ask is there a better assumption about the process under study than assuming $x_n = 0$ outside the record? In practice, the question of the best assumption for a particular experiment is usually difficult to answer, but it is relatively easy to formulate *data models* that do lead to readily applied spectral estimates.

One such model is to assume that the time series under question is the output of an MA filter S of length N with white random noise n at its input. White noise is an abstract concept of random data whose power spectrum is constant over all frequencies; its phase spectrum must then vary randomly to produce the innovational component. The output x of our MA operator S is

$$x = s * n \qquad (11.10)$$

and

$$X(\omega) = S(\omega)N(\omega)$$

therefore,

$$|X(\omega)|^2 = |S(\omega)|^2 \qquad (11.11)$$

because the power spectrum of the noise is a constant in ω, taken here to be unity. Thus, PSDA of this process attempts to measure the magnitude response of the MA filter. This process is frequently used in practical applications: broadband noise is used to excite an LTI device, and PSDA of the output yields the device's frequency response.

The MA Model: The Approach of Blackman and Tukey

```
     DIM A(1024), B(M)
     DO 10 J = 1, N
     DO 10 I = 1, M - J + 1
10   A(J) = A(J) + B(J + I - 1) * B(I)
     DO 20 I = 1, N - 1
20   A(1024 - I + 1) = A(I + 1)
     CALL FFT(A, 1024)
```

Figure 11.7 Program to compute the Blackman–Tukey power spectral estimate of lag N. The array A is zero padded in the middle to 1024 points for the subsequent DFT. The input is the data B. The resulting power spectrum is the squared magnitude of A.

What now is the proper assumption to place on our data x if it results from this MA model? The data cannot be assumed to be zero outside of the record as the periodogram assumes, for the data is generated from a stationary time series that is infinitely long. By the very nature of the MA operator, it produces correlations among the output data, but only over its length N. Elsewhere, the data is uncorrelated random noise. Thus, we know that, not the data itself is zero, but its autocorrelation is zero beyond N lags. This is a less strict assumption than the periodogram makes.

It now becomes clear how to estimate the PSD of data produced by this MA process. Suppose that we have 64 data points, and we have reason to believe that the MA order is 16. Then, we use just the central 16 lags of the autocorrelation function to compute the power spectrum. The assumption is that the higher lags should really be zero, but when they are computed from the available record, erroneous nonzero values are produced for lack of the complete data stream. This is the Blackman–Tukey approach to power spectral estimation. A simple computer program that implements the Blackman–Tukey method is shown in Fig. 11.7 with results for a 64-long data record compared in Fig. 11.8 with the periodogram of the same data.

Basically, we are saying that the reliability of the autocorrelation estimate decreases with increasing lags because of the increasingly poor overlap in the autocorrelation sum due to the finite record. This observation leads to windowing the autocorrelation function so as to place more weight on the central lags. Then, the PSD estimate becomes

$$\hat{P}(\omega) = \sum p_k w_k e^{-i\omega k}$$

where the window function ω_k is $2n - 1$ long. In this framework, the truncated example discussed earlier is the special case of a rectangular window. Many of the windows common to digital filter design, such as the

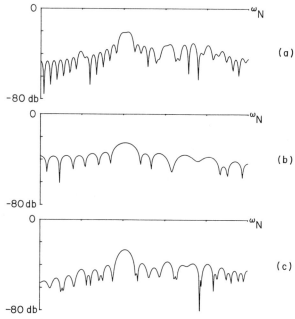

Figure 11.8 (a) Periodogram of a 64-point data sample of unknown origin. (b) The Blackman–Tukey power spectrum using 8 lags and in (c) 16 lags. As the Blackman–Tukey lags approach the length of the data, the estimate approaches the periodogram. At 64 lags, this estimation is identical to the periodogram because there is no truncation of the autocorrelation coefficients.

Bartlett, Hamming, and Hanning windows, have been used in this Blackman–Tukey approach to PSDA, but unfortunately seldom, if ever, does their selection relate to any known properties of the data or the process actually being analyzed. Additionally, the selection of the order of the MA process produces radically different PSD estimates, as shown in Fig. 11.8; in practical cases, there is seldom any decent way of selecting an order based on available knowledge of the data.

This windowing process increases the statistical resolution (i.e., reduces the bias and the variance of the spectral estimate at each frequency) at the expense of frequency resolution. Indeed, multiplying the autocorrelation in the time domain by the window function is equivalent to convolving the periodogram with the Fourier transform of the window function. For all reasonable windows, this is a smoothing operation applied to the periodogram, reducing the statistical fluctuations.

An additional effect of windowing concerns the sign of the PSD estimate. Certainly, any real power spectrum is a nonnegative function of frequency. So is the periodogram. The Fourier transform of the Bartlett window (a triangle) is the sinc-squared function, which is nonnegative.

Thus, it must necessarily produce nonnegative results when convolved with the periodogram. But, the rectangle, the Hamming, and the Hanning windows all have Fourier transforms that go negative. These windows produce better frequency resolution than the others, but they can also produce a revolting estimate having a negative power spectrum. Put another way, the trouble with windowing is that it can produce an autocorrelation estimate which, in fact, cannot be a valid autocorrelation function. Recall from our discussion on the root method of spectral factorization that not every symmetric sequence can be an autocorrelation.

Many variations on windowing can be included within the MA model in PSDA. Combined with the segmented averaging that we discussed in the previous section, a bewildering collection of PSD estimates can be made. A data record can be segmented and windowed and periodograms computed, and then these periodograms averaged. Alternately, the data can be segmented, windowed or not, and autocorrelations estimated, and then these autocorrelations averaged before taking the Fourier transform. Also, the autocorrelations from the segmented data may be windowed. When segmenting, the individual data segments may or may not overlap. Finally, the spectral averaging effect of windowing can be applied directly to the periodogram in the frequency domain.

Uncertainties in the selection of windows, segmental averages, and MA order, in the face of any relevance to the process actually under study, reflect the strong empirical guidance required in applying PSDA in practice.

The AR Model: The Approach of Yule–Walker and Burg

After discussing the MA data model, it is natural to consider next a model in which we assume our time series under study has been generated by an autoregressive process. In this model, we assume that an AR filter $1/S$ of order m generates our data by acting on a random time series of white noise:

$$X(Z) = N(Z)/S(Z) \qquad (11.12a)$$

or equivalently,

$$X(Z)S(Z) = N(Z) \qquad (11.12b)$$

In the frequency domain,

$$X(\omega) = N(\omega)/S(\omega)$$

```
DIM P(M,M), B(N), C(M), X(M)
C(1) = 1
DO 10 J = 1, M
DO 10 I = 1, N - J + 1
10  P(1, J) = P(1, J) + B(J + I - 1) * B(I)
DO 20 J = 1, M
DO 20 I = J, M
P(J, I) = P(1, I - J + 1)
20  P(I, J) = P(J, I)
X = INV(P) MPY(C)
```

Figure 11.9 Program to compute the Yule–Walker spectral estimation of order N. The program is similar to the Wiener filter program of Fig. 9.6, using a brute force matrix inverse to solve the normal equations. For greater speed, the Levinson recursion should be used. The input is the data B and the output is the AR operator coefficients X.

giving the PSD estimate

$$\hat{P}(\omega) = |X(\omega)|^2 = \frac{1}{|S(\omega)|^2} \quad (11.13)$$

Certainly, we would only consider a model where the order of the AR process m is less, normally much less, than the length of our data record n. Thus, the plan is to solve Eq. (11.12b) via the least-squares solution for S, then evaluate $S(\omega)$ using the DFT, and finally form $P(\omega)$ by Eq. (11.13). Using our previous Wiener least-squares inverse program from Chapter 9 and an FFT algorithm yields a quite simple program code as diagrammed in Fig. 11.9. The results of this program acting on the same data used in the MA model of Fig. 11.8 are shown in Fig. 11.10.

This AR approach was first used by Yule in 1927 in a search for periodicities in sunspots. Walker published the same type of approach in 1931. An essential feature of this Yule–Walker method is that it necessarily requires an estimate of the data's autocorrelation for use in the least-squares inverse. Because of the possibility of using different autocorrelation estimates, the AR model shares the same large number of such possibilities with the MA model. Also, there is the choice of the AR order, which again has no simple or direct relationship to what is usually known about the process or the data under study. As would be expected, the all-pole structure of the AR model, in general, produces more detail in the estimated power spectrum as demonstrated in Fig. 11.10. Looking at the AR model in another way, it continues the data outside of the recorded window in, hopefully, some sensible way. Even more, it continues the data's autocorrelation function beyond the recorded window to infinity.

The AR Model: The Approach of Yule–Walker and Burg

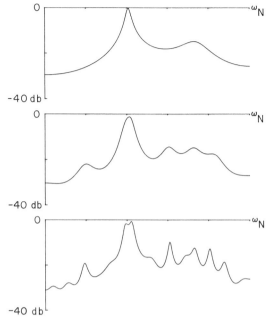

Figure 11.10 The Yule–Walker spectral estimation of the same 64-point data sample used in the periodogram and Blackman–Tukey estimates of Fig. 11.8. Top to bottom, the Yule–Walker order is 8, 16, and 32. Compared to the all-zero model of Blackman–Tukey, the Yule–Walker estimate displays the distinctive character of an all-pole model with increased resolution.

So, we observe that while the MA model is less presumptuous about the data's behavior outside the record than is the periodogram, the AR model is one more level less presumptuous than the MA model.

The problem with all of these methods of spectrum estimation is the low reliability of computing the autocorrelation coefficients from a finite sample, a problem that is aggravated for short records. In 1967, John P. Burg presented a paper at the 37th Meeting of the Society of Exploration Geophysicists in Oklahoma City, Oklahoma, that had a revolutionary impact on spectral analysis. He showed how to estimate the autocorrelation without running computations off of the end of the data, thereby avoiding end effects and assumptions about the signal beyond the available data. Burg's procedure is recursive, starting out with a 2×2 least-squares solution for a prediction error filter F that will convert the input data to white noise. Through the recursion, this filter is extended to some order m; then, the estimated power spectrum of the data is the power spectrum of this filter's inverse.

Before describing the Burg recursive algorithm, we will first consider some properties of the AR process. As in the Yule–Walker approach, we first focus on the inverse problem, involving an MA operator. That is, we seek an MA operator that filters the data into white noise as expressed in Eq. (11.12b). Such an operator will recursively filter $\{x\}$ into an unpredictable sequence

$$x_t + f_1 x_{t-1} + \cdots + f_m x_{t-m} = n_t$$

that is,

$$x * F = n \qquad (11.14)$$

The operator F is a prediction error filter, trying to predict the time series $\{x\}$ from the m preceding values. When m is large enough, n loses all of its predictability and it becomes white noise. We would anticipate then, on developing F's of ever increasing order, that if the AR model of $\{x\}$ is really valid, the coefficient f_m eventually will approach zero, indicating that the order of the AR process has been reached.

In Chapter 9, we developed the normal equations for the solution to the linear least-squares prediction error filter; now, it will be insightful to pursue a different approach with the problem of power spectral estimation foremost in mind. The Z transform of Eq. (11.14) is

$$x(Z)(1 + f_1 Z + f_2 Z^2 + \cdots + f_m Z^m) = N(Z)$$

or

$$X(Z)F(Z) = N(Z)$$

and the power spectrum of this equation is

$$X(Z)X^*(1/Z)F(Z)F^*(1/Z) = N(Z)N^*(1/Z)$$

which can be written

$$P(Z)F(Z) = \frac{K}{1 + f_1^* Z^{-1} + f_2^* Z^{-2} + \cdots f_m^* Z^{-m}} \qquad (11.15)$$

Now comes an important point. For the noise to be perfectly white—that is, its autocorrelation to be the unit impulse at zero lag—the sequence $\{x\}$ must be infinitely long, not just a finite sample $\{x_n\}$ of the hypothesized AR process. Given this observation, $P(Z)$ in Eq. (11.15) becomes unknown autocorrelations of this idealized data, and K is the power of the noise sequence N. Because the process by which the $\{x\}$ are generated is AR, F must be minimum phase. Hence, the denominator in Eq. (11.15) can be expanded in negative powers of Z. So, we write the equation as

$$(\cdots + \phi_m Z^{-m} + \cdots + \phi_1 Z^{-1} + \phi_0 + \phi_1 Z + \cdots + \phi_m Z^m + \cdots)$$
$$(1 + f_1 Z + f_2 Z^2 + \cdots + f_m Z^m) = K(1 + C_1 Z^{-1} + C_2 Z^{-2} + \cdots) \qquad (11.16)$$

The AR Model: The Approach of Yule–Walker and Burg   275

By simply equating powers of Z on each side of this equation, we find (1) there are only a finite number of positive powers of Z, from zero to m, that involve the autocorrelations up to lag m, and (2) there are an infinite number of negative powers of Z involving autocorrelations beyond m, plus the unknown coefficients C. It is easy to see that the positive powers of Z lead to the same normal equations of Chapter 9:

$$\begin{pmatrix} \phi_0 & \phi_1 & \cdots & \phi_m \\ \phi_1 & & & \\ \vdots & & & \\ \phi_m & & & \phi_0 \end{pmatrix} \begin{pmatrix} 1 \\ f_1 \\ \vdots \\ f_m \end{pmatrix} = \begin{pmatrix} K \\ 0 \\ \vdots \\ 0 \end{pmatrix} \qquad (11.17)$$

This interesting exercise shows that the same normal equations arise whether we start from the viewpoint of solving the least-squares prediction error filter problem or start from the viewpoint of solving the PSDA problem assuming an AR data model. In either case, the development of these normal equations used the full autocorrelations computed from untruncated data; hence, the ϕ are really unknown, as is F. On the other hand, our development suggests that we should be able to cast the problem into a prediction error filter problem working directly on the data, thereby avoiding the involvement of unknown autocorrelations. Burg's approach is exactly this; he estimates both the first m lags of the autocorrelation and the m coefficients of F directly from the available data. To see how this is possible, we first start with an example for $m = 2$. The normal equations read

$$\begin{pmatrix} \phi_0 & \phi_1 \\ \phi_1 & \phi_0 \end{pmatrix} \begin{pmatrix} 1 \\ F \end{pmatrix} = \begin{pmatrix} K \\ 0 \end{pmatrix} \qquad (11.18)$$

To begin the autocorrelation estimation, the zero-lag autocorrelation is taken from the data in the most obvious fashion:

$$\phi_0 = \sum_1^N x_i x_i$$

Now, for Burg's twist: he estimates the remaining lags of the autocorrelation from F itself. By Eq. (11.18), the estimate of ϕ and K are

$$\phi_1 = -\phi_0 F \qquad (11.19a)$$

$$K = \phi_0(1 - F^2) \qquad (11.19b)$$

leaving only F to be determined by a separate means. This 2-long prediction error filter $(1, F)$ operates on the data, never running off the end, to produce a mean squared error in the forward direction:

$$E^+ = \frac{1}{N-1} \sum_1^{N-1} |x_{i+1} + Fx_i|^2 \tag{11.20}$$

Because a time series and its conjugate time-reversed version have the same autocorrelation, the discussion starting with Eq. (11.18) is also valid for the same prediction error filter F operating on $X(-)$. However, the mean squared error will be different for the time-reversed operation because of end effects. This suggests minimizing the total errors in both directions:

$$E = E^+ + E^-$$

where

$$E^- = \frac{1}{N-1} \sum_1^{N-1} |x_i + Fx_{i+1}|^2 \tag{11.21}$$

By straightforward differentiation, it is easily seen that this minimization produces

$$F = -\frac{2 \sum_1^{N-1} x_{i+1} x_i}{\sum_1^N x_i^2 + \sum_1^{N-1} x_{i+1}^2} \tag{11.22}$$

completing the estimation of the autocorrelation of F and, hence, of the power spectrum for the assumed single-pole AR process. Thus, the original intention of estimating the autocorrelation using only the available data has resulted in directly giving us the AR filter of interest $S = (1, F)$.

Fundamental to the AR data model is the notion that Eq. (11.12a) is actually implemented in the AR mode (as opposed to imagining that the data were generated by an IIR MA operator). This autoregressive operation demands that $1/S$ be causal and stable; that is, S must be minimum phase. That $|F| \leq 1$ is easily seen (Problem 11.3), as must also be true from the implications of Eq. (11.18).

This minimum phase requirement of the AR model can be subtle. In the Yule–Walker treatment, the autocorrelation is computed from the data by

$$\phi_j = \sum x_i x_{i+j}$$

so that ϕ is a genuine autocorrelation, although certainly not that of the actual innovational time series $\{x\}$. Thus, by the property of the solution to the Wiener least-squares problem in Eq. (11.17), the Yule–Walker S will necessarily be minimum phase. However, in the Burg algorithm, the prediction error filter is estimated in quite a different manner. Thus, Burg

The AR Model: The Approach of Yule–Walker and Burg

had to take special care to assure that his algorithm generates an F with minimum phase. Otherwise the spectrum, computed from the inverse of F, could diverge.

By extending this discussion from a 2-long prediction filter to a 3-long filter, we can arrive at the recursive algorithm for finding the filter. We start by looking at the forward and backward prediction error filters. The 2-long forward filter gives an error

$$E^+ = x_{i+1} + Fx_i$$

and the 3-long forward filter error is

$$E^+_{new} = x_{i+2} + F_1 x_{i+2} + F_2 x_i$$

Next, Burg used the Levinson recursion, which always assures that the new F generated at each stage of the recursion is minimum phase. Then, the 3-long F becomes

$$(1, F_1, F_2) = (1, F - CF, -C) \tag{11.23}$$

or

$$E^+_{new} = x_{i+2} + (F - CF)x_{i+1} - Cx_i$$

This can be written

$$E^+_{new} = (x_{i+2} + Fx_{i+1}) - C(x_i + Fx_{i+1})$$

which provides a recursion

$$E^+_{new} = E^+_{old} - CE^-_{old} \tag{11.24a}$$

It is equally easy to see that also

$$E^-_{new} = E^-_{old} - CE^+_{old} \tag{11.24b}$$

The factor C is determined by minimizing the total summed squared forward and backward errors:

$$E = \sum (E^+)^2 + \sum (E^-)^2$$

Minimizing this relative to C gives

$$C = \frac{2 \sum E^+ E^-}{\sum (E^+)^2 + \sum (E^-)^2} \tag{11.25}$$

We now have the Burg recursive algorithm expressed in Eqs. (11.23)–(11.25). The order of business is to compute E^+ and E^- from Eqs. (11.24), and then to compute C from Eq. (11.25) for the subsequent determination of F by Eq. (11.23). Since at each stage of the Levinson recursion a new operator F is formed in the same manner as Eq. (11.23), always incorporating factor C [review Eq. (9.23)], the minimization of the squared error

```
      REAL A(N), C(N), B(M), E1(M), E2(M), S(1024)
      A(1) = 1
      DO 10 I = 1, M
      E1(I) = B(I)
   10 E2(I) = B(I)
      DO 60 J = 2, N
      T = 0
      B = 0
      DO 20 I = J, M
      B = B + E2(I)**2 + E1(I - J + 1)**2
   20 T = T + E2(I)*E1(I - J + 1)
      C(J) = 2*T/B
      DO 30 I = J, M
      E3 = E2(I)
      E2(I) = E2(I) - C(J)*E1(I - J + 1)
   30 E1(I - J + 1) = E1(I - J + 1) - E3*C(J)
      A(J) = 0
      DO 40 I = 1, J
   40 S(I) = A(I) - C(J)*A(J - I + 1)
      DO 60 I = 1, J
   60 A(I) = S(I)
```

Figure 11.11 Program to compute the Burg spectral estimation AR coefficients. The input data is in array B; the output coefficients are in array S.

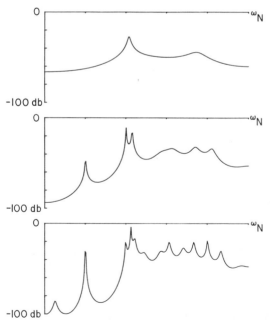

Figure 11.12 The Burg spectral estimation of the same 64-point data sample used in Figs. 11.8 and 11.10. Top to bottom, the Burg order is 8, 16, and 32. At each order, the resonances in the Burg and the Yule–Walker estimates correlate well, but the Burg estimate tends to have poles closer to the unit circle, resulting in sharper spectral peaks. Note the vertical scale difference in Figs. 11.8, 11.10, and 11.12.

gives the same result for C at each step. Thus, it is clear how to extend the recursion from two or three to any order. A simple computer program that does this is shown in Fig. 11.11; the last remaining step is to evaluate the power spectrum $1/F(\omega)$ by a zero-padded DFT computation. Figure 11.12 shows the results of this Burg algorithm acting on the same data used to generate Figs. 11.8 and 11.10. Note the greatly enhanced resolution of the Burg results compared to the Yule–Walker estimation.

Strangely, in the Burg approach, we set out to estimate the autocorrelation coefficients by a method that makes no assumptions about the data beyond the available values, but in the end, the algorithm does not explicitly contain these autocorrelation estimates.

The Maximum Entropy Principle

Much ink has been spilt over the significance and the properties of the maximum entropy spectral estimation—much more than we can afford here. Nonetheless, we cannot stop with the mere presentation of the Yule–Walker and Burg algorithms. The first item of old business is the ado about maximizing the entropy.

As every college physics student knows, entropy is a measure of disorder. In information theory, this disorder is equated to the information content because the more disorder in a signal, the greater the number of messages are possible and hence the higher the information content of the received message. To focus in on this concept, let us consider a time-invariant source of digital random numbers. Each number produced by the source is treated as a symbol or as a message. Each such message is a binary number consisting of N_i binary digits (bits). The number of possible results for a given message is 2^{N_i}. However, the only information required to describe this result consists of the N_i bits. Hence, we argue that the proper measure of information is proportional to the number of bits rather than the number of possible outcomes. That is, the information I is proportional to $\ln_2 2^{N_i} = N_i$. Stated in terms of the probability P of occurrence of a given number, the information is

$$I = K \ln 2^{N_i} = K \ln(1/P_i)$$

It may well be that not all symbols (our binary numbers) are transmitted with the same probability of occurrence. In this case, $P_i \neq 1/2^{N_i}$. But we would still insist that the information is $I = -K \ln P_i$. (The constant K is really a matter of units only, and not of much concern here.)

Next, we seek an information rate. Imagine that we have recorded a transmission of the source over a long period T. Then, if T is very large, on the average, we expect to find a particular result P_iT times since P_i is the relative probability of occurrence of a given result. It is then natural to associate the information provided by a given symbol with

$$I_1 = -KP_1T \ln P_1$$

since the symbol will occur P_iT times. After a time T, the total information received from all symbols would be

$$I_{\text{total}} = -K(P_iT \ln P_1 + P_2T \ln P_2 + \cdots)$$

Then, the average information per unit time, called the *entropy* by Shannon (1949) is

$$h = -K\sum_i^n P_i \ln P_i \qquad (11.26)$$

This basic mathematical form for the measure of uncertainty in a distribution $\{P_1 \cdots P_n\}$ has been shown by Shannon to follow directly from certain elementary consistency and additivity requirements. Our approach to the concept of entropy was intended merely to suggest that Eq. (11.26) is a reasonable definition. In fact, this definition is identical to that of entropy in statistical mechanics, not exactly a coincidence. In the final analysis, a definition is only justified by its usefulness.

As an example of Eq. (11.26), consider a message source consisting of the toss of a loaded coin that turns up heads $\frac{3}{4}$ of the time and tails $\frac{1}{4}$ of the time. If we use the base two logarithm, $K = 1$ and the units of entropy are bits/toss. So, the resulting entropy is

$$h = -(\tfrac{1}{4}\ln\tfrac{1}{4} + \tfrac{3}{4}\ln\tfrac{3}{4})$$

$$h = 0.811 \text{ bits/toss}$$

Whereas, if the coin were honest, the entropy of the source would be 1 bit/toss. It seems reasonable that the loaded coin source produces less information than the honest coin source because its bias has robbed it of the flexibility inherent in the message of the honest coin source. This association of information content of a source to its number of possible alternative messages is fundamental to information theory. A certain message received out of only two possible results is not nearly as informative as the same message received out of a possible 10,000 alternative messages. It is apparent then that the unpredictability of the source is related to its information content and to its entropy. The more unpredictable, or innovative, a time series, the higher its entropy.

So far, this discussion seems to have little to do with spectral estimation. Clearly, there must be a relationship between the innovational character of a time series and its power spectrum. As an example, consider a spectrum

The Maximum Entropy Principle 281

composed of only sharp spectral lines with no continuum. Each spectral line represents a completely predictable sinusoid and hence the time series itself is completely predictable, or deterministic. It can carry no information. Its entropy is negative infinity.

This relationship between the predictability of a time series and its spectrum was first developed by N. Wiener (1949) in his famous work during World War II. Building on his previous work in generalized harmonic analysis, which we discussed briefly in Chapter 6 in connection with the Wiener–Khintchine theorem, Wiener solved the problem of predicting stationary time series using frequency domain methods. Recall that a stationary time series has no Fourier transform, but it does have a power spectrum. Later in 1949 Shannon, influenced by Wiener's work, related his newly defined measure of predictability, that is, entropy, to a time series' power spectrum. Because Shannon's discussion would lead us too far away from our study of PSDA into his information theory, it will suffice to just quote Shannon's result. He showed that the entropy is proportional to the logarithm of the power spectrum $\Phi(\omega)$ integrated over the Nyquist interval:

$$h \propto \int_{-\pi}^{\pi} \ln \Phi(\omega)\, d\omega \qquad (11.27)$$

This result has an immediate consequence. If the power spectrum is zero over any finite bandwidth, the above integral will diverge, giving a negatively infinite entropy. From information theory, the interpretation must be that such a time series contains no randomness; it is entirely predictable.

The condition that

$$\int_{-\pi}^{\pi} \ln \Phi(\omega)\, d\omega > -\infty$$

for a time series to be nondeterministic is known as the *Paley–Wiener criterion* (Paley and Wiener, 1934). It is only for these time series that prediction theory should be applied. It is perhaps worth emphasizing that prediction theory only applies to stochastic processes where the time series under consideration is one member of an ensemble generated by a stationary ergodic source. Only then can meaningful error criteria be formulated by comparing the predicted results with the statistics of the ensemble.

Now, with the result of Eq. (11.27), we are finally ready to see what entropy has to do with power spectral estimation. The idea is to be as noncommittal as possible concerning data that has not been measured. The approach is to assume that we have known autocorrelation coefficients of the data out to some lag. Within this constraint, we maximize the uncertainty of the unknown information. That is, we maximize the entropy given by Eq. (11.27), subject to the known autocorrelations ϕ_n, $|n| \le p$:

$$\frac{\partial h}{\partial \phi_n} = 0 = \int_{-\pi}^{\pi} \frac{1}{\Phi(\omega)} \frac{\partial \Phi(\omega)}{\partial \phi_n} d\omega \qquad |n| > p \qquad (11.28)$$

and we know that

$$\Phi(\omega) = \sum_{-\infty}^{\infty} \phi_n e^{-i\omega n} \qquad -\pi \le \omega \le \pi$$

so that

$$\frac{\partial \Phi(\omega)}{\partial \phi_n} = e^{-i\omega n}$$

Thus, the maximization procedure immediately gives

$$\int_{-\pi}^{\pi} \Phi^{-1}(\omega) e^{-i\omega n} d\omega = 0 \qquad \text{for} \qquad |n| > p$$

Because of the symmetry in n, this can also be written as

$$\int_{-\pi}^{\pi} \Phi^{-1}(\omega) e^{i\omega n} d\omega = 0 \qquad \text{for} \qquad |n| > p \qquad (11.29)$$

Since the process under consideration produces an innovational time series, its power spectrum cannot be zero over any finite bandwidth, and its inverse, whose power spectrum is $\phi^{-1}(\omega)$, must exist. Equation (11.29) says that the autocorrelation of this inverse is zero beyond lag p. It follows that the inverse process is an MA of order p. This result shows that the maximum entropy process itself is AR of order p.

Apparently then, both the Yule–Walker and the Burg algorithms, which use the AR model to estimate spectra, are maximum entropy procedures. For more details of the relationship among these, as well as other spectral estimates, see, for example, Jaynes (1982) and Smith and Grandy (1985), which give further discussion of this fascinating subject. We can make several important observations here however.

First, the Yule–Walker method estimates the autocorrelation coefficients directly from the data, while the Burg method estimates a minimum phase prediction error filter from the data, never running off the end of the data. These estimates are different, and indeed, they are both at odds with the assumption used in Eq. (11.28). This assumption was that autocorrelations are known exactly out to lag p. Presumably, the estimated power spectrum would be forced to conform to these known autocorrelation coefficients, while the remaining ones for $|n| > p$ would be determined by the max-entropy principle. In fact, however, having an exact knowledge of the first p autocorrelations is an unusual state of affairs experimentally, perhaps occurring only in some coherency experiments (such as optical interferometry).

The maximum entropy principle has then not solved anything, but it has merely recast the problem of estimating the spectrum into one of estimating the AR coefficients. The Yule–Walker and Burg approaches are just two different ways of doing this.

Obviously, both methods estimate all of the autocorrelation coefficients, just as any method must, since these coefficients are uniquely determined from the final estimated power spectrum. Both methods, as any AR process must, lead to satisfying the same normal equations of Eq. (11.17). Significantly, these equations have the form of the convolution operation $\phi * F$; they thus show that the same one-step prediction error filter that filters the data into white noise also annihilates autocorrelation coefficients for lags 1 through p. Likewise, because of the symmetry of the equations (recall the Levinson recursion of Chapter 9), the backward prediction operator annihilates the autocorrelations for negative lags -1 to $-p$.

Our minds quicken. Our insight deepens. Burg's approach assumes no given autocorrelations. Rather it estimates all of them via the prediction error filter in such a way that maximizes the entropy. The prediction error filter destroys order, permitting as much chaos as possible, giving way to maximum entropy. In particular, the inverse filter F whitens the data as much as possible, yielding a result that is completely uncorrelated for lags $|n| < p$. For greater lags, the chips fall where they may. Their control is lost because there is no data. Ironically, the situation is quite different from our early conception of power spectral estimation. It is not the autocorrelations for large lags of the data that are being directly controlled by the estimation procedure; rather the central ones of the inverse operator are.

Furthermore, the normal equations immediately show the difference between the Yule–Walker approach and Burg's method. In their own ways, both the Yule–Walker and the Burg methods satisfy these normal equations exactly; the difference is the problem they are solving. In the Yule–Walker case, the problem that is solved is the max-entropy problem given the exact autocorrelation coefficients ϕ_0 to ϕ_n. But, in most actual applications of the Yule–Walker method, the exact autocorrelation coefficients are not used; they are not known. Rather, they are estimated from a direct computation from available data, a particularly poor estimate for short data runs. In the Burg case, the problem that is solved is the max-entropy problem given the data x_1 to x_n. The autocorrelation coefficients ϕ_0 to ϕ_p are determined, along with the prediction error filter, so that the normal equations are satisfied. So, again, the autocorrelations do not in fact provide a constraint under which the entropy is maximized; rather they are adjusted to satisfy the normal equations. If we insist on equating these two algorithms to the solution of a max-entropy problem, then we are forced to wait to discover exactly what the problem is until its solution is obtained. Only then do we know the exact autocorrelation coefficients that were to be constrained in the original statement of the max-entropy problem. In the Yule–Walker case, it is almost certain that these autocorrelations used are not the correct ones for the process under study. In the Burg case, these estimated autocorrelations of lag $|n| < p$ are in fact optimal in terms of inverse filtering the data into white noise.

The Burg method has been criticized because the order p seems to be undetermined, open to the discretion of the analyst. Usually, adding poles to the spectrum (by increasing p) will increase the resolution of the spectrum and give it more structure. Frequently, spectral lines in a Burg spectrum of a given order divide or split when p is further increased. This increased structure due to line splitting is caused by information present in the data. If $2\frac{1}{4}$ cycles of a 1-kHz sinusoidal component are discovered in a 2.25 msec data run (which is supposed to be a segment of a stationary time series), it is highly unlikely that the true spectrum of this process actually contains a spectral line at 1 kHz. If we know that there is indeed a 1-kHz spectral line in the data, then we would need to maximize the entropy including this constraint. The Burg algorithm includes no such constraint. Its prediction filter can only reflect information that is actually contained in the data. In this sense, the Burg procedure is actually fail-safe, producing the smoothest spectrum consistent with the data at hand. Indeed, we can even conclude that the Burg method dictates the order. An order equal to the data run is the only order that whitens the output spectrum of the inverse filtering the most. If the resulting structure seen in the Burg spectrum is objectionable, it is not because of a failure of max-entropy, but because the analyst either has prior knowledge that he failed to include in max-entropy, or he has an unfounded prejudice. Interestingly, we criticize the periodogram on similar grounds, that is, it does not look right. When we make this judgment, we are using additional information that we apparently have at our disposal.

In some cases, increasing the order of the spectral estimate may not increase the structure of the spectrum. This means that some poles are migrating far away from the unit circle and are becoming ineffective and redundant. This is a sign that the order of the prediction error filter has become larger than the actual order of the stationary process. The extra poles do no harm; max-entropy is fail-safe.

Other Data Models

At first it may seem that the Blackman–Tukey spectral estimation assumes that an MA system generated the given data, and that max-entropy assumes that an AR process did. But, we know that technically these two

Other Data Models   285

processes can be represented by one another. That is to say, any MA operator can be represented by an AR operator of high order (infinite theoretically), and an AR operator can be likewise represented by a high-order MA operator. So, it seems that the question of MA versus AR is more one of degree rather than kind. Even so, it would be reasonable to insist on a parsimonious description of the process, using as few parameters as possible. It follows quite naturally then to consider the ARMA model that incorporates that descriptive power of both models.

The ARMA model arose quite naturally in Chapter 10 in the discussion of spectral factorization. There we developed a least-squares iterative procedure for factoring a sequence into an MA numerator and a minimum phase AR denominator. If the original data were in fact actually generated by true ARMA processing, this iteration will converge, and a power spectrum estimate can be formed by evaluating the rational fraction on the unit circle using the DFT.

Numerical experiments using data actually generated by MA, AR, and ARMA processes are easily carried out (see Problems 11.10–11.14). Not unexpectedly, they show that the best results occur when the estimate assumes the same generating process as actually used to generate the data. Although these experiments are educational amusement in themselves, there is a catch in their interpretation.

First, seldom in any real spectral estimation problem does one have knowledge of the generating mechanism of the data. Hence, the conclusion to match the data model to the generating mechanism is of little practical value. Second, in arriving at the conclusion of these numerical experiments, the spectrum of the data-generating operator has been interpreted as the true spectrum. In fact, there is no such true spectrum available in any spectral estimation problem. Let us recall that the problem at hand is to estimate the power spectrum of a stationary process, producing an infinite run (or at least a very long run) of random numbers. Any finite run (or a very short run) of data could have been produced by an infinity of different generating mechansims; there is no correct one. Therefore, simple comparisons of spectral estimates to spectra that are supposedly correct do not properly fit within the framework of estimation theory. Rather, one needs to compare the estimated spectrum to the most desirable spectrum, as determined by some criteria.

The max-entropy principle is one way of proceeding that is firmly based in estimation theory. However, the true max-entropy solution to spectral estimation problems under a variety of different constraints has never been given. Rather, other data models have been developed that include, each in their own way, assumptions of prior knowledge.

One method assumes that the data is a superposition of anharmonic sinusoids (the Pisarenko method). In this method, an AR model of the same order is used for both the sinusoids and noise, which leads to an eigenvalue problem. Since the sinusoids are represented by a second-order process, the order of the eigenvalue problem is twice the number of sinusoids included in the model. The poles of the AR process are on the unit circle, which results in zero-width spectral lines for the estimated power spectrum. From the Paley–Wiener criterion, such a spectrum only occurs in a deterministic process. Hence, in this case, the prior knowledge of the assumed sinusoids contradicts the precept of a stationary random process. Another method assumes a superposition of damped sinusoids (the Prony method), which results in a power spectrum that is nonzero everywhere.

To be completely satisfactory, all of these data models would need to be justified by prior knowledge of the process under study. Generally, it is not clear in an actual experiment just how such prior knowledge would arise. One exception is the ARMA model of marine seismic reverberations discussed in Chapter 10, where it was argued that the water bottom multiples represent an AR process, and the reflections are the result of an MA process. Still, this model is quite an idealization of the actual physical process that really occurs.

The point is that given prior knowledge of the process, as in these examples, the application of the true max-entropy principle could then be sought, using this prior knowledge as a constraint. In most cases, this presents a challenging problem—a yet unsolved problem.

Even within the framework of estimation theory, the max-entropy principle does not reign supreme. There are other approaches. One is to maximize the probability that the estimated spectrum is the correct one. Such an approach, called *maximum likelihood*, develops an operator based on passing power in a narrow band about the signal frequency of interest while minimizing the power in interfering spectral components. Both the max-entropy and the max-likelihood methods can resolve spectral peaks arbitrarily close together for any nonzero time–bandwidth product. Furthermore, the estimates of the two methods are related: the average of the inverse spectra of the max-entropy procedure, for all orders 1 through p, is equal to the inverse of the max-likelihood spectrum (Burg, 1972). This interesting result implies that at the expense of spectral resolution, the max-likelihood procedure achieves greater statistical stability than does the max-entropy method.

By now it should be clear that the subject of spectral estimation is indeed an interesting and a complex one—complex enough that we cannot present an in-depth survey of all its aspects in an introductory treatment. We have, nonetheless, given a logical and historical development that began with the periodogram. Only the periodogram gives the exact spectrum of the actual data at hand. But, we have additional information; we know (or think we know) that we are estimating the spectrum of a stationary process and that the data is not zero outside of the given data sample. We feel that if we had some other data sample of the process, its periodogram would be different than the first result. We somehow wish to preserve those stationary characteristics that are in common with each periodogram and suppress those variable characteristics that are not indicative of the generating process's statistics. Furthermore, we have shown that just extending the data run does not solve this problem; it only increases the frequency resolution, not the statistical resolution.

We were thus brought to various ideas designed to smooth the spectral estimate: ensemble averaging, windowing, and AR models. All of these seemed to be ad hoc and plagued with objections of one kind or another. All except one—the Burg algorithm. This one method, which has done more to revolutionize the field of spectral estimation than any other approach, has been shown to estimate all of the autocorrelation coefficients of the process while making the most noncommittal assumptions consistent with the actual data at hand.

The concept of maximizing the entropy has a strong appeal because of its firm base in information and estimation theory. However, arguments can be made for other criteria, such as max-likelihood. Furthermore, the full max-entropy principle is not easy to apply in most cases, given prior information in various forms. So far, only the Burg solution has surfaced. It would be extremely useful, if only as guiding light, to have one or two other max-entropy solutions available. For those interested, here is a challenge worthy of consideration.

Problems

11.1 State and prove the form of the convolution theorem appropriate for discrete time and continuous frequency. Then, adapt your result to the autocorrelation function, finally showing that Eq. (11.9a) follows from Eq. (11.4).

11.2 Explain the difference between the periodogram and the discrete Fourier transform of a given sample of a stationary time series.

11.3 Using the result of Eq. (11.22), show that $|F| \leq 1$.

11.4 Show that each recursion arising from Eq. (11.23), forming an $n + 1$ term filter from an n term filter, can be written as

$$A'(Z) \to A(Z) - CZ^n A(1/Z)$$

Then show that if $|C| < 1$, $A'(Z)$ is necessarily minimum phase if $A(Z)$ is.

11.5 An information source consisted of the results of tossing one loaded die that has probabilities (0.054, 0.788, 0.1142, 0.1654, 0.2398, 0.3475) for the numbers 1 through 6. Compute the entropy per unit time (or per toss) of this source and compare it to the entropy of a similar source using an honest die.

11.6 An experimental study of the relative frequency of the 8727 most frequently used words in the English language has shown that their frequency of occurrence can be represented empirically by

$$P_n \simeq 0.1/n$$

Compute the entropy of common English text using this information. Furthermore, compute the entropy in bits/letter assuming that there are an average of 5.5 letters/word (this includes spaces). Discuss your result compared to the ANSCII encoding of 8 bits/character.

11.7 Using the definition of entropy from Eq. (11.27), show that the output of any LSI operator with a stationary time series at its input has less entropy than its input signal does.

11.8 Relate the Paley–Wiener criterion to sinc interpolation performed by zero padding in the frequency domain.

11.9 Write out the terms of Eq. (11.16), showing that the normal equations of Eq. (11.17) follow for positive powers of Z up through Z^m. What can be said of powers greater than m, and of negative powers in Z?

11.10 The zero-phase Spencer smoothing operator is a pretty poor low-pass filter by many standards, but its spectrum does offer character (recall Problem 7.21). Using synthetic data generated by passing white noise through the Spencer MA operator, experiment with spectral estimation by applying the following different PSDA methods to this data: (1) periodogram, (2) Blackman–Tukey, (3) Yule–Walker, and (4) Burg. Compare these results, including different orders, with the exact spectrum of the Spencer operator.

11.11 In the same spirit of the previous problem, generate synthetic AR data by passing white noise through the operator shown below. Then, study the results of applying the same four PSDA methods to this data, experimenting with different orders in the Blackman–Tukey, Yule–Walker, and Burg estimates. Discuss your results, comparing them to the actual spectrum of the AR operator.

$$H(Z) = \frac{1}{1 - 2.7607Z + 3.8160Z^2 - 2.6535Z^3 + 0.9238Z^4}$$

11.12 Apply the method of the previous two problems to synthetic data generated by passing white noise through the ARMA elliptic filter operator shown here:

$$H(Z) = \frac{0.05634(1 + Z)(1 - 1.0166Z + Z^2)}{(1 - 0.6830Z)(1 - 1.4461Z + 0.7957Z^2)}$$

11.13 Repeat Problems 11.10 through 11.12 using the iterative least-squares method of Fig. 10.7 to form power spectral estimates of the MA, AR, and ARMA data samples. Discuss your results, being sure to address questions of order and data sample length.

11.14 Repeat Problems 11.10 through 11.12 using ensemble averaging of periodograms. Discuss your results and compare them with Eq. (11.8).

11.15 Investigate the so-called line splitting phenomena by applying the Burg algorithm to synthetic data. Use three sinusoids imbedded in colored noise. Choose two of the sinusoids with incommensurate frequencies close together, forming a spectral doublet. Study the resolution of the Burg method for high and low signal-to-noise ratios. Observe line splitting at orders approaching the data length. Compare the Burg resolution with the uncertainty principle. Discuss the significance of this numerical experiment.

11.16 A standard method of finding the extrema of a function subject to a constraint is to use Lagrange multipliers. Show that taking the variation of the entropy

$$\delta \int_{-\omega_N}^{\omega_N} \ln \Phi(\omega) \, d\omega = 0$$

subject to the constraint

$$\phi_K = \int_{-\omega_N}^{\omega_N} \phi(\omega) e^{i\omega k} \, d\omega \quad p \le k \le p$$

gives

$$\phi(\omega) = \frac{1}{\sum_{-P}^{P} \lambda_K e^{i\omega k}}$$

where the λ's are the Lagrange multipliers.

11.17 Compare Eq. (11.22) for the second coefficient of the 2-long Burg prediction error filter with the same coefficient from the Yule–Walker method. Show that the Burg pole is necessarily closer to the unit circle than the Yule–Walker pole, giving greater spectral

290   11/ Power Spectral Estimation

resolution. Show that for large data runs this difference becomes insignificant.

11.18 In the Pisarenko PSDA, the data is modeled as a superposition of sinusoids y_t and additive noise $N_t: x_t = y_t + N_t$. The $p/2$ sinusoids are represented by a second-order difference equation with its zeros on the unit circle (see Problem 9.15):

$$y_t = \sum_{1}^{p} a_K y_{t-k}$$

Show that this data model leads to the eigenvalue problem:

$$\mathbf{\Phi A} = \sigma^2 \mathbf{A}$$

where $\mathbf{\Phi}$ is the autocorrelation matrix of the data x, σ^2 is the power in the noise spectrum, and A is the operator $(1 - a_1 Z - a_2 Z^2 - \cdots - a_p Z^p)$.

11.19 Show that the spectral lines of the Pisarenko method in the previous problem lie at the frequencies where the spectrum of the operator A vanishes.

11.20 The solutions to the previous two problems only give the frequencies of the sinusoids present in the Pisarenko estimate. The power contained at these frequencies can be estimated by a least-squares fit to the autocorrelation coefficients of the process. Show that the Pisarenko spectral lines of power P_k have autocorrelation coefficients

$$2\pi \phi_n = P_0 + 2 \sum_{1}^{p/2} P_k \cos(n\omega_k) \qquad n = 0, N$$

where the ω_k are the known spectral line frequencies. Write this equation as a matrix relation:

$$\mathbf{WP} = \mathbf{\phi}$$

where **W** is the nonsquare $(N+1) \times (p/2 + 1)$ matrix:

$$w_{nk} = \begin{cases} 2\cos(n\omega_k) & k = 0 \\ 1 & k = 0 \end{cases}$$

Finally, write a computer algorithm for fitting the P to the given ϕ_n, thereby producing the Pisarenko spectral estimate.

11.21 Show quantitatively that the results of Figs. 11.4 and 11.5 are consistent with the time–frequency statistical resolution principle expressed in Eq. (11.8).

11.22 An X-band police moving radar unit operates on a carrier frequency of 10,525 MHz with a 24° beam width. Random noise is returned from incoherent scatters in various targets. Some of these scatters are from vehicles moving toward and away from the patrol car, others are from the stationary environment. The Doppler effect shifts the frequency of this scattered energy an amount $\Delta f/f = \pm 2(v/c)$. But, because of the beam width, scatters have a range of apparent velocities, producing a frequency spread in the returned radar signal. Describe the power spectrum of the returned signal as quantitatively as you can. Suggest methods of determining the speeds of various targets to a given precision, for example, one mile per hour.

12

Multidimensional DSP

We have seen that many important problems can be addressed within the one-dimensional context of the preceding chapters. Admittedly, some interesting questions, such as issues surrounding optimal inverses and estimation of power spectra, have not been fully dealt with; they, in fact, promise research fodder into the foreseeable future. Since our goal is to provide an introduction to digital signal processing (DSP), we leave these more advanced topics in favor of an introduction to another arena—problems in two or more dimensions.

While many practical situations can indeed be treated within the 1-D concept, many cannot; we do not live in a 1-D world. In addition to the obvious three spatial variables of our world, there are also many cases in which other signals interact. In meteorology, for example, one may wish to describe pressure as a function of three coordinates, time, temperature, and density. In magnetohydrodynamics, one has the complexity of vector fields of both electrodynamics and fluid mechanics. Even in some 1-D cases, it turns out to be useful to cast the problem into a multivariate format. For example, the motion of a particle in one dimension can be described in terms of its position, velocity, and acceleration, all three treated as independent variables that interact via the equation of motion.

Again, our treatment will be introductory, providing information sufficient for both insight and practical applications. We will concentrate on problems described in terms of a function of two variables. Generally, the extension to three dimensions requires less edification; nothing would be gained by so complicating the discussion. Once the 2-D ideas are understood, application to higher dimensions is usually straightforward tedium.

Even the extension from 1-D to 2-D is frequently obvious, made in the most natural manner. Examples are the extension of the DFT and FIR filter design to multidimensions. However, there are some crucial differences that demand understanding. Perhaps at the most fundamental level stands the Z transform. Its extension to multidimensions is obvious and natural enough, but a basic algebraic fact thwarts one of its most important applications: polynomials in multidimensions cannot, in general, be factored. Thus, we are unable to perform pole–zero analysis of multidimensional systems. We can, of course, synthesize Z transforms from couplets in more than one variable. Such polynomials are factorable, but they are special cases.

This problem of factoring Z transforms impacts just about every topic in 2-D DSP. To emphasize the significance of the factoring problem, we will briefly discuss its effect in three applications of DSP: digital filter design, stability (which in turn effects many other applications, such as difference equations and IIR filter design), and power spectral estimation.

First, we consider the digital filter design problem. In the 1-D case, the design can proceed independently of the implementation scheme. Then, the filter's transfer function can be implemented in any fashion we please. Not so in multidimensions. If, for example, we wanted to filter a 2-D function by cascading parts of the transfer function, the design would be required to produce a factorable Z transform; this is a constraint on the design problem not present in one dimension. Thus, the design of the filter and its implementation become interwoven, a new phenomena compared to the 1-D case.

Second, we present an example concerning stability that has far reaching consequences. The three transfer functions

$$F(Z_1, Z_2) = \frac{1}{1 - (1/2)Z_1 - (1/2)Z_2}$$

$$G(Z_1, Z_2) = \frac{(1 - Z_1)^8(1 - Z_2)^8}{1 - (1/2)Z_1 - (1/2)Z_2}$$

$$H(Z_1, Z_2) = \frac{(1 - Z_1)(1 - Z_2)}{1 - (1/2)Z_1 - (1/2)Z_2}$$

all have the same denominator. It turns out that a function of the form $(1 - az_1 - bz_2)^{-1}$ is stable providing that $|a| + |b| < 1$. Therefore, F, above, is unstable. Additionally, Goodman (1977) has shown that G is stable while H is unstable. Thus, the numerator has the ability to make an unstable denominator, such as F's, stable. Clearly, the stability question in 2-D is complicated. Even the numerator has an effect, completely unlike the 1-D situation. Although, as of yet, there is no completely general approach for determining the requirements for stability for 2-D and higher dimension systems in general, sufficient conditions are known for certain subclasses of systems. Unfortunately, frequently even these are difficult to employ in practice.

Last, we comment on the 2-D power spectral estimation problem. In the 1-D case, we can estimate a power spectrum whose inverse Fourier transform has exactly the given autocorrelation coefficients. This is not true in higher dimensions because there is insufficient freedom. In the 2-D case, for example, the normal equations would determine an $N \times M$ prediction error filter, but there are $2 \times M \times N - M - N - 1$ independent specified autocorrelation coefficients that the estimated spectrum would have to recover. Thus, the all-pole model cannot match a given set of autocorrelation coefficients. If one wishes to formulate a max-entropy problem with prior knowledge of given autocorrelation coefficients, the all-pole model is not the answer in the 2-D case. It is the factorability of 1-D polynomials that allows an all-pole spectral estimate to be written as a product of a minimum phase prediction error filter and its complex conjugate [recall Eq. (11.15), which is essentially this statement].

So the nonfactorability of multivariate system functions prohibits a *carte blanche* extension of the 1-D discussions in the preceding chapters to higher dimensions. We shall review selected topics that are free from the factorability limitation, are easy to apply, and are extremely important in practical applications.

Our first topic is difference equations in multidimensions. We shall see that many partial differential equations that are important in areas of physics and engineering lead to difference equations with separable convolutional operators. This does allow for a satisfactory understanding of their stability properties.

Multidimensional Difference Equations

Partial differential equations arise in many branches of science, engineering, and mathematical physics. Compared to the large number of applications, only a relatively few analytical solutions to these equations can be found. Generally, these solutions are only possible in restricted circumstances of highly geometrical boundary conditions or simplistic initial conditions. Even so, such solutions are, of course, of great importance for providing insight to problems and for guiding analysis of more complex situations.

In cases where the boundary and initial conditions prohibit analytical solutions, we turn to numerical methods of some kind. The constraints, even if expressible analytically, may still defy solution; or, as is quite

Multidimensional Difference Equations 295

frequently the case in applied problems, the constraints may be expressed numerically. Even in cases where analytical solutions are known, numerical solutions sometimes can be just as practical as evaluating infinite sums of complicated terms of an analytical solution. Numerical solutions to partial differential equations are thus of great practical value. Because of this value to many fields of science and engineering, much effort has been expended in studying numerical methods of solving partial differential equations.

In practical applications, usually boundary conditions are so specified that the solution is uniquely determined by them. The unknown function depends on two or more variables. Where possible, the variables are selected so that the boundary conditions are simply expressed as surfaces in that coordinate system. For example, the unknown function might be $\psi(x, y, z, t)$. One boundary condition might be $\psi(x, y, 0, 0) = f(x, y)$; another might specify the normal derivative to the surface:

$$f'(x, y) = \frac{\partial \psi(x, y, z, t)}{\partial z}\bigg|_{\substack{z=0 \\ t=0}}$$

The specifications of only the values on the boundary are called *Dirichlet conditions*, while the specifications of normal gradients are called *Neumann conditions*. In some cases, the boundary conditions may be mixed, being Dirichlet along a portion of a surface and Neumann along the remainder. Some equations may require both the Dirichlet and the Neumann conditions for a unique solution; when both are given, they are called *Cauchy conditions*.

Finally, the boundaries themselves may be open or closed. A *closed* boundary encloses the solution, while an *open* boundary extends to infinity with no conditions imposed there. *Initial value problems* are a type of open boundary value problem where one boundary is the surface specified at $t = 0$ and no specification is given for the infinite future.

The exact boundary conditions required for fixing a unique solution depend on the partial differential equation. It turns out that these conditions really depend on categories of partial differential equations defined by the geometry of certain characteristic curves. Three such categories are elliptic, hyperbolic, and parabolic. Poisson's equation, the wave equation, and the diffusion equation are the archetypal members of these respective categories. In general, questions of existence and uniqueness are quite difficult to answer, but Fig. 12.1 summarizes well-known results for the three categories mentioned. We will concentrate on the initial value problem. Since this has an open boundary surface (specified by the disturbance at $t = 0$), Fig. 12.1 shows that the diffusion equation and the wave equation will provide unique solutions for our following examples.

	TYPE OF EQUATION		
TYPE OF BOUNDARY CONDITION	ELLIPTIC (POISSON'S EQ.)	HYPERBOLIC (WAVE EQ.)	PARABOLIC (DIFFUSION EQ.)
DIRICHLET OPEN SURFACE	INSUFFICIENT	INSUFFICIENT	UNIQUE, STABLE SOLUTION IN ONE DIRECTION
CLOSED SURFACE	UNIQUE, STABLE SOLUTION	OVER SPECIFIED	OVER SPECIFIED
NEUMANN OPEN SURFACE	INSUFFICIENT	INSUFFICIENT	UNIQUE, STABLE SOLUTION IN ONE DIRECTION
CLOSED SURFACE	UNIQUE, STABLE SOLUTION IN GENERAL	OVER SPECIFIED	OVER SPECIFIED
CAUCHY OPEN SURFACE	UNPHYSICAL RESULTS	UNIQUE, STABLE SOLUTION	OVER SPECIFIED
CLOSED SURFACE	OVER SPECIFIED	OVER SPECIFIED	OVER SPECIFIED

Figure 12.1 Effect of boundary conditions on three categories of partial differential equations.

Our approach will use the familiar difference operators, including the bilinear transform, that we employed in studying ordinary differential equations. At times, we will also take advantage of the Z-transform, but now it will be multidimensional. Again stability will be an important consideration, but now the problem is much more complex. Nonetheless, for the equations under study, we will be able to give general stability criteria. In the strict sense, they will apply only to boundary conditions imposed on periodic solutions. But, they will still provide theoretical insight and practical guidance in the use of finite difference methods.

Even though the numerical solution to partial differential equations is a vast field, we will see that the methods of digital signal processing will serve us well in gaining a working knowledge of this important application.

The diffusion equation

Of all the partial differential equations that arise in mathematical physics, perhaps the one that is best suited for our introductory treatment is the diffusion equation. It is not too difficult to solve, yet it is not a trivial equation either. It arises in many applications: heat flow, diffusion, and highly viscously damped waves. The equation in one spatial dimension is

Multidimensional Difference Equations

$$\frac{\partial \psi}{\partial t} = \alpha \frac{\partial^2 \psi}{\partial x^2} \qquad (12.1)$$

In the diffusion problem, ψ is a concentration and α is the diffusion constant. In the heat flow case, ψ is the temperature and $\alpha = T/C$, where T is the thermal conductivity and C is the heat capacity; α is assumed to be independent of x and t.

To convert this equation to a difference equation, we proceed as in Chapter 3: we simply substitute finite difference operators for the partial derivatives occurring in the equation. For the time derivative, we use the first forward difference operator

$$\frac{\partial \psi}{\partial t} \to \frac{\psi_k^{n+1} - \psi_k^n}{\Delta t} \qquad (12.2)$$

Writing the time indices as superscripts and the spatial indices as subscripts is conventional in the theory of multivariate difference equations. Similarly, the required second-order partial derivative in x is written

$$\frac{\partial^2 \psi}{\partial x^2} \to \frac{\psi_{k+1}^n - 2\psi_k^n + \psi_{k-1}^n}{(\Delta x)^2} \qquad (12.3)$$

So that the finite difference equation representing Eq. (12.1) becomes

$$\psi_k^{n+1} - \psi_k^n = a(\psi_{k+1}^n - 2\psi_k^n + \psi_{k-1}^n) \qquad (12.4)$$

after setting $a = \alpha \, \Delta t/(\Delta x)^2$.

Solving for ψ at the latest time gives

$$\psi_k^{n+1} = a\psi_{k+1}^n + (1 - 2a)\psi_k^n + a\psi_{k-1}^n \qquad (12.5)$$

The right-hand side of this equation is an MA operation:

$$\psi^{n+1} = (a, 1 - 2a, a) * \psi^n \qquad (12.6a)$$

$$\psi^{n+1} = Q * \psi^n \qquad (12.6b)$$

So, an initial value problem can be solved by the straightforward application of Eq. (12.6): given an initial ψ_k^0 defined on the grid k, we can compute ψ_k^n at each new time step by an MA operation on the previous values of ψ.

Before such an iteration is begun, we must stipulate the boundary conditions at the two extremes of the x coordinate. According to Fig. 12.1, we have a choice between Dirichlet and Neumann conditions. Having properly specified one of these conditions, the simple iteration implied by Eq. (12.6) is easy to carry out, giving good results in practice. The convolution in the spatial coordinate is stable since it is an MA operation. But what about stability of the time recursion? In general, this is a difficult question to answer. However, in the case of Eq. (12.6), we can draw some conclusions. We propose that the sufficiency condition for stability is

$$a \leq \tfrac{1}{2} \qquad (12.7)$$

Under this condition, all the coefficients of Q are positive. They also sum to one. Thus, by looking at maximum values across the spatial grid, we can write

$$\text{Max}|\psi^{n+1}| \leq |\textstyle\sum q_i| \quad \text{Max}|\psi^n| \leq \text{Max}|\psi^n|$$
$$\leq \text{Max}|\psi^{n-1}| \leq \cdots \leq \text{Max}|\psi^0|$$

Therefore, ψ^{n+1} is bounded if ψ^0 is.

At this point, an example of using Eq. (12.6) will be instructive. Figure 12.2(a) shows an initial distribution of ψ^0. We have selected a ψ^0 with discontinuities because it seems intuitively that such an initial distribution is likely to provide the best education on the subject of stability. This initial condition could represent the temperature of a rod that has had three parts separated and held at fixed temperatures. At $t = 0$, the three parts are brought into contact forming the initial temperature distribution.

At the ends of the rod, $x = 0$ and $x = L$, we can either specify the temperature (Dirichlet conditions) or the heat flow expressed by $\partial \psi / \partial x$

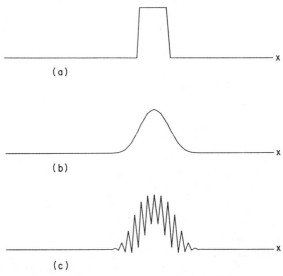

Figure 12.2 Solutions to the diffusion equation using the finite difference equation, Eq. (12.6). (a) The initial distribution defined over 71 grid points. (b) The distribution evolved through 10 steps in time with $a = 0.45$. (c) The distribution after 10 steps in time with $a = 0.55$, showing instability because $a > \tfrac{1}{2}$.

(Neumann conditions). We can have one type of condition at $x = 0$ and the other at $x = L$. The conditions could even vary in time. To specify constant temperature at each end, linear convolution would be used running the operator onto the constant values at each end. The Neumann condition of zero heat flow at the ends (insulated ends) can be emulated by imagining a periodic solution, one rod repeated after another. Circular convolution appropriately simulates this boundary condition.

We have selected this periodic boundary condition for our example. In Fig. 12.2(b), the temperature distribution is seen after 10 iterations using the solution of Eq. (12.6) with $a = 0.45$. The discontinuity has been removed by the subsequent heat flow. But with $a = 0.55$, Fig. 12.1(c) displays an instability after the same 10 time steps. In 100 steps, the amplitudes of these oscillations grow to the order of 10^6. Apparently then, instability sets in at $a = \frac{1}{2}$, suggesting that the criterion of Eq. (12.7) is also a necessary condition for stability.

Instability of this kind is a major concern in the study and use of finite difference techniques. It is a fundamental phenomenon of difference equations, having no analog in the one-dimensional case. It has nothing to do with numerical limitations, such as sound-off errors, but depends entirely on the Δt, Δx grid size. The constant $a = \alpha \Delta t / (\Delta x)^2$ scales the proportion of this grid, and its size, to the physical parameter α. Therefore, the $a \leq \frac{1}{2}$ criterion tells us that given α, the relationship between the grid size in each coordinate direction is crucial. Simply using a finer grid in the x direction, for example, makes the stability situation even worse, a circumstance that might have surprised us without a knowledge of the $a \leq \frac{1}{2}$ criterion. A finer grid size in x actually requires a finer time step to maintain stability.

The difference between the finite difference solution and the exact solution of the partial differential equation could also be studied. Two questions could be addressed: (1) what is the behavior of their difference as $t \to \infty$ for fixed Δx and Δt? and (2) what is the behavior of their difference as the grid Δx and $\Delta t \to 0$ for fixed time? The second question is the more important of the two because it addresses the accuracy of a given solution—a question of great practical value. It turns out that both questions lead to similar conclusions; namely, that as long as the relative rates at which Δx and Δt go to zero satisfy the stability requirements, the errors in the difference equation solutions will be small. For closed boundary value problems, the situation is more complicated. Therefore, for our purpose of relating digital signal processing techniques to finite difference methods, we will limit outselves to a few examples of initial value problems, concentrating on their stability requirements.

For our next example, we employ the backward difference operator for the time derivative in the diffusion equation, Eq. (12.1). This time we work in the Z domain, writing Z for the time Z transform and z for the spatial Z transform. Then, using the same second spatial derivative as before, Eq. (12.1) becomes

$$(1 - Z)\psi = a(z^{-1} - 2 + z)\psi$$

$$\psi = \frac{Z\psi}{-az^{-1} + (1 + 2a) - az}$$

Thus, a time recursion results using an AR operator on the previous spatial distribution:

$$\psi^n = (-a, 1 + 2a, -a)^{-1} * \psi^{n-1} \tag{12.8a}$$

$$\psi^n = Q * \psi^{n-1} \tag{12.8b}$$

The operator Q can be written in terms of its poles

$$Q = \frac{z}{-a(z - r_1)(z - r_2)}$$

where

$$r_{1,2} = \frac{(1 + 2a) \pm \sqrt{1 + 4a}}{2a} \tag{12.9}$$

Both poles are real since the constant a is necessarily positive. Because of the symmetric nature of Q, its poles must occur in reciprocals pairs. Therefore, one must be inside the unit circle, we will call that one r, and the other, $1/r$, must lie outside the unit circle.

Since Q does not have a minimum phase denominator, it cannot be used in the AR mode; the convolution in Eq. (12.8) can only be implemented in the MA mode. The required MA coefficients are computed using partial fractions as discussed in Chapter 2, expanding each pole as appropriate in positive and negative powers of z. Because of the symmetry of Q under an interchange of z and z^{-1}, the coefficients must likewise be symmetric in z. Carrying out this expansion gives

$$Q = \frac{1}{-az^{-1} + (1 + 2a) - az} = \frac{1}{a(1 - r^2)} \sum_{-\infty}^{\infty} r^{|n|+1} z^n \tag{12.10}$$

Since r is the pole that is inside the unit circle, this expansion converges for all values of a. Therefore, if the grid is selected to keep r small, relatively few terms of Q will be required in the convolution.

Having decided the question of how to handle the IIR operator in the spatial coordinate, the next question concerns the stability of the time recursion. The same discussion as before, following Eq. (12.7), applies. Each coefficient of the operator Q is positive independent of the constant

Multidimensional Difference Equations

a, and they sum to one (Problem 12.5). Therefore, ψ^n is bounded if ψ^0 is. This conclusion is valid for all values of a, perhaps a surprising result. Thus, an intriguing trade-off has been made. For the price of the longer spatial convolution, we reap unconditional stability in the time recursion. This is particularly nice because it means that Δx and Δt can be selected independently, based on considerations such as desired resolution and accuracy, rather than based on stability.

Although this discovery is rewarding, it gives us no insight into why the backward difference operator leads to such a different result than the forward operator. In our discussion of difference operators in Chapter 3, we saw that one key for maintaining the proper stability characteristics of an ordinary differential equation lies in centering the finite differences on the grid points. Both the forward and backward difference operators resulted in a mismatch of the centers of the differences. Figure 12.3 shows the locations of ψ on the n–k grid and the mismatch of the locations of the first- and second-order differences. In Chapter 3, we saw how the bilinear transform centered the first-order difference on the grid points, matching the grid location used with other values in the difference equation. Without this match, proper stability behavior was not guaranteed. The two-step central difference operator did not work either (the unnecessarily large step size, in effect, creates a second-order operator). The same considerations apply to Eq. (12.1). But now, because of the two dimensions we can think of centering in either the temporal dimension or the spatial dimension. If we pick time, we could substitute the bilinear transform for

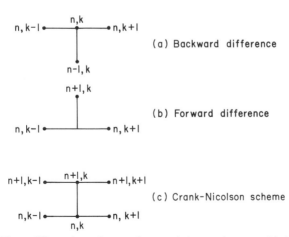

Figure 12.3 Three differencing schemes shown relative to the x–t grid. Time is plotted vertically, increasing upward. The x coordinate is plotted horizontally, increasing to the right. The Crank–Nicolson scheme centers the difference equation on the point $(n + \frac{1}{2}, k)$. It gives results identical to the bilinear transform.

Eq. (12.2). If we chose to center in x, we could substitute for Eq. (12.3) the average

$$\partial^2 \psi / \partial x^2 \rightarrow \tfrac{1}{2}(\psi_{k+1}^{n+1} - 2\psi_k^{n+1} + \psi_{k-1}^{n+1}) + \tfrac{1}{2}(\psi_{k+1}^n - 2\psi_k^n + \psi_{k-1}^n) \quad (12.11)$$

This latter method is called the Crank–Nicolson scheme. It gives the same result as the bilinear transform (Problem 12.7). To perform our centering, we will select the bilinear transform. Again using transform variables Z and z for time and space, the diffusion equation now becomes in the transform domain

$$2\frac{(1-Z)}{(1+Z)}\psi = a(z^{-1} - 2 + z)\psi$$

$$\frac{(1-Z)}{(1+Z)}\psi = bT\psi \quad (12.12)$$

where we have written T for the operator $(1, -2, 1)$ and $b = a/2$. Using just algebraic manipulation, Eq. (12.12) can be solved for ψ in terms of its time delayed values and an ARMA spatial operator:

$$\psi = \frac{1 + bT}{1 - bT} Z\psi \quad (12.13a)$$

$$\psi^n = \frac{(b, 1 - 2b, b)}{(-b, 1 + 2b, -b)} \psi^{n-1} \quad (12.13b)$$

$$\psi^n = U * \psi^{n-1} \quad (12.13c)$$

This equation has, by now, a familiar form. Aside from the factor of 2 in $b = a/2$, the numerator of the ARMA operator U is the same as the operator that arose from the use of the forward time difference, and the denominator is the same as the operator that arose from the use of the backward time difference. The implementation of Eq. (12.13) is similar to that of Eq. (12.8) for the AR case: expand the denominator in positive and negative powers of z and then convolve with the numerator. We already have the denominator expansion described in Eq. (12.10). Carrying out the convolution (see Problem 12.8) gives the coefficients of U:

$$u_0 = \frac{(1 - 2b + 2br)r}{b(1 - r^2)} \quad (12.14a)$$

$$u_n = \frac{2r^{|n|+1}}{b(1 - r^2)} \quad (12.14b)$$

Multidimensional Difference Equations 303

where

$$r = \frac{(1 + 2b) - \sqrt{1 + 4b}}{2b} \qquad (12.14c)$$

is the pole that is located inside the unit circle.

Because we have taken care to center the differences properly, we expect that Eq. (12.13) will give stable time recursions for all values of b. This is indeed true, as we shall see later. It is easy to carry out solutions according to Eqs. (12.13) and (12.14) for a large range of b, experimentally suggesting that they are, in fact, stable for all b (see Problem 12.9).

Such a mathematical experiment will, however, yield another surprise. For values of b greater than about 0.8, the central coefficient u_0 becomes negative in Eq. (12.14a), and large errors occur in the solution. A plot of solutions for the first eight time steps using $b = 2.0$ is shown in Fig. 12.4. The discontinuity in ψ^0 produces large deviations from the exact solution, but these departures do diminish with increasing time. Thus, the time

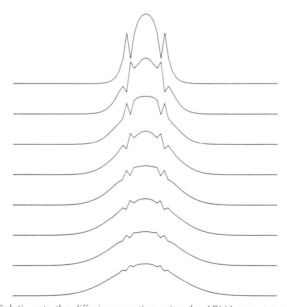

Figure 12.4 Solutions to the diffusion equation using the ARMA operator according to the finite difference equation, Eq. (12.13). The initial conditions are the same as those in Fig. 12.2. Time increases from top to bottom in 8 time steps. In Eq. (12.13b), the coefficient $b = 2$, producing poor behavior in the solution, which does decay in time; hence, the solution is stable. For $b < 0.8$, the solution is smooth for all time steps. The ARMA operator has been truncated to 19 terms in the MA mode.

recursion is stable. Such behavior is dependent on the initial conditions, discontinuities being the toughest for the finite differences to handle. Nonetheless, the behavior shown is not instability, and as the time steps are made smaller (that is b made smaller), the solution does approach the exact solution. For b less than about 0.8, the nonphysical temperature gradients seen in Fig. 12.4 suddenly disappear.

The wave equations

These finite difference methods that we have applied to the diffusion equation can be equally well applied to any partial differential equation. Stability and accuracy must, however, remain a primary concern. To further demonstrate this procedure, we will give examples from three forms of the wave equation. In its simplest form, the wave equation represents wave motion in one dimension, such as waves on a string. The equation is

$$\frac{\partial^2 \psi}{\partial t^2} = v^2 \frac{\partial^2 \psi}{\partial x^2} \qquad (12.15)$$

where ψ is the disturbance (such as the small transverse deflection of a string or the pressure of an acoustic plane wave), and v is the propagation velocity. Applying the central second difference operator to both sides gives

$$\psi_k^{n+1} - 2\psi_k^n + \psi_k^{n-1} = \frac{v^2 (\Delta t)^2}{(\Delta x)^2} (\psi_{k+1}^n - 2\psi_k^n + \psi_{k-1}^n)$$

This equation can be written in terms of a three-term MA operator acting on the x coordinate

$$\psi^n = [a, 2(1-a), a] * \psi^{n-1} - \psi^{n-2} \qquad (12.16)$$

where $a = v^2 (\Delta t)^2 / (\Delta x)^2$. An interesting variation to the simple one-dimensional wave equation occurs when one imagines springs attached at right angles to the string (or alternately imagines immersing the string in an elastic medium). The result is the Klein–Gordon equation in one dimension:

$$\frac{\partial^2 \psi}{\partial t^2} = v^2 \frac{\partial^2 \psi}{\partial x^2} - \alpha \psi \qquad (12.17)$$

where α is related to the transverse elastic constant. The Klein–Gordon equation also arises in quantum mechanics when the Schrödinger equation is applied to a relativistic particle. In that case, α is related to the rest mass of the particle. The same procedure used above yields only a slightly

Multidimensional Difference Equations 305

modified difference equation:

$$\psi^n = [a, 2(1 - a - b/2), a] * \psi^{n-1} - \psi^{n-2} \qquad (12.18)$$

where $b = \alpha(\Delta t)^2$.

Equation (12.6a), representing the diffusion equation, Eq. (12.16), representing the wave equation, and Eq. (12.18), representing the Klein–Gordon equation, all appear quite similar. But the wave equations, being second order in time, require both ψ^{n-1} and ψ^{n-2} for the computation of each time step. Therefore, they also require a knowledge of the disturbance at $t = 0$ and its first difference (the Cauchy conditions on an open surface).

The three-term x coordinate convolutional operator is also quite similar for all three equations, differing only in the relative weight of the central coefficient compared to its two neighbors. The stability lesson that we have learned from the diffusion equation applies here as well. As long as this central coefficient is positive, the time recursion will be stable. Figure 12.5 shows the solution of the wave equation for an initial triangular displacement with zero initial velocity. The disturbance propagates in both directions and is reflected back from the end of the x grid where the disturbance

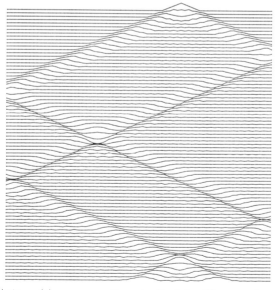

Figure 12.5 Solutions of the wave equation using the finite difference scheme of Eq. (12.16). The initial condition (at top of figure) is a triangular disturbance with zero initial velocity. Subsequent traces are the evolution in time (every two steps) as the disturbance propagates outward, left and right. The 71-point X direction grid acts like a perfect reflector at the end points, producing disturbances that return to the original form after two reflections. In Eq. (12.16), the coefficient $a = 0.9$.

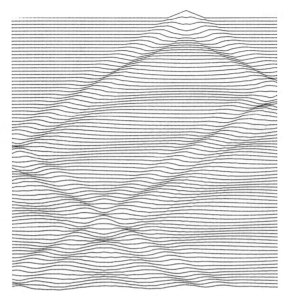

Figure 12.6 Solutions of the wave equation showing dispersion. The finite difference solution, Eq. (12.18), of the Klein–Gordon equation was used with $a = 0.5$ and $b = 0.1$. Other conditions are the same as in Fig. 12.5. It is the relationship between the central coefficient and its two neighbors of the MA operator in Eq. (12.18) that produces dispersion.

is held at zero deflection. When the effects of the constant b are brought into play via the Klein–Gordon equation, dispersive propagation occurs. Figure 12.6 shows this dispersion whereby the pulse changes shape as it propagates. In both cases, the solutions are stable because the selection of a and b maintained a positive central coefficient. Otherwise, instability sets in. For example, Fig. 12.7 shows growing oscillations in the solution of the wave equation with $a = 1.0005$. For larger values of a, the oscillations grow quite rapidly.

The second-order derivatives in these wave equations have permitted the use of the centered second-order difference operators. Their use has

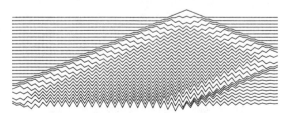

Figure 12.7 Solution to the wave equation using the finite difference scheme of Eq. (12.16). The coefficient $a = 1.0005$, producing instability. For $a < 1$, the solution is stable.

Multidimensional Difference Equations

led to quite simple difference equations with easily understood stability properties and easily applied recursions.

As our last example of finite difference methods, we select a wave equation that contains first-order derivatives, suggesting a retreat back to the bilinear transform. We have seen that the bilinear transform is not necessary for successful use of finite difference methods, but it has superior stability properties compared to first-order difference operators. The equation in question is

$$\frac{\partial^2 \psi}{\partial z \partial t} = -\frac{v}{2} \frac{\partial^2 \psi}{\partial x^2} \qquad (12.19)$$

This equation, called the 15° Claerbout equation, arises in seismic wave imaging. It is an approximation to the wave equation, describing waves in the x–z plane moving parallel to the z axis within about 15°. Furthermore, it only includes waves traveling in one direction along the z axis, as suggested by its first-order derivative in z. The equation attempts to describe an equivalent model used in reflection seismology whereby, after preliminary processing, data are treated as though a coincident source and receiver were deployed on the earth's surface. The receiver records the energy returned from subsurface reflectors along unknown ray paths. The pair is then moved to a new x coordinate and a new record is made there, eventually obtaining the gridded data $\psi(x, 0, t)$. Because two-way travel time is used, the velocity is halved in Eq. (12.19). Thus, one imagines that the energy left the reflector at $t = 0$. The model then implies that $\psi(x, z, 0)$ is the location of the seismic wavefront when it was on the reflectors; hence, $\psi(x, z, 0)$ is a picture of the subsurface reflectors. So, the seismic imaging problem is: given the sampled wavefield $\psi(x, 0, t)$ measured over a large range of x and t, determine $\psi(x, z, 0)$, the picture of the subsurface. An alternate problem is: given $\psi(x, z, 0)$, determine $\psi(x, 0, t)$. For more background on this seismic application, see Claerbout (1976, 1985).

To solve Claerbout's equation using finite differences, we employ the bilinear transform for both of the first-order derivatives and the centered second-order difference operator for the x coordinate derivatives. Using, Z, z, and X for the Z-transformed variables of the time, z coordinate, and x coordinate, respectively, Eq. (12.19) becomes

$$4\left(\frac{1-Z}{1+Z}\right)\left(\frac{1-z}{1+z}\right)\psi = -\frac{v \Delta t \Delta z}{2(\Delta x)^2}(X^{-1} - 2 + X)\psi$$

Writing $a = v \Delta t \Delta z / 8 (\Delta x)^2$, $T = (X^{-1} - 2 + X)$ and solving for ψ in terms of recursions on z and Z gives

$$(z + Z)\psi = \left(\frac{1 + aT}{1 - aT}\right)(1 + zZ)\psi \qquad (12.20a)$$

or

$$(z+Z)\psi = \frac{(a, 1-2a, a)}{(-a, 1+2a, -a)}(1+zZ)\psi \qquad (12.20\text{b})$$

This is the appropriate form to solve the two imaging problems mentioned previously because it relates differences in depth (the z coordinate) and time through an operator on the x coordinate grid (see Problem 12.12). Because this x coordinate operator arises from the same algebraic manipulations used in developing the ARMA difference equation for diffusion, it is identical to the one in Eq. (12.13). Thus, we have already computed the MA mode coefficients. They are given by Eqs. (12.14). Again, both the depth and time recursions are stable for all values of the parameter a.

Stability of difference equations

The stability of difference equations is an important consideration in their use. It not only is important for avoiding bad solutions, but it is related to understanding their convergence rate. Now that we have seen the application of various differencing techniques to several partial differential equations, representing a fairly wide class of equations of mathematical physics, we are in a position to gain some insight into the stability question. For convenience of the discussion, we list some of our previous results, using an abbreviated description:

fwd, diffusion: $\qquad \psi = (a, 1-2a, a)\psi Z \qquad (12.21\text{a})$

bwd, diffusion: $\qquad \psi = \dfrac{\psi Z}{(-a, 1+2a, -a)} \qquad (12.21\text{b})$

bilinear, diffusion: $\qquad \psi = \dfrac{(a, 2-2a, a)}{(-a, 2+2a, -a)}\psi Z \qquad (12.21\text{c})$

bilinear, Claerbout: $\qquad (z+Z)\psi = \dfrac{(a, 1-2a, a)}{(-a, 1+2a, -a)}(1+zZ)\psi \qquad (12.21\text{d})$

These equations present a glaring occurrence of similar operators $A(a)$ and their inverses with the value of a negated, $A^{-1}(-a)$, a fact of great significance. To see why, we first consider the first two entries. Recall that the forward differencing version of the diffusion equation is conditionally stable, while the backward differencing form is unconditionally stable. This backward version can be used to work backward in time:

$$\psi^{n-1} = (-a, 1+2a, -a)\psi^n \qquad (12.22\text{a})$$

$$\psi^{n-N} = (-a, 1+2a, -a)^N \psi^n \qquad (12.22\text{b})$$

So, we can start at a time n and work backward to a time N, arbitrarily far in the past. This requires the operator to the Nth power. Similarly, starting at time n and working to the future requires the inverse of the operator to the Nth power. The forward working time steps are stable. Therefore, the backward ones are unstable.

A little thought shows that backward stepping in time should be unstable. If we start with some temperature distribution and ask what distribution in the infinite past did it diffuse from, the solution should diverge. Any roughness in the current temperature came from an infinitely rough distribution in the distant past. Thus, physical considerations require that Eq. (12.22) be unstable. Therefore, its forward stepping in time given by Eq. (12.21b), using the inverse operator, must be stable. This discussion points out the relationship among the operator $(-a, 1+2a, -a)$, its inverse, and time reversal. Equation (12.22) works the time steps backward. But changing $a = k\,\Delta t/(\Delta x)^2$ to $-a$ reverses the flow of time again, transforming Eq. (12.22) back to Eq. (12.21a).

So, we have interestingly related the consistency of the stability properties of Eqs. (12.21a) and (12.21b) to time reversal and physics. But we would like to know specifically what controls their individual stability properties. Additionally, we would also like to know why both of the ARMA operators in Eqs. (12.21) happen to be unconditionally stable. To this question, we provide mathematical, rather than physical, insight.

This insight comes from Sylvester's matrix theorem. This theorem allows us to write powers of a matrix A as

$$A^N = \lambda_1^N B_1 + \lambda_2^N B_2 + \cdots + \lambda_n^N B_n \qquad (12.23)$$

The λ's are the eigenvalues of the diagonalizable matrix A and the B's are certain basic matrices composed of the row and column eigenvectors of A (see Problem 12.16). The recursions of Eqs. (12.21) involve just such powers of the operators in question. Recall that the convolutions can be written as Toeplitz matrices. For circular convolution, appropriate for periodic solutions, the matrix is circulant. It follows then that these recursions in time will be stable if and only if the largest eigenvalue of A (sometimes called the *spectral radius* of A) lies inside the unit circle (the eigenvalues can be complex).

It is easy to compute the eigenvalues of matrices corresponding to operators of the form (a,b,a); see Problem 12.17. We do not need to worry about matrix inverses because Eq. (12.23) also applies for negative N. Using this approach, we find that $(a, 1-2a, a)$ has all its eigenvalues inside the unit circle for $a \leq \frac{1}{2}$, while $(-a, 1+2a, -a)$ has all of them outside the unit circle for all values of a. Hence, by Eq. (12.23), the inverse of $(-a, 1+2a, -a)$ is unconditionally stable.

Furthermore, the two ARMA operators of Eqs. (12.21) are of the form $A(a)A^{-1}(-a)$, where $A(a) = (a, b - 2a, a)$. Applying Eq. (12.23) and the properties of the B's to this special form gives

$$\frac{A^N(a)}{A^N(-a)} = \frac{\lambda_1(a)}{\lambda_2(-a)} B_1 + \frac{\lambda_2(a)}{\lambda_2(-a)} B_2 + \cdots + \frac{\lambda_n(a)}{\lambda_n(-a)} B_n \quad (12.24)$$

The eigenvalues of such ARMA operators are seen to lie inside the unit circle independent of the value of a providing that b is positive (Problem 12.18). Incidentally, by Eq. (12.24), the inverse of these ARMA operators will therefore be unconditionally unstable—and this result makes good physical sense.

Another implication of Eq. (12.23) is that the rate of convergence in time is controlled by the eigenvalues. In fact, the division by the $\lambda(-a)$ in Eq. (12.24) for the ARMA operators not only serves to make them unconditionally stable, but also accelerates their convergence.

We have seen that the study of the stability and convergent rate of difference equations is both an interesting and challenging business. Because of the application to a wide variety of practical problems, this business is also profitable with much research currently being conducted. We have only scratched the surface. Yet, we have shown how to use digital signal processing methods to construct, analyze, and use difference equations arising from common partial differential equations. The principle moral—which experience, not mathematical proof, bears out—is that centering differences leads to difference equations that display the same stability properties of the parent partial differential equation. Other differencing can produce difference equations that have the incorrect stability properties for certain values of the grid parameter. In either case, the prudent worker will investigate the region of proper behavior of the difference equations under study.

Fourier Transform Methods in Multidimensions

Extending our previous discussion of Fourier transforms in two or more dimensions is rather straightforward with no particular pitfalls in store. All of the four types of Fourier transforms summarized in Fig. 7.2 can be so extended. In just 2 dimensions, the possible combinations of the 4 types for each coordinate add up to 16 different possibilities. We need not discuss all combinations, or even very many of them, because their extension will be obvious from just looking at 2 archetypical possibilities. The best 2 for this purpose are the 2-D DFT and the 2-D Fourier integral transform. We look at the latter first.

Fourier Transform Methods in Multidimensions 311

The 2-D Fourier integral transform simply treats the 2 variables alike:

$$F(k, \omega) = \int_{-\infty}^{\infty} \int_{\infty}^{\infty} f(x, t) e^{-i(kx+\omega t)} \, dx \, dt \tag{12.25a}$$

$$f(x, t) = \frac{1}{4\pi^2} \int_{-\infty}^{\infty} \int_{\infty}^{\infty} F(k, \omega) e^{i(kx+\omega t)} \, dk \, d\omega \tag{12.25b}$$

Anticipating applications using time and space as variables, we have written x for the spatial coordinate and k for its frequency. This variable k is called the x wave number. It is measured in radians/length. Both x and t, and k and ω, are on equal mathematical footings; only the physical significance that we give them make them different. In some applications, these variables may represent something else physically; then, the physical meaning we are using here would change accordingly.

All properties, such as the special values and symmetry properties of the 2-D Fourier transform, follow directly from the corresponding 1-D properties. One important example is the symmetry property for real $f(x, t)$. By computing the real and imaginary parts of F from Eq. (12.25), it follows that

$$\text{Re } F(-k, -\omega) = \text{Re } F(k, \omega) \tag{12.26a}$$

$$\text{Im } F(-k, -\omega) = -\text{Im } F(k, \omega) \tag{12.26b}$$

That is for real $f(x, t)$, the Fourier transform's real part is symmetric about the origin in $k-\omega$ space, and its imaginary part is antisymmetric about the origin. Thus, only one-half of the $k-\omega$ plane is required to display the 2-D spectrum.

To gain some insight to the 2-D Fourier transforms, we will look at two special examples, the transform of sinusoids and of a line. We first look at the sinusoid

$$f(x, t) = e^{i(k_0 x + \omega_0 t)} \tag{12.27}$$

Substituting this into the definition of Eq. (12.25a) and using our knowledge of 1-D transforms gives

$$F(k, \omega) = 4\pi^2 \, \delta(k_0 - k) \delta(\omega_0 - \omega) \tag{12.28}$$

This result is a delta function located at (k_0, ω_0) as shown in Fig. 12.8(a). This delta function is eminently reasonable because $f(x, t)$ contains only pure sinusoids at the two frequencies k_0 and ω_0. The function $f(x, t)$ in Eq. (12.27) is called a plane wave because the locations of constant phase $\phi = k_0 x + \omega_0 t$ form a plane in three spatial dimensions. [In Eq. (12.27), $f(x, t)$ is independent of y and z; thus, the surfaces of constant phase are planes.] Since we can write $x = \phi_0/k_0 - \omega_0 t/k_0$, these constant phase surfaces move with velocity $v = \omega_0/k_0$, an interpretation only valid when we

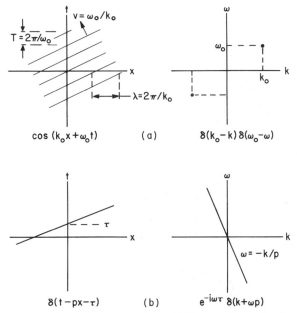

Figure 12.8 Two pairs of 2-D Fourier transform. (a) The transform of a plane wave is a δ function located at (δ_0, k_0). (b) The transform of a δ function line is another δ-function line, with reciprocally related slopes. Only the magnitude spectrum is sketched. The phase spectrum is linear with slope $-\tau$.

ascribe the physical properties of time and distance to t and x. These plane waves in 2-D are analogous to sinusoids in 1-D. The plane wave has a sinusoidal variation in both x and t. It also has a delta function spectrum in both k and ω. Thus, any reasonable function $f(x,t)$ can be built up of plane waves according to Eq. (12.25b) or analyzed in terms of them by Eq. (12.25a). This latter decomposition is sometimes called plane wave decomposition. Just about any 2-D image can be thought of as a superposition of plane waves.

The other 2-D Fourier transform that we wish to consider is the delta-function line. Let

$$f(x,t) = \delta(t - px - \tau) \qquad (12.29)$$

This describes a line in x–t space. It is zero everywhere except along $t = px + \tau$, where it is infinite in the delta-function fashion. Substituting this $f(x,t)$ into Eq. (12.25a) and performing the easy integration over the delta function gives

$$F(k,\omega) = 2\pi e^{-i\omega\tau}\delta(k + \omega p) \qquad (12.30)$$

This result, shown in Fig. 12.8(b), shows that the Fourier transform takes a line into another line. Interestingly though, the intercept of the line in the x–t domain is wholly contained in the phase spectrum; all lines of the same x–t slope have the identical magnitude spectra. Note that lines of positive slope in x–t have negative slopes in k–ω, and that the slopes are reciprocally related.

The extension of the 1-D DFT to 2-D DFT is also straightforward. The 2-D DFT is given by

$$F_{\mu,\nu} = \sum_{mn} f_{m,n} e^{-i2\pi\mu n/N - i2\pi\nu m/M} \tag{12.31a}$$

$$f_{m,n} = \frac{1}{NM} \sum_{\mu\nu} F_{\mu,\nu} e^{i2\pi\mu n/N + i2\pi\nu m/M} \tag{12.31b}$$

These transforms are easily related to the 1-D DFT. For example, Eq. (12.31a) can be written as

$$F_{\mu,\nu} = \sum_{m} \left(\sum_{n} f_{m,n} e^{-i2\pi\mu n/N} \right) e^{-i2\pi\nu m/M}$$

$$F_{\mu,\nu} = \sum_{m} \mathrm{DFT}_n(f_{m,n}) e^{-i2\pi\nu m/M} \tag{12.32}$$

This shows that the 2-D DFT is obtained by first performing a 1-D DFT on each row of the $M \times N$ array f, and then performing a 1-D DFT on the columns of that result. The same is true for the inverse 2-D DFT. Thus, $M + N$ 1-D DFT's are required to perform the 2-D DFT. Given 1-D FFT routines, the programming of 2-D (or higher dimension) routines is then straightforward. Of course, computation time jumps up in moving to multidimensional transforms, but then the inherent speed in FFT algorithms frequently makes FFT processing even more attractive in multidimensional applications compared to the alternatives. For example, in 2-D low-order convolutions may be faster than FFT processing. But the FFT processing time is independent of the filter order; so for filters of high order, the FFT processing will win out. Furthermore, there are logical extensions of the 1-D FFT to multidimensions that are faster than the row–column 2-D DFT algorithm just discussed; see Dudgeon and Mersereau (1984).

All of the precautions discussed in Chapter 7 on applying the 1-D DFT to continuous data also apply to the 2-D DFT with the expected extensions: (1) First, of course, is aliasing. Two-dimensional data can be aliased in x alone, t alone, or in both coordinates. (2) The 2-D DFT spectrum is periodic in both dimensions, requiring appropriate precautions in applying the convolution theorem. (3) In using the 2-D DFT to approximate the spectrum of continuous signals, zero padding in both coordinates is required to reduce leakage.

This last requirement can be particularly demanding of computer time. Since the padding is in two dimensions, arrays may become undesirably large just to accommodate the zero padding. The sampling rate required to avoid aliasing frequently becomes a problem in one dimension but not the other. For example, in earthquake studies it is relatively easy to record unaliased data in time at each site, but it is quite another economic problem to have sites sufficiently close together to avoid aliasing in the spatial dimension. This limitation frequently eliminates the potential of using 2-D frequency domain techniques in such studies. In other applications, such as image processing, dense sampling in all coordinates is rather easy to achieve, allowing the full advantage of multidimensional Fourier methods. We will now give an interesting example, one from seismic reflection work where sampling is quite dense in all coordinate directions.

The seismic migration problem

Fourier transforms have been central to our study of 1-D LSI systems. Their utility is directly connected to the observation that sinusoids are eigenfunctions of LSI systems. This, in turn, leads directly to solving the problem of finding the output to an LSI system, given its input; in the Fourier transform space, the output is simply the product of the system's impulse response and the input function.

In higher dimensions, Fourier transform methods are again extremely useful for understanding and for solving problems. (In fact, Joseph Fourier, 1768–1830, first developed his theory while working on the heat flow equation.) In the 2-D case, however, the operation required in the transformed space can be quite different from the multiplication in the 1-D case. This is because the open temporal boundary condition yields recursions working from the initial disturbance, rather than from an independent driving function [Eq. (12.1) is not of the form $y = h * x$]. In essence, the frequency domain operation is data dependent. Because of this difference, we will present an example of using Fourier methods to solve a problem in three variables. In the end, the DFT is used to transform to the Fourier domain, an operation is performed, and the IDFT brings the solution back to x–y–t space.

The example selected is the so-called seismic migration problem discussed in the preceding section: given a wave field $\psi(x, 0, t)$, find $\psi(x, z, 0)$. So, in our example, we will be seeking a certain kind of solution to the wave equation—the same one that we found using finite differences, but now frequency domain methods will be used.

Let us first think of the fundamental property of waves. For simplicity,

Fourier Transform Methods in Multidimensions

let us think of a one-dimensional wave, such as a wave on a string. If we take a picture at one time, $t = 0$, then we see the disturbance for all x. If we know both the propagation velocity and the propagation direction, then we know the motion for all time from these facts. This is the forward problem in geophysics. Conversely, if we know the motion of our string for all t at one given location x, then we could find the shape of the string at $t = 0$ for all x. This is called the inverse or migration problem in reflection seismology.

Thus, we are motivated to think that from a knowledge of the surface seismogram (that is, a knowledge of the disturbance at all x and t for $z = 0$), we may be able to reconstruct the disturbance at $t = 0$ for all x and z (that is, the migrated data). [The word migration is used because events in (x, t) appear to have moved, or migrated, when viewed in a (x, z) display.] We know in one dimension that all functions that we may be interested in can be synthesized from a summation of sinusoids. The 2-D analogous statement would be expanded to include 2-D sinusoids. We are thus led to consider a superposition of sinusoids (plane waves) of the type

$$e^{i(k_x x + k_z z + \omega t)}$$

because these are upgoing solutions to the wave equation (you can verify this). A general superposition of these functions could be written as

$$f(x, z, t) = \frac{1}{4\pi^2} \int_{-\infty}^{\infty} \int_{-\infty}^{\infty} A(k_x, \omega) e^{i(k_x x + k_z z + \omega t)} \, dk_x \, d\omega$$

We do not superimpose plane waves for various k_z because k_z is not independent of k_x and ω:

$$k_x^2 + k_z^2 = 4\omega^2 / v^2$$

Clearly, we could use any pair of k_x, k_z, and ω in our superposition, but not all three. If $f(x, z, t)$ is going to be a valid representation of the disturbance for all x, z, and t, then it must match the observed seismogram at $z = 0$:

$$f(x, 0, t) = \frac{1}{4\pi^2} \int_{-\infty}^{\infty} \int_{-\infty}^{\infty} A(k_x, \omega) e^{i(k_x x + \omega t)} \, dk_x \, d\omega$$

or

$$\text{FT } f(x, 0, t) = A(k_x, \omega)_{z=0} = F(k_x, \omega)$$

Thus,

$$f(x, z, t) = \frac{1}{4\pi^2} \int_{-\infty}^{\infty} \int_{-\infty}^{\infty} F(k_x, \omega) e^{i(k_x x + k_z z + \omega t)} \, dk_x \, d\omega \quad (12.33)$$

where $F(k_x, \omega)$ is the 2-D Fourier transform of the surface seismogram.

Equation (12.33) has two important properties: (1) it satisfies the wave equation, and (2) it reduces to $f(x,t)$ at $z = 0$. An immediate important question is: Is Eq. (12.33) the only function that has these two properties, that is, is Eq. (12.33) unique? The answer is no. But, it is an excellent start to the migration problem. If we do assume Eq. (12.33) gives us $f(x,z,t)$, then the migrated section is given by

$$f(x,z,0) = \frac{1}{4\pi^2} \int_{-\infty}^{\infty} \int_{-\infty}^{\infty} F(k_x, \omega) e^{i(k_x x + k_z z)} \, dk_x \, d\omega$$

Note that this is not the IFT of F because $ik_z z$ is the variable in the exponential, not $i\omega t$. However, it can be converted to a standard 2-D IFT by a simple change of variables from ω to k_z:

$$k_x^2 + k_z^2 = 4\omega^2/v^2$$

$$k_z \, dk_z = \frac{4\omega \, d\omega}{v^2}$$

$$d\omega = \frac{v}{2\sqrt{1 + k_x^2/k_z^2}} \, dk_z$$

and we get

$$f(x,z,0) = \frac{v}{8\pi^2} \int_{-\infty}^{\infty} \int_{-\infty}^{\infty} \frac{F(k_x, v/2\sqrt{k_x^2 + k_z^2})}{\sqrt{1 + k_x^2/k_z^2}} e^{i(k_x x + k_z z)} \, dk_x \, dk_z$$

This is a standard inverse 2-D Fourier transform. To see this, let us relabel variables back to a new angular frequency: Let $vk_z/2 = \omega'$ be the new variable. The second argument of F, which is ω, is then related to ω' by

$$\sqrt{\frac{v^2 k_x^2}{4} + \frac{v^2 k_z^2}{4}} = \sqrt{\frac{v^2 k_x^2}{4} + \omega'^2} = \omega$$

or

$$\omega' = \sqrt{\omega^2 - (vk_x/2)^2}$$

and the migrated result is

$$f(x, 2z/v, 0) = \frac{1}{4\pi^2} \int_{-\infty}^{\infty} \int_{-\infty}^{\infty} \frac{F(k_x, \omega')}{\sqrt{1 + (vk_x/2\omega')^2}} e^{i[k_x x + \omega'(2z/v)]} \, dk_x \, d\omega' \quad (12.34)$$

The desired migration process, expressed in the ω–k plane, is rather simple, as diagrammed in Fig. 12.9. In the ω–k plane, each value $F(k, \omega)$ is moved to a new location $F(k, \omega')$ and multiplied by a factor depending on k_x and ω'. After all values receive this treatment, the IFT brings our

Fourier Transform Methods in Multidimensions 317

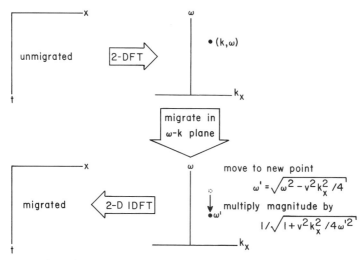

Figure 12.9 Solving the seismic migration problem by finding $\psi(x,z,0)$, given $\psi(x,0,t)$, using Fourier methods. In the $\omega-k$ plane, each datum is moved down in ω to a new frequency ω' and scaled by a factor. Transforming back via the IDFT gives $\psi(x,z,0)$. This result is scaled in the z direction by $t = 2z/v$, where t is the two-way reflection travel time.

answer back to the desired function $f(x,z,0)$. Clearly, computer code to perform this algorithm consists of only a few simple lines, given a 2-D FFT subroutine.

In spite of its simplicity (or perhaps because of it), several questions surround the validity of results obtained by this suggested procedure. First, of course, is the question of whether Eq. (12.33) represents a valid continuous theory approximation that is appropriate to the seismic problem under question. Much of this answer belongs to the domain of seismic reflection theory and lies outside our purpose here. We have noted that this is not a unique solution as indicated by Fig. 12.1. The signal processing questions of interest here relate to approximating the 2-D Fourier transforms of Eq. (12.34) by the 2-D DFTs. These questions and answers are completely analogous to the 1-D case. They concern record length and sample rate. The record length in the x direction, called the migration aperture in seismic terminology, controls the spectral resolution in that direction just as the record length in the t coordinate controls the frequency resolution. Typically, these record lengths might be 10 miles and 5 sec, respectively. Sampling controls aliasing; typically, the sample intervals are the order of 20 ft and 2 msec, respectively. The required approximation of the continuous Fourier transform by the DFT is achieved by adequate sampling of the band-limited data and zero padding to prevent spectral leakage.

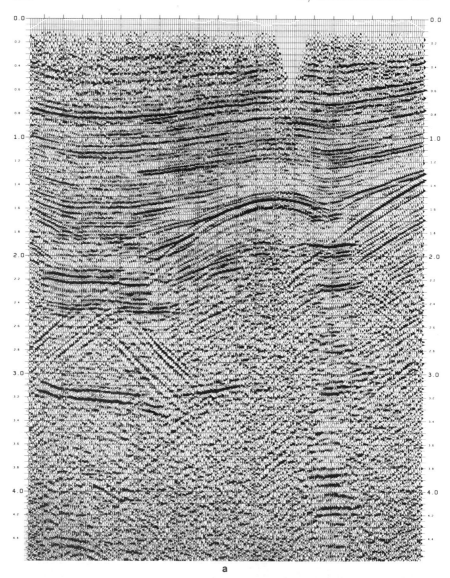

Figure 12.10 An example of seismic migration, that is, solving for $\psi(x, z, 0)$ given $\psi(x, 0, t)$. (a) Seismic data approximating $\psi(x, 0, t)$ from a coincident source/receiver experiment. The display is 5.0 sec by approximately 13.5 km. (b) Finite difference migration, the result of using Eq. (12.20b) to solve Eq. (12.19). (c) Migration in the $\omega-k$ plane, the result of using the scheme of Fig. 12.9 to solve the wave equation. Note how migration has changed the shape and location of reflectors. (From R. H. Stolt, 1978, Migration by Fourier transform, Geophysics **43**, 23–48. Reprinted with permission of the Society of Exploration Geophysicists.)

Fourier Transform Methods in Multidimensions

Within these limitations, the $\omega-k$ migration algorithm achieves excellent results and is widely used in the seismic data processing industry. Figure 12.10 shows results of $\omega-k$ migration of high quality seismic reflection data that has been carefully migrated.

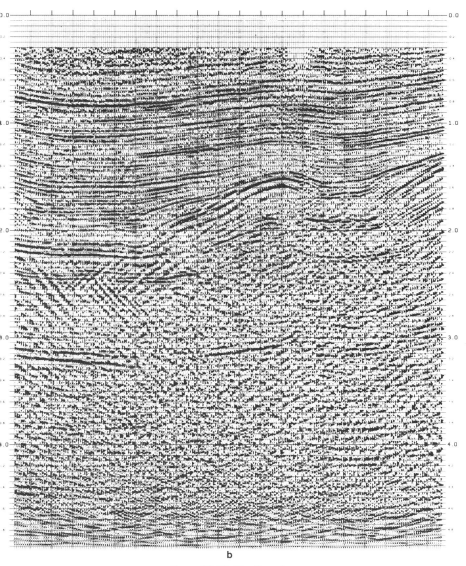

Figure 12.10 (Continued)

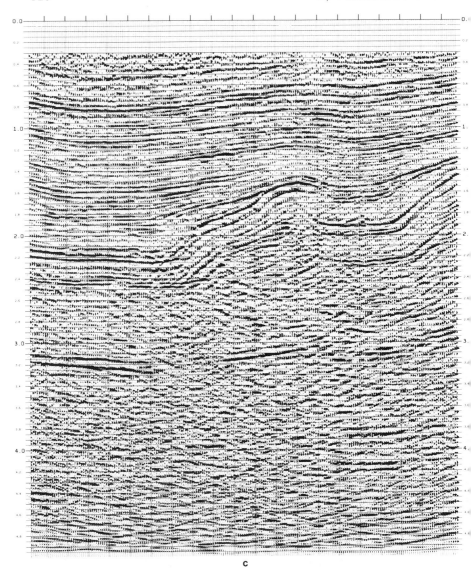

c

Figure 12.10 (Continued)

Two-Dimensional FIR Frequency Filter Design

We know that any LSI operation can be performed by a convolution in the time domain or as a multiplication in the frequency domain. The same is true for multidimensional LSI operations. The terminology frequency filter implies that the frequency domain function is rather simple, such as in lowpass filtering, bandpass filtering, or highpass filtering. In contrast, an output energy filter, for example, normally would have a complicated spectrum and, therefore, we would not call it a frequency filter. As in the 1-D case, the frequency response of 2-D digital filters is specified by considerations arising from its application. The design problem is to determine the convolutional coefficients that yield this desired frequency response.

While there are many similarities between 1-D and 2-D digital filter design, there are some important differences, such as the nonfactorability of the Z transform. Another important difference is the implementation of 2-D frequency filters. Frequency filtering in 2-D is much more likely to be actually carried out in the frequency domain than in one dimension. One reason is speed. As we know, to achieve sharp transition zones, long convolutional operators are required. As we have already noted, when the spatial domain operators become long, frequency domain filtering becomes more attractive because of the speed advantage of the FFT algorithms. In two or more dimensions this effect is even more pronounced.

Furthermore, in many 2-D applications, extreme precision of the frequency response is not required and naive approaches are satisfactory, such as simply taking the DFT, setting the appropriate frequency components to zero, and then doing an IDFT. However, careful implementation would include windowing the data, smoothing transition zones to prevent ringing in the spatial coordinates, and zero padding to provide a reasonable approximation to the 2-D spectrum. Recall that the true frequency response of this operation is not the frequency domain filter function used, but that function convolved with the data window.

Because of the additional dimensions, a richer array of filter functions is possible in two dimensions. A common one, useful in the array processing found in radar and sonar technologies and seismology, is the so-called fan, or pie-slice, filter. Its passband is pie shaped, passing frequency components of a given slope. As we have seen in the previous section, this also corresponds to the energy of a given slope in the spatial coordinates, which means that beam forming can be affected by digital filters. Another possibility in the 2-D case, with no counterpart in the 1-D case, is a filter that is FIR in one coordinate and IIR in the other. The operators we introduced earlier in this chapter concerning finite difference equations are just this type of 2-D filter. In spite of these obvious possibilities, which result from the addition of another coordinate, many similarities remain that we need to formalize before proceeding. The fundamentals of 2-D LSI systems for the most part are just the extensions that we might expect. The convolution is again the most general operation possible by a 2-D LSI system. Again, the output is the convolution of the system impulse response h with the input x:

$$y_{n_1,n_2} = \sum_{k_1} \sum_{k_2} h_{n_1-k_1, n_2-k_2} x_{k_1, k_2} \tag{12.35}$$

The double summation includes all nonzero values of h and x. There can be linear or circular convolution. For linear convolution, the impulse response can be finite or infinite. For FIR operators and finite x extent, the resulting region of nonzero y is not simply stated as in the 1-D case (see Problem 12.35).

In the special case that h is factorable, the convolution sum becomes particularly simple. If $h = fg$, then Eq. (12.35) can be written

$$y_{n_1,n_2} = \sum_{k_1}\sum_{k_2} f_{n_1-k_1} g_{n_2-k_2} x_{k_1,k_2}$$

$$y_{n_1,n_2} = \sum_{k_1} f_{n_1-k_1} \sum_{k_2} g_{n_2-k_2} x_{k_1,k_2}$$

$$y_{n_1,n_2} = \sum_{k_1} f_{n_1-k_1} g'_{k_1,n_2}$$

where

$$g'_{k_1,n_2} = \sum_{k_2} g_{n_2-k_2} x_{k_1,k_2}$$

Thus, the intermediate array g' is formed by taking the 1-D convolution of g with the rows of x. Next, the columns of g' are convolved with the f, so we see that the 2-D convolution is obtained by a series of 1-D convolutions in this special case.

Factorable system functions offer considerable savings in computation time over nonfactorable ones. If f were a nonfactorable $n \times m$ array and x were $p \times q$, then about $n \times m \times p \times q$ multiplications and additions are required to compute their 2-D convolution (we are neglecting end effects in this estimate). While, if f were factorable, only about $n \times p + m \times q$ multiplications and additions are required. The price paid for this dramatic savings is the rather limited structure of f.

Next, we document that, as expected, 2-D LSI systems have 2-D sinusoids as their eigenfunctions. Let the input be the complex sinusoid of the form

$$x(n_1, n_2) = e^{i\omega_1 n_1 + i\omega_2 n_2}$$

Then, Eq. (12.35) gives

$$y_{n_1,n_2} = \sum_{k_1}\sum_{k_2} h_{k_1,k_2} e^{i\omega_1(n_1-k_1) + i\omega_2(n_2-k_2)}$$

$$y_{n_1,n_2} = e^{i(\omega_1 n_1 + \omega_2 n_2)} H(\omega_1, \omega_2) \qquad (12.36)$$

where

$$H(\omega_1, \omega_2) = \sum_{k_1}\sum_{k_2} h_{k_1,k_2} e^{-i(\omega_1 k_1 + \omega_2 k_2)} \qquad (12.37)$$

Two-Dimensional FIR Frequency Filter Design

showing that output is indeed the same complex sinusoid, but with the complex amplitude (the eigenvalue) $H(\omega_1, \omega_2)$, which is the frequency response of the system (its spectrum). Again, we emphasize that H is a continuous and repetitive function of frequency.

We are now in a position to entertain a simple convolutional filter, just as we did in Chapter 3. There we saw that the simpler filter $\frac{1}{4}(1, 2, 1)$ had a reasonable frequency response for a lowpass filter $H(\omega) = \frac{1}{2}(1 - \cos \omega)$. This 1-D filter motivates the 2-D extension, which uses these coefficients in both the rows and columns:

$$h = \frac{1}{16} \begin{pmatrix} 1 & 2 & 1 \\ 2 & 4 & 2 \\ 1 & 2 & 1 \end{pmatrix} \quad (12.38)$$

where the center element is taken as the origin, making the filter symmetric in both directions; hence, it is zero phase. The factor of $\frac{1}{16}$ normalizes the gain to unity at zero frequency (the filter acting on a constant array will return that array unaffected). Using Eq. (12.37) to compute the frequency response gives

$$H(\omega_1, \omega_2) = \tfrac{1}{16}(e^{-i\omega_1 - i\omega_2} + 2e^{-i\omega_1} + e^{-i\omega_1 + i\omega_2}$$
$$+ 2e^{-i\omega_2} + 4 + 2e^{i\omega_2} + e^{i\omega_1 - i\omega_2} + 2e^{i\omega_1} + e^{i\omega_1 + i\omega_2})$$

Factoring and collecting terms of the same ω allows us to write it as

$$H(\omega_1, \omega_2) = \tfrac{1}{4}(1 + \cos \omega_1)(1 + \cos \omega_2) \quad (12.39)$$

This does seem like an entirely reasonable result considering its 1-D parentage; it is plotted in Fig. 12.11. This particular form of H in Eq. (12.39) is factorable, suggesting that h itself is. Indeed, h is factorable, a fact that leads us to a general statement: factorable impulse responses have factorable frequency responses (see Problem 12.36). This is the imposition placed on the frequency response by a factorable impulse response and its associated computational efficiency.

We, of course, recognize the frequency response of Eq. (12.37) as the logical extension to 2-D of one of the four kinds of Fourier transforms considered in Chapter 7; recall Fig. 7.2. Therefore, we expect that it can also be inverted recovering h from $F(\omega_1, \omega_2)$. By multiplying both sides of Eq. (12.37) by a 2-D complex sinusoid, and integrating using the orthogonality of sinusoids, it is straightforward to show that

$$h_{n_1, n_2} = \frac{1}{4\pi^2} \int_{-\pi}^{\pi} \int_{-\pi}^{\pi} H(\omega_1, \omega_2) e^{i(\omega_1 n_1 + \omega_2 n_2)} d\omega_1 d\omega_2 \quad (12.40)$$

So Eqs. (12.37) and (12.40) are in complete analogy with the 1-D case of Fig. 7.2(a). The other three Fourier transforms of Fig. 7.2 can likewise be

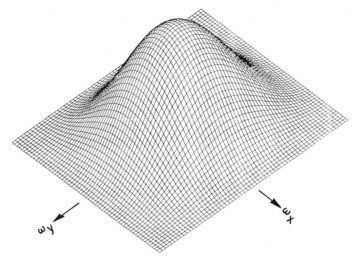

Figure 12.11 Plot of the frequency response of the simple 2-D, lowpass filter of Eq. (12.38). The frequency response is given by Eq. (12.39).

extended to two dimensions or higher. However, having spent some detail on the 1-D case, we can afford to proceed more directly now, using the preceding filter as an example of a 2-D FIR filter.

Two-dimensional FIR filter design is based on Eqs. (12.37) and (12.40) in a fashion analogous to 1-D design. The brute force approach is frequency sampling, using the 2-D DFT. A trial impulse response is computed by emulating Eq. (12.40) with the 2-D IDFT. The desired frequency response is specified at points in (ω_1, ω_2) space. Common specifications of $H(\omega_1, \omega_2)$ result in an h whose extent is infinite, causing aliasing in the spatial domain. To minimize this aliasing, the frequency sampling should be made on a dense grid, perhaps several times larger than the desired extent of h. Then, h is brought down to the desired size by windowing. The common 1-D windows can be used, the rectangular, the Hamming, and the Kaiser being among the most popular.

There is, however, the question of how to extend these 1-D windows to two dimensions. Two obvious possibilities immediately come to mind. The first is the outer product of the 1-D windows:

$$W(n_1, n_2) = W_1(n_1)W_2(n_2) \qquad (12.41)$$

This approach allows for different windows in each direction and for different extents of h in each direction. The filter performance may then be adjusted in each direction according to need. The other obvious window

Two-Dimensional FIR Frequency Filter Design 325

extension to two dimensions is the circular window:

$$W(n_1, n_2) = W(\sqrt{n_1^2 + n_2^2}) \tag{12.42}$$

As in the 1-D case, truncating the impulse response with these windows convolves the specified frequency response with the Fourier transform of the selected window. Before the window is applied, the actual frequency response is a sinc interpolated version of the response specified on the frequency sampling grid. Windowing smoothes this interpolated version according to the Fourier transform of the window. The final design is always checked with Eq. (12.37) by using a zero-padded, 2-D DFT. The result of this check is the true frequency response.

Examples of results from this frequency sampling procedure are shown in Fig. 12.12. The desired frequency response was specified on a 100 × 100

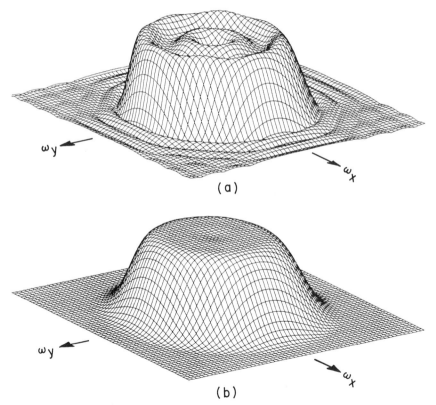

Figure 12.12 Frequency responses (power spectra) of 2-D, lowpass FIR filters designed by frequency sampling and windowing. Rectangular (a) and Hamming (b) windows truncated the operator to a circle with a radius of seven coefficients. As in the 1-D case, the Hamming window reduces ripple at the expense of broadening the transition zone.

grid. Then, the impulse response was truncated to a circle of coefficients seven points in radius, using the rectangular and the Hamming windows for comparison. In both cases, the circular forms of these windows were used according to Eq. (12.42).

Many advanced design methods cannot be easily extended from one dimension to two dimensions. For example, the Chebyshev optimal design of equiripple filters is thwarted when attempted in two dimensions. It turns out that the alternation property of the extremal points is not carried over the 2-D case because of the lack of factorization of the Z transform. This makes the problem difficult to handle and algorithms slow to converge. A different approach that has been proven useful is a transform method, called the *McClellan frequency transform*, which substitutes Chebyshev polynomials for the $\cos(n\omega)$ factors appearing in the frequency response of a zero-phase FIR filter. Then, a transformation is made in the frequency domain that depends on the filter design, making the procedure design dependent. The 2-D equiripple and McClellan transform design methods are sufficiently complicated to be beyond our introductory level. Those interested should start with Dudgeon and Mersereau (1984).

Problems

12.1 Write a computer program to implement Eq. (12.6) for given initial conditions. Reproduce the results of Fig. 12.2 and then try a suite of initial conditions, investigating whether instability depends on the initial conditions.

12.2 Show that the two poles given in Eq. (12.9) are reciprocals of each other.

12.3 Show that the smaller of the two poles of Eq. (12.9) lies between zero and one.

12.4 Carrying out the partial fraction expansion of the two-pole AR operator in Eq. (12.8), verify Eq. 12.10.

12.5 Prove that the infinite series comprised of the sum of the IIR time domain coefficients of the operator Q in Eq. (12.10) sum to one.

12.6 Show that Eqs. (12.13) do follow from using the bilinear transform in Eq. (12.12).

12.7 Show that Eqs. (12.13) also follow from the Crank–Nicolson scheme suggested in Eq. (12.11).

12.8 Carry out the required convolution to show that Eqs. (12.14) are indeed the MA mode IIR coefficients of the spatial operator U in Eqs. (12.13).

12.9 Write a computer program to solve the diffusion equation, using a truncated version of the spatial IIR operator of Eqs. (12.13) and

Problems

(12.14). By solving the diffusion equation for a range of values of b and initial conditions, convince yourself that this method is stable under all conditions. Verify that the physically impossible conditions of Fig. 12.4 suddenly disappear for small enough values of b.

12.10 Derive Eq. (12.18), the finite difference approximation to the Klein–Gordon equation, using central second differences.

12.11 Derive Eq. (12.20), the finite difference approximation to the 15° Claerbout equation, using the bilinear transform for the first-order derivatives in Eq. (12.19).

12.12 Using Eq. (12.20) write out explicit solutions for the two imaging problems mentioned in the text: (1) given $\psi(x, 0, t)$, find $\psi(x, z, 0)$ and (2) given $\psi(x, z, 0)$, find $\psi(x, 0, t)$. Discuss the significance of these solutions in terms of seismic wave imaging.

12.13 Using the results of Problem 12.12 and the expression in Eqs. (12.14) for the coefficients of the IIR operator, write a computer program to solve the first imaging problem of Problem 12.12. This is called *migration* in seismology; sometimes it is also called the inverse problem. As a test of your program use a point source for $\psi(x, 0, t)$. Devise other simple wave fields and use them to test your program.

12.14 Proceeding in similar fashion to the suggestion in Problem 12.13, write a computer program to solve the forward problem, given $\psi(x, z, 0)$, find $\psi(x, 0, t)$. As a test of your program, see if you can recover the point source using the output of the migration routine of Problem 12.13.

12.15 Of the four possible solutions of Eq. (12.20) for one of the ψ, Problem 12.12 requests two of them. Show that the remaining two solutions result in unstable operators.

12.16 Sylvester's theorem is frequently very handy in problems involving matrices. The theorem states that any given function of a diagonalizable matrix $\mathbf{A}$ can be written as a superposition of matrices constructed from the row and column eigenvectors of $\mathbf{A}$, each scaled according to the same given function of the corresponding eigenvalue:

$$f(\mathbf{A}) = f(\lambda_1)\mathbf{c}_1\mathbf{r}_1 + f(\lambda_2)\mathbf{c}_2\mathbf{r}_2 + \cdots + f(\lambda_n)\mathbf{c}_n\mathbf{r}_n$$

where

$$\mathbf{A}\mathbf{c}_n = \lambda_n \mathbf{c}_n$$

and

$$\mathbf{r}_n \mathbf{A} = \lambda_n \mathbf{r}_n$$

Also,

$$\vec{r}_i \cdot \vec{c}_j = \delta_{ij}$$

$$(\vec{c}_i \vec{h}_i)(\vec{c}_i \vec{r}_i) = \vec{c}_i \vec{r}_i$$

$$(\vec{c}_i \vec{h}_i)(\vec{c}_j \vec{h}_j) = 0$$

Show that repeated operations of a matrix **A** on itself produce

$$\mathbf{A}^N = \lambda_1^N \mathbf{c}_1 \mathbf{r}_1 + \lambda_2^N \mathbf{c}_2 \mathbf{r}_2 + \cdots + \lambda_n^N \mathbf{c}_n \mathbf{r}_n$$

12.17 Convolution can be expressed by matrix multiplication where the matrix has each row shifted one step from the row above. For circular convolution, the rows rap around as they are shifted through the matrix. Thus, circular convolution of the operator (a, b, a) requires the so-called circulant matrix:

$$\begin{pmatrix} b & a & 0 & 0 & \cdots & & & a \\ a & b & a & 0 & \cdots & & & 0 \\ 0 & a & b & a & 0 & \cdots & & 0 \\ \vdots & & & & & & & \\ 0 & & \cdots & & & a & b & a \\ a & 0 & \cdots & & & 0 & a & b \end{pmatrix} \begin{pmatrix} x_1 \\ x_2 \\ \cdot \\ \cdot \\ \cdot \\ x_n \end{pmatrix} \to (a, b, a) * x$$

In Problem 5.12, you showed that the eigenvalues of such a circulant matrix are the DFT of the first row. Using this result, compute the eigenvalues of the above matrix where $(a, b, a) = (a, 1 - 2a, a)$ and show that $|\lambda| \leq 1$ only when $0 \leq a \leq \frac{1}{2}$.

12.18 Consider the operator $A(a) = (a, b - 2a, a)$, where b is any positive number. Next form the operator $Q(a) = A(a) A^{-1}(-a)$. Using the method of Problem 12.17 and Sylvester's matrix theorem, show that Q is bounded for all values of a and n.

12.19 The operator $(a, 2 - 2a, a)$ occurs in the wave equation difference equation, Eq. (12.16). Show that the spectral radius of the circulant corresponding to this operator is greater than one. Reconcile this result with the experimental fact that Eq. (12.16) is stable for $a < 1$.

12.20 We want to plot a circle by using finite differences. The equations $x = R \cos \theta$ and $y = R \sin \theta$ define the circle. Using forward difference operators, show that the iterative scheme given here follows:

$$\begin{pmatrix} x_{i+1} \\ y_{i+1} \end{pmatrix} = \begin{pmatrix} 1 & -\Delta\theta \\ \Delta\theta & 1 \end{pmatrix} \begin{pmatrix} x_i \\ y_i \end{pmatrix}$$

Write a computer program that implements the iteration and observe that it is unstable.

Problems

12.21 Explain the instability of the previous iteration in terms of Sylvester's matrix theorem.

12.22 Intuitively, we feel that using one forward difference operator and one backward difference operator in the scheme of Problem 12.20 will result in proper centering of the differences, leading to a stable iteration. Find the matrix that results from this approach and show that the iteration is indeed stable.

12.23 Show that a function satisfying Laplace's equation in two dimensions,

$$\frac{\partial^2 \psi}{\partial x^2} + \frac{\partial^2 \psi}{\partial y^2} = 0$$

leads to the difference expression that expresses the fact that a potential function at a point is the average of its values over a circle centered at that point:

$$\psi_{i,j} = \tfrac{1}{4}(\psi_{i+1,j} + \psi_{i-1,j} + \psi_{i,j+1} + \psi_{i,j-1})$$

12.24 Using a small demonstration grid, for example about 3 × 4 or so, show that the result of the previous problem applied to the Laplace equation Dirichlet boundary value problem leads to the linear algebra formulation:

$$\mathbf{A}\psi = \psi_0$$

where **A** is a sparse symmetric matrix of known coefficients, ψ is a vector representing the unknown interior points, and ψ^0 is a vector representing the known boundary values.

12.25 In the spirit of our discussion of 2-D Fourier transforms, write down equations that extend all of the 1-D transforms of Fig. 7.2 to 2 dimensions.

12.26 In the section on Fourier transform methods, we mentioned 16 possibilities for the 4 different kinds of Fourier transforms when extended to two dimensions. Write down one of the hybrids suggested by this statement, and give an example where it will apply in a practical situation.

12.27 Assuming that you have a computer subroutine that performs 1-D DFTs, write a computer program that does 2-D DFTs by rows and columns as suggested by Eq. (12.32). Test the program by transforming several obvious examples.

12.28 Next, use your 2-D DFT program to study aliasing and zero padding in two dimensions. First, construct synthetic plane wave data with aliasing in only one direction and examine the effect on its 2-D spectrum. Next, study the effect of zero padding by investigating the 2-D spectrum of a plane wave that has one frequency placed exactly on a DFT computation frequency and the other between DFT computation points.

12.29 Verify that the plane wave $\psi = \exp[i(k_x x + k_z z + \omega t)]$ is traveling along the z axis in the negative direction. What angle do the wave fronts (surfaces of constant amplitude) make with the surface $z = 0$?

12.30 Some seismologists write the wave equation as

$$\frac{\partial^2 \psi}{\partial x^2} + \frac{\partial^2 \psi}{\partial y^2} = \left(\frac{2}{v}\right)^2 \frac{\partial^2 \psi}{\partial t^2}$$

halving the velocity because t represents two-way travel time in reflection seismology. Show that the plane wave $\psi = \exp[i(k_x x + k_z z + \omega t)]$ satisfies this wave equation only if

$$k_x^2 + k_z^2 = 4\omega^2/v^2$$

12.31 Show that Eq. (12.33) satisfies the wave equation.

12.32 Using Fig. 12.9 as a guide, write a $\omega-k$ migration program. As test input data, use a single unit impulse located midway in the $x-t$ plane. It should migrate to a hyperbola.

12.33 According to the $\omega-k$ migration scheme of Fig. 12.9, spectral energy of frequency below the line $\omega = v^2 k_x^2/4$ does not participate in forming the migrated spectrum. Explain this.

12.34 Write a computer program that inverts the migration process. That is, given $\psi(x, z, 0)$, find $\psi(x, 0, t)$.

12.35 Consider the convolution of the two 2-D arrays that have the extent shown by the dots in the figure below. Find the extent (the region of nonzero values) of their convolution.

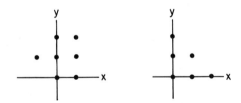

12.36 Show that the impulse response of Eq. (12.38) is factorable, and that in general all factorable impulse responses $h(n_1, n_2) = f(n_1)g(n_2)$ have factorable spectra $H(\omega_1, \omega_2) = F(\omega_1)G(\omega_2)$.

12.37 Consider a function $f(x, y)$. By digitizing f on an $N \times M$ grid, show that the one extension of the 1-D trapezoid concept of integration leads to the following Z transform expression for the volume under f over the grid:

$$V = \frac{1}{6}(\Delta x \, \Delta y) \sum_{i=1}^{N} \sum_{j=1}^{M} (1 + 2Z_1 + 2Z_2 + Z_1 Z_2) f_{i,j}$$

12.38 Frequency domain ω–k migration was developed as a solution to the LSI wave equation. Is the ω–k migration of Fig. 12.9 an LSI operation? Explain your answer.

12.39 A simple 2-D integral transform that arises in many fields, such as tomography, is an integration over straight lines of slope p and intercept τ of an image f in the x–y plane:

$$F(\tau, p) = \int_{-\infty}^{\infty} f(x, \tau + px) dx \, dy$$

This transform is called variously the Radon transform, the Hough transform, or a slant stack (by seismologists). Because the 2-D Fourier transform takes lines into lines, there is an intimate connection between the Radon transform and the 2-D Fourier transform. Show that 1-D Fourier transforms along straight lines of slope $-1/p$ in the k_x–k_y plane yield the Radon transform; that is, show that

$$F(\tau, p) = \int_{-\infty}^{\infty} F\left(k_x, -\frac{k_x}{p}\right) e^{-ik_x \tau/p} dk_x$$

12.40 Approximating the continuous Radon transform in the previous problem by direct integration of gridded data in the x–y plane immediately runs into problems because the line $y = px + \tau$ will almost never fall on the 2-D grid points. Therefore, some kind of interpolation will be required. Based on the results of the previous problem, devise a computer algorithm using the 2-D DFT to evaluate the Radon transform. Discuss digital methods of inverting the Radon transform.

12.41 Is the Radon transform of Problem 12.39 an LSI operation? If so, what is its x–y plane convolutional operator? What is its k_x–k_y plane multiplication operator? What is its impulse response?

References

Box, E. P.G., and Jenkins, G. M. (1976). "Time series analysis: Forecasting and Control." Holden-Day, San Francisco.
Bracewell, R. N. (1978). "The Fourier Transform and Its Applications," 2nd ed. McGraw-Hill, New York.
Burg, J. P. (1972). The relationship between maximum entropy spectra and maximum likelihood spectra, *Geophysics* **37**, 375–376.
Champeney, D. C. (1973). "Fourier Transforms and Their Physical Applications." Academic Press, New York.
Claerbout, J. F. (1976). "Fundamentals of Geophysical Data Processing with Applications to Petroleum Prospecting." McGraw-Hill, New York.
Claerbout, J. F. (1985). "Imaging the Earth's Interior." Blackwell, Palo Alto, California.
Dudgeon, D. E., and Mersereau, R. M. (1984). "Multidimensional Digital Signal Processing." Prentice-Hall, Englewood Cliffs, New Jersey.
Dym, H., and McKean, H. P. (1972). "Fourier Series and Integrals." Academic Press, New York.
Elliott, D. F., ed. (1988). "Handbook of Digital Signal Processing." Academic Press, New York.
Goodman, D. (1977). Some stability properties of two-dimensional linear shift-invariant filters, *IEEE Trans, Circuits Syst.* **CAS-24** (4), 201–208.
Hamming, R. W. (1983). "Digital Filters." Prentice-Hall, Englewood Cliffs, New Jersey.
Jaynes, E. T. (1982). On the rationale of maximum-entropy methods, *Proc. IEEE* **70**, 939–952.
Lighthill, M. J. (1958). "An Introduction to Fourier Analysis and Generalised Functions." Cambridge Univ. Press, London.
Oppenheim, A. V., and Schafer, R. W. (1975). "Digital Signal Processing." Prentice-Hall, Englewood Cliffs, New Jersey.
Paley, R. E. A. C., and Wiener, N. (1934). "Fourier Transforms in the Complex Domain." American Mathematical Society Colloquium Publication, Vol. 19. Providence, Rhode Island.

Press, W. H., Flannery, B. P., Teukolsky, S. A., and Vetterling, W. T. (1986). "Numerical Recipes." Cambridge Univ. Press, Cambridge, England.
"Programs for Digital Signal Prócessing." (1979). IEEE Press, New York.
Rabiner, R. L., and Gold, B. (1975). "Theory and Application of Digital Signal Processing." Prentice-Hall, Englewood Cliffs, New Jersey.
Shannon, C. E., and Weaver, W. (1949). "The Mathematical Theory of Communication." Univ. of Illinois Press, Urbana, Illinois.
Smith, C. R., and Grandy, W. T. eds. (1985.) "Maximum-Entropy and Bayesian Methods in Inverse Problems." Reidel, Boston.
Wiener, N. (1949). "Extrapolation, Interpolation, and Smoothing of Stationary Time Series." MIT Press, Cambridge, Massachusetts.

Index

Acausal inverse, 32, 300
Accumulated energy, 253, 254
Advance operator, 22, 28
Aliasing
 frequency domain, 7, 135, 162
 spatial, 191, 324
Allied function, 122
Allpass filter, 71, 238
Analog signal, 2
 quantization in sampling, 1-8
 sampling of, 134
Analog systems, 2, 62, 63
Analog-digital conversion, 5, 113
Analog-digital filter transformations, *see* Infinite response filter design
Analytic signal, 122
Anti-Hermitian, 84
Aperture, 317
Aperture error, 113
Apodization, 115
Autocorrelation, 92
Autocovariance, 262
Autoregressive (AR), 18
 mode, 20
Autoregressive moving average (ARMA), 19

Band-limited signal, 106
Bandpass filter, 70, 75, 77, 166, 256

Bartlett's spectral estimation, 270
Bartlett's window, 172, 270
Beam forming, 321
Bilinear transformation, 53
 analog-to-digital, 62, 63, 184
 s-plane relation, 62
Blackman-Tukey PSDA, 269
Block diagrams, 21
Boundary conditions, 295
Boundary value problem, 295, 296, 329
Burg, J. P., 273, 286
Burg PSDA, 273
Butterworth filter, 185
 bilinear transformation, 187-190
 design definition, 185
 frequency-domain filtering, 191
 iterative least-squares design, 251

Canis Major, 115
Canonical form networks, 23
Cascade form networks, 23
Cauchy conditions, 295, 296, 305
Causal allpass filter, 71
Causal inverse, 33
Causality, 16
 Hilbert transform and, 240, 243
Central limit theorem, 111, 125

335

Chebyshev criterion, 178
Chebyshev filter, 185, 326
Chebyshev polynomials, 185, 326
Circulant matrix, 95, 309, 328
Circular convolution, 89, 199
Claerbout, J. F., 307
Claerbout equation, 307
Coherent signal averaging, 258
Compensator, 150, 159
Complex logarithm, 77, 78
Computer programs, see FORTRAN programs
Constant filter, 76
Constant-coefficient difference equations, 18, 57, 294–310
Continuous signal, see Analog signal
Continuous-time systems, see Analog systems
Controller, see Compensator
Convolution, 14
 circular, 89, 199
 computer program, 15
 linear, 89
 periodic, see Circular convolution
 theorem, 88
Correlation sequence, 91–93, 208
Correlator, 220
Couplet, 29
 maximum phase, 32
 minimum phase, 32
Crank–Nicolson scheme, 302, 326
Cross-correlation, 91, 208
Cross-power spectrum, 104

Data models, 49, 143, 247, 284
 AR, 285
 ARMA, 34, 246, 250, 285
 MA, 267
Deadbeat response, 156
Decimation, 145
Deconvolution, 16, see also Inverse
Delay operator, 22, 28
Delta function, 99, 124
Demodulation, see Demultiplexing
Demultiplexing, 147
Desired output signal, 209
Deterministic approach, 4
Deterministic signals, 3, 214
DFT, see Discrete Fourier transform

Difference equations, 18, see also Constant-coefficient difference equations
Difference operators, 19, 52, 297–308
Diffusion equation, 296–304
Digital filters, see also Finite impulse response filter design; Infinite impulse response filter design
 delay properties, 211
 frequency responses, 42–53, 138
 narrow-band, 70
 pole-on-pedestal, 76
 single-zero, single-pole, 74–76
Digitization, 3
Dirac delta function, see Delta function
Direct current value, 86
Direct form networks, 23
Dirichlet conditions, 295, 296
Discrete Fourier series, see Fourier series
Discrete Fourier transform (DFT), 79–93
 computation of, 95, 96
 filtering, 190–192
 inverse, 198–203
 linear convolution using, 90
 matrix representation of, 93
 two-dimensional, 313
Discrete Hilbert transform, 242
Discrete random signal, see Random process
Discrete-time impulse, 11
Discrete-time sequence, 3, 10
 causality, 16
 convolution, 14
 invertibility, 16
 stability, 15
Dispersion, 73
Downsampling, see Decimation
Duality, 81, 94
Duhamel's integral, 25

Eigenfunction, 42, 95
Eigenvalue, 42, 95, 309, 328
Elliptic filter, 185
Energy, in sequences, 105
Energy distribution in wavelets, 253
Ensemble, 261
Entropy, 279
Envelope function, 122
Equiripple approximation, 179
Ergodic, 265

Index

Error in least-squares filtering, 209, 229
Error signal, 150
Extremal point, 179

Fan filter, 321
Fast Fourier transform (FFT), 81
 computation of, 95
 in two dimensions, 313
Feedback, 18, 150
Feedback stability, see Stability
Feedforward, 23
Filter design, see Finite impulse response
 filter design; Infinite impulse response
 filter design
Filter performance, 230
Finite energy, 105
Finite impulse response (FIR) filter design
 165, 182
 Chebyshev, see Parks–McClellan
  algorithm
 equiripple, see Parks–McClellan
  algorithm
 minimax, see Parks–McClellan algorithm
 minimum phase, 257
 Parks–McClellan algorithm, 175
 windowing, 171
First value, 86
Flow graphs, 23
Folding frequency, see Nyquist frequency
Folding operation, see Convolution
FORTRAN programs, 15, 95, 96, 177, 181,
 210, 215, 220, 224, 227, 242, 245, 269,
 277, 278
Fourier, J., 314
Fourier coefficients, 129
Fourier integral representation, 102
Fourier integral transform, 98–123, 130
Fourier series, 44, 128
 convergence, 131
Frequency division multiplexing (FDM), 147
Frequency domain filtering, 190
Frequency response, 43
 of differential and integral operators,
  50–60
 of digital filters, 44–49, 166, 168, 174,
  175, 182, 189, 192, 241, 252
 of simple systems, 44–49
Frequency sampling, 173
Frequency transformations, 190, 326

Frequency warping, 55, 190
Frequency wave-number domain, 311, 317
Fundamental theorem of algebra, 29

Gauss method of least squares, see Least-
 mean-square error criterion
Generalized eigenvalue problem, 222
Generalized Fourier integrals, 106
Geometric series, 31
Gibbs, J. W., 133, 261
Gibbs phenomenon, 133
Goodman, D., 293
Group delay, 72, 77, 78
Guard bands, 148

Heat-flow equation, see Diffusion equation
Hermitian, 84
Hilbert transform, 119, 125,
 generalized, 123
Hough transform, see Radon transform

IIR, see Infinite impulse response
 filter design
Imaging, 307, 327
Impulse, discrete-time, 10, 80
Impulse invariance design, 39, 183
Impulse response, 11
Incoherent signal averaging, 260
Infinite impulse response (IIR) filter design,
 182–192
 all-pole, 233–234
 ARMA, 233–234
 bilinear transform, 62, 63, 184, 187
 Butterworth, 185, 187–190, 233–234
 Chebyshev, 185
 differential mapping, 183
 direct pole–zero, 77
 elliptic, 185
 equiripple, 185
 hold invariance, 184
 impulse invariance, 39
 iterative least-squares, 250–253
 Jacobi elliptic, 185
 least-squares, 233–234
 single-zero, 65

Index

single-zero, single-pole allpass, 71
step invariance, 40, 184
Infinite-energy signal, see Wiener-Khintchine theorem
Information, 279
Initial value problem, 295
Integrated spectrum, 281
Interpolation, 142, 143
  midpoint, 145
  polynomial, 61
  sinc, 145
Inverse
  DFT, 198
  least-squares, see Wiener filter
  physically unrealizable, 32
  stability of, 31
Inverse filtering, see Inverse
Inverse Z-domain operator, 30–37
  partial fraction expansion, 33
Invertibility, 16

Jacobi elliptic filter, 185
Jaynes, E. T., 282

Kernel, 128
Klein–Gordon equation, 304
Kolmogoroff spectral factorization, 242
Kronecker delta function, 10, 80

Lag angle, 65
Lagrange multiplier, 289
Lag-window (MA) spectral analysis, 267–271
Laplace transform, 56, 62
Laplace's equation, 329
Leakage, 137
Least-mean-square-error criterion, 46, 131, 207, 209, 228
Least-squares filter, see Wiener filter
Levinson recursion, 225
Line splitting, 284, 289
Linear constant-coefficient difference equations, see Constant-coefficient difference equations
Linear convolution, 89
Linear phase, 72, 74, 87, 172

Linear prediction, see Prediction operator
Linear time-invariant systems, 8–18
Loran, 77

Magnitude spectrum, 65
  of allpass filter, 71
Marine seismic trace, 255, 286
Matched filter, 217
Matrix representation of convolutions, 13, 205, 206, 283
Maximum delay, see Maximum phase sequence
Maximum entropy method (MEM), 279–284
Maximum likelihood PSDA, 286
Maximum phase sequence, 32, 34
McClellan frequency transform, 326
Mean-square error, minimization of, see Least-mean-square-error criterion
Michelson, A., 133
Midpoint interpolation, 144
Migration, 327
  finite difference, 307
  frequency domain, 314, 330
Minimax criterion, 178
Minimum delay, see Minimum phase sequence
Minimum group delay, 73
Minimum phase FIR filter, 257
Minimum phase lag, see Minimum phase sequence
Minimum phase sequence, 32, 34
  accumulated energy in, 253
Minimum phase spectrum, 66
Mixed phase, 32, 34
Mode, MA and AR, 20
Modulation, see Multiplexing
Modulation theorem, 118
Moving average (MA), 18
  mode, 20
Multidimensional Fourier transforms, 310–314
Multiple reflections and reverberations, 217, 247, 255
Multiplexing, 148

Networks, see Canonical form networks
Neumann conditions, 295, 296
Ninety-degree phase shifter, see Hilbert transform

Index

Noise-like, 4, 261
Normal equations, 207, 212, 216, 275, 283
Numerical solution of differential equation, *see* Constant-coefficient difference equations
Nyquist frequency, 7
Nyquist value, 86

Optimum prediction distance, 215
Orthogonal matrix, 222
Orthogonality relation, 80
Output energy filter, 221

Paley-Wiener criterion, 281, 286
Parks-McClellan algorithm, 175-182
Parseval's theorem, 86
Partial differential equations, 294-296
Partial energy, *see* Accumulated energy
Partial fraction expansion, 33
Periodic sequence, 82, 90
Periodic convolution, *see* Circular convolution
Periodogram, 264
  averaging, 265
  variance of, 265
Phase delay, *see* Phase spectrum
Phase lag, *see* Phase spectrum
Phase spectrum, 44, 65, *see also* Minimum phase sequence
Phase unwrapping, 77, 88
Phase-locked loop, 71
Phase-shift filter, 71, *see also* Hilbert transform
Phase-shift theorem, 87
PID controller, 159
Pie-slice filter, *see* Fan filter
Pisarenko PSDA, 285, 290
Plancherel, 101
Plancherel's theorem, 103
Plane wave, 311
Plane wave decomposition, 312
Plant, 150
Poisson's equation, 296
Pole-on-pedestal filter, 76
Poles, defined, 34
Pole-zero plots, 65-76
Polya, G., 111
Polynomial interpolation, 48, 61

Power spectral density, 105, 262, 267
Power spectral estimation
  Blackman-Tukey, 267-271
  Burg, 275-279
  maximum entropy, 279-284
  maximum likelihood, 286
  periodogram, 264
    variance of, 263-265
  Pisarenko method, 285
  Prony method, 286
  smoothed spectrum estimators, 269-271
    Bartlett window, 271
    windowing, 269
  Yule-Walker, 271-275
Power spectrum, 69, 72
Power theorem, 103
Prediction distance, 216
Prediction error operator, 216, 233
  gapped, 216
Prediction operator, 213, 231
Prony PSDA, 286

Quadrature function, 122

Radar, 199, 204, 220, 247, 290
Radiotelegraphy, 70, 125, 147
Radon tranform, 331
Random process, 261
Rational Z transform, 34, 246
Rayleigh's theorem, 103
Realization, 261
Real-time processing, 190
Recursive, *see* Autoregressive
Recursively computable, 253
Reflection coefficient, 213, 255
Region of convergence, Z transforms and, 31
Resolution, 137
Reverberation, 255
Roots, 29, *see also* Zeros, defined

Sampled data, 3
Sampling and aliasing, 4-8
Sampling function, 133
Sampling theorem, 145
Seismic trace model, 212, 255, 286

Sequence
  acausal, 16
  autocorrelation, 92
  causal, 16, 238
  cross-correlation, 91
  maximum phase, 32, 34
  minimum phase, 32, 34
  mixed phase, 32, 34
  periodic, 82, 90
  sampling, 134
Shannon, C. E., 280, 281
Shaping filter, 228
Shift invariance, 13
Shift-invariant system, *see* Linear time-invariant systems
Signal
  analytic, 122
  band-limited, 106
  discrete-time, *see* Discrete-time sequence
  quadrature, 122
  random, 261
  time-limited, 106
Signal averaging
  coherent, 258
  incoherent, 260
Signal flow graph, *see* Flow graphs
Signal-like, 4, 261
Signal-to-noise ratio enhancement filters
  least-squares filter, 232, 234
  matched filter, 217, 232
  output energy filter, 221, 232
Similarity theorem, 103
Simpson integration rule, 51
Sinc interpolation, 143
Slant stack, *see* Radon transform
Spectral estimation, *see* Power spectral estimation
Spectral factorization, 198, 235–253
  IIR filter design, 250
  iterative least-squares, 246
  Kolmogoroff method, 242
  least squares, 245
  root method, 236
Spectral radius, 309
Spectral resolution, *see* Resolution
Spectral response, *see* Frequency response
Spectrum, 43, *see also* Frequency response
  of causal function, 238
Speech signals, 201
Spencer smoothing operator, 163
Spiking filters, *see* Wiener filter

Stability, 15, 30–37
  of difference equations, 56–60
  of difference equations in two dimensions, 296–310
  of IIR filters, 187
Stationary process, 218, 261
Stationary time series, 261
Statistical approach, 4
Step response, 40, 57, 154
Step-invariance procedure, 40, 184
Sylvester's matrix theorem, 309, 327, 328
Symmetry properties
  of discrete Fourier transform, 84, 85
  of Fourier integral transform, 102, 103
  of two-dimensional Fourier transform, 311
System, defined, 8
System transfer function, *see* Transfer function

Tellegen's theorem, 23
Tick's integration rule, 51
Time averages, 258–259, 261
Time division multiplexing (TDM), 148
Time invariance, 12
Time series, *see* Stationary time series
Time–frequency resolution, 109
Time–frequency-statistical resolution, 265
Time-limited band-limited theorem, 106
Titchmarsh, 101
Toeplitz matrix, 14
Tomography, 331
Transfer function, 21, 50
  of allpass filter, 72
  of dispersive filter, 305
  of rational fraction, 64
  of RC circuit, 62, 63
  of single-pole couplet, 69
  of single-zero couplet, 65
Two-dimensional discrete Fourier transform, 313
Two-dimensional Z transform, 293

Uncertainty principle, 109
Uncorrelated noise, 219
Unit circle, 32, 65, 66
Unit-delay operator, *see* Delay operator

Index

Unrealizable filter, 32
Upsampling, *see* Interpolation

Van Der Monde matrix, 93

Wave equation, 304–308
  Claerbout, 307
  Klein–Gordon, 304
Wave number, 311
Wave-equation migration, *see* Migration
Wave-field reconstruction, *see* Migration
Wavelets, 212, 213
Wave-shaping filter, *see* Wiener filter
White noise, *see* Uncorrelated noise
Wiener, N., 101, 207, 225, 281
Wiener filter, 206
  frequency domain, 234
  IIR filter design, 233
  least-squares property, 228
  shaping filter, 228
  signal-to-noise enhancement, 232, 234

Wiener–Khintchine theorem, 104
Wilbraham, H., 133
Window, 112, 269
Windowing, 138, 169, 172

Yule–Walker PSDA, 271

Z transform, 27–40
  convolution theorem, 29
  cross-correlation and autocorrelation sequences, 91–93
  rational, 34, 246
Zero interlacing, 148
Zero padding, 138
Zero phase, 176
Zero-lag value, 208
Zero-order hold, 152
Zeros, defined, 32
Zoom processing, 148

ISBN 0-12-398420-3